# *Cleaner Production and Energy Conservation for Sustainable Development*

# *Cleaner Production and Energy Conservation for Sustainable Development*

*- Editors -*

*Jurgis Staniškis*
C. *Jayaraman*

CENTRE FOR SCIENCE & TECHNOLOGY OF THE
NON-ALIGNED AND OTHER DEVELOPING COUNTRIES
(NAM S&T CENTRE)

2010
DAYA PUBLISHING HOUSE
Delhi - 110 035

ISBN 9788170359838



**Centre for Science and Technology of the Non-Aligned and Other Developing Countries (NAM S&T Centre)**
Core-6A, 2nd Floor, India Habitat Centre, Lodhi Road,
New Delhi-110 003 (India)
Phone: +91-11-24644974, 24645134, Fax: +91-11-24644973
E-mail: namstct@gmail.com
Website: www.namstct.org

*Published by* : **Daya Publishing House**
**A Division of**
**Astral International Pvt. Ltd.**
**– ISO 9001:2008 Certified Company –**
4760-61/23, Ansari Road, Darya Ganj
New Delhi-110 002
Ph. 011-43549197, 23278134
E-mail: info@astralint.com
Website: www.astralint.com

*Laser Typesetting* : **Classic Computer Services**
Delhi - 110 035

*Printed at* : **Chawla Offset Printers**
Delhi - 110 052

Printed in India

# Foreword

As a supporter of and participant in the 3-day international workshop on "Cleaner Production and Energy Conservation for Sustainability" held in Kochi, the Renewable Energy and Energy Efficiency Partnership (REEEP) is extremely gratified to see the lessons of this workshop being shared further through this publication.

The Kochi Declaration rightly pointed out that the challenge we all face is to pass this earth on to the next generation in a condition that is both more green and more clean, without reducing the level of energy services, and expanding those services in a sustainable manner.

REEEP views renewable energy and energy efficiency as two sides of the sustainable energy development coin. In line with this, the insights in this book address both the supply and demand side of the energy equation. The geographical spread also mirrors the workshop's diversity, with interesting contributions from an extremely wide range of emerging, transition and developing economies from Lithuania to Zambia.

The width and breath of perspectives from the unique mix of scientists, academics, technocrats, diplomats, engineers, consultants and managers who made the Kochi workshop such a success, are also guaranteed to make this book very compelling reading.

**Dr. Marianne Osterkorn**

*Director-General, REEEP*

# *From the Editors' Desk*

## Background

Modernisation, industrialisation and development often appear to be synonyms. Hence, when we talk about "developing countries" it implies that they are going to be more developed in the future and hence more industrialised and more energy intensive. It is been pointed out that a great majority of world's energy demand growth in the coming decade will come from Asian developing countries which are more oil intensive and much less energy efficient than their industrial counterparts. Industrialisation very often adopted a strategy "Pollution First and Control Afterwards" which focused on the end-of-pipe treatment. Many pollution control systems have been constructed, with large amounts of investment, to treat the wastes generated from industrial production processes. Years of experience, however, have shown that such end-of-pipe control systems can result in not only tremendous waste of resources and high costs, but they also have complicated technical requirements. Based on the lessons learned from traditional industrial pollution control practices, a new concept–Cleaner Production (CP)–was proposed and advocated. It is a pro-active and integrated solution to pollution problems by eliminating or reducing pollutants at the source during the course of production processes.

Cleaner production aims at reducing environmental impact, reducing wastes from processes, products and services by using better management strategies, methods and tools and has become a fundamental policy in many countries around the world to achieve the objective of sustainable development. The most common definition of CP coined by United Nations Environment Program calls for "the continuous application of an integrated preventive environmental strategy applied to processes, products, and services to increase eco-efficiency and reduce risks to humans and the environment." CP can yield both environmental and economic gains both in the short term as well as in the future.

Adoption of appropriate policies and implementation measures to lower the energy as well as lowering dependency on fossil fuels through policy changes, energy efficiency, and fuel diversification thus assumes extreme importance today for the developing countries. With this understanding an international workshop was organised at Kochi, in India to bring in different stakeholders in the NAM and other developing countries, working in the fields of Energy Efficiency (EE), Renewable Energy, Industrial Productivity and Waste Minimisation to understand the concept of cleaner production. The workshop had an expected out come to entice policy makers in the respective locations to provide the required patronage, and to make policy level interventions to accelerate the penetration of CP/ EE benefits into all the sectors of economy. A group of professionals working in the field of Cleaner Production techniques and Energy Conservation methodologies in the NAM and other developing countries including the host country – India, thus assembled in the workshop to deliberate on the issue and the deliberations are consolidated in this publication as a useful experience sharing tool. The articles in this publications have touched upon various aspects of cleaner production, energy needs and impacts, the challenges to science and technology, energy policies established in their respective countries with an analysis of its prospects and challenges, present status, achievements, Government's initiatives and future trends towards energy conservation for sustainability.

## Purpose

The basic purpose of this publication is to provide a platform for knowledge and awareness enhancement in the field of Cleaner Production and to provide guidelines for policy makers, management and technical personnel on developing and implementing CP solutions. It also aims to promote and disseminate the achievements and experience of developing countries.

## The Organization

The articles are categorized into FIVE sections viz;

- Energy Policies and Strategies,
- Cleaner Production,
- Renewable Energy,
- Energy Efficiency, and
- Case Studies.

Policy Intervention strategies for developing Asia, Government's initiatives in Tanzania and in The Republic of South Africa, the challenge to Science and Technology in Egypt, future Trends towards energy conservation for sustainability in Sri Lanka and analysis of energy policy for Mauritius are covered in the first section.

In the next section innovation, development and implementation of Cleaner Production strategies are covered in detail with inputs from Lithuania, Zambia and Kenya.

Section 3 deals with Renewable Energy and it covers various applications of renewable energy sources such as biogas production from distillery slops, solar traffic lighting system by way of using multi colour LED lamps and also on the issues of sourcing local raw material for sustainable manufacturing of solar energy powered components. A proposed action for solving the Indonesian's problems in energy development through Bio Energy is also included in this section.

Section 4 is on energy efficiency and it attempts to give a global overview of energy efficiency in buildings through a paper which attempts to highlight the importance of energy efficiency in buildings and lists the key barriers to development of this opportunity. It also lists various approaches to promote the development of this market and lists some of the key initiatives in this space.

In the last section various case studies from India are included.

## Remarks

Effective awareness-raising programs can involve different groups in the wider society–from government representatives and business stakeholders to the public who are impacted by industrial pollution. Region wide programs such as conferences are useful for both raising awareness among firms, as well as sharing information among agencies promoting CP and providers of CP expertise.

Many national approaches have been based on the implementation of demonstration projects in the hope of generating chain reactions among those who are exposed to those projects. In many instances, demonstration projects remain as just that, not even being replicated in the very companies where they are demonstrated. Experiences suggest that it is better to use a combination of strategies such as training, and case studies by industrial sector in addition to demonstration projects. Where possible, case studies of projects from other locations with similar industrial processes can also be used to supplement the demonstration projects.

Developing an effective mechanism for communication between the directors of a CP program and industry is a key element for the success or failure of any program. A variety of media can be used to disseminate information about CP, including printed materials (guidance manuals, fact sheets, reports, brochures, case studies), videos, electronic systems including the Internet and databases. An important consideration however, is that these materials be presented in language that is familiar to and easy to understand by the target audience. We hope that this publication has served its purpose in meeting this important consideration.

**Jurgis Staniškis**

**C. Jayaraman**

# Introduction

Cleaner production is a continuous application of an integrated environmental strategy to processes, products and services in order to increase the efficiency and reduce risks to humans and environment. It aims at minimizing the waste and emissions and maximizing the product output by analysing the flow of materials and energy inside a system. In the past, polluting companies concentrated on treating the waste generated by an industrial process in an attempt to reduce its impact on the environment, but often no attempt was made at source to reduce the overall level of waste. With cleaner production, industrial processes can often be improved so that not only the amount of waste, and consequently the pollution, is reduced, but money is also saved or generated for the company, which in other words becomes a win-win situation for the company.

Cleaner production is most often applied to manufacturing processes, but it is also relevant to other sectors of the economy, including agriculture, mining and the provision of services. Instead of relying on penalties that force companies to treat their waste, cleaner production offers rewards in the form of increased profits and an enhanced environmental image to those who can reduce their overall level of waste. Cleaner Production has been pursued in many countries for several years under various names such as pollution prevention, waste minimization, clean technology, environmental management, materials productivity, global competitiveness, natural resource conservation, green productivity etc. However, all these approaches represent different facets of Cleaner Production and collectively they address the triple objectives of sustainable development, namely, economy, environment and society.

Increasing energy demand exerts pressure on the energy producers to over exploitation of the limited natural sources for producing more energy. In developing countries, larger part of mechanical and electrical power is generated by burning of coal and petroleum products, which dramatically gives birth to the second major problem, *i.e.* pollution. In this way, it is much significant to club energy conservation and cleaner production together as the balancing parts of a single combo-solution.

Energy conservation is an important element of energy planning and policy as it leads to reduction in energy consumption and energy demand per capita and offsets the growth in energy supply needed to keep up with population growth. Up to some extent this strategy reduces the rise in energy costs and also the need for new power plants and energy imports. The reduced energy demand can provide more flexibility in choosing the most preferred methods of energy production. Energy conservation facilitates the replacement of non-renewable resources with renewable energy. It is often the most economical solution to energy shortages, and is a more environmentally benign alternative to increase energy production. It represents a cost-effective approach to raising profitability, enhancing competitiveness and improving environmental performance ensuring a sustainable development. The possibilities range from relatively simple and low-cost process modifications to sophisticated and more costly investments in pollution prevention technologies.

Enabling developing countries to take full advantage of cleaner production options is particularly challenging. Many of these countries and their enterprises lack the human, institutional, technical and financial capacities needed to apply cleaner production methods. Development assistance and technology cooperation are therefore important policy tools in helping developing countries in their efforts to manage technological change for cleaner production. In order to address the issues relevant to what has been stated above, the Centre for Science and Technology of the Non-Aligned and other Developing Countries (NAM S&T Centre) organised an international workshop on Cleaner Production and Energy Conservation during 24x26 June 2008 at Riviera Suites in Cochin, India jointly with the Society of Energy Engineers and Managers (SEEM). The workshop was also supported by the Renewable Energy and Energy Efficiency Partnership (REEEP) and the French Agency for Environmental and Energy Management (ADEME).

The Cochin workshop was inaugurated by Mr. L. Radhakrishnan, IAS, Secretary to the Government of Kerala in the Department of Power followed by Welcome Address by Mr. C. Jayaraman, President, SEEM and Senior Manager, BPCL-Kochi Refinery; Introductory Remarks by Prof. Arun P. Kulshreshtha, Director, NAM S&T Centre; Brief Remarks by Mr. V. M. Trehan, Chairman, Mekaster Engineering and Equipments (P) Ltd., and Vote of Thanks by Mr. R. Harikumar, General Secretary, SEEM and Project Director, Agency for Non-Conventional Energy and Rural Technology (ANERT) of Kerala.

The workshop was attended by 85 experts and professionals, including 20 foreign delegates from 16 countries and UNIDO. Besides 16 participants from India, the overseas participants were from Austria, Egypt, France, Indonesia, Kenya, Lithuania, Malaysia, Malawi, Mauritius, Myanmar, Nepal, Nigeria, South Africa, Sri Lanka, Tanzania, Uganda and Zambia. The overall programme was carried out in six technical sessions broadly categorised under 'Key Note Addresses', 'Energy Efficiency Policy Intervention', 'Cleaner Production', 'Integrating CP/EE Tools – Case Studies and CDM', 'Energy Efficiency in Buildings, Energy Needs and Access' and 'Renewable Energy'. These sessions were chaired/co-chaired respectively by Mr. C. Jayaraman, President, SEEM and Prof. H.J. Staniškis, Director, Institute of Environmental Engineering, Kaunas of Lithuania; Mr. R. Sudhir Kumar of Central

Power Research Institute (CPRI), Bangalore and Mr. Binu Parthan, Deputy Director, REEEP, Vienna, Austria; Mr. C.R. Surendranathan, Dy. Chief Controller of Explosives, Government of India and Mr. G. Sivakumar, Director, Kerala State Productivity Council; Prof. V.K. Damodaran, UNIDO International Energy Consultant and Vice-Chairman, Energy Management Centre, Kerala and Mr. R.K.R. Pillai, Chief Manager, BPCL Kochi Refineries; Mr. A.K. Asthana of Indo-German Energy Program, German Technical Corporation (GTZ) and Mr. A.M. Narayanan, Head, Energy Efficiency Division, Energy Management Centre, Kerala; and Prof. K.K. Sasi of Amritha Viswavidyapeetham and Co-ordinator, SEEM Tamil Nadu Chapter and Dr. M.P. Sukumaran Nair, Managing Director, Travancore Cochin Chemicals Ltd. The participants also had an opportunity to visit the National Exhibition on Energy Efficient and Renewable Energy Products and Technologies, also organised by SEEM jointly with the Centre for Innovation in Science and Social Action (CISSA) at the Cochin Town Hall, on its inaugural and valedictory sessions, which were respectively chaired by Mr. L. Radhakrishnan, IAS, Power Secretary of Kerala Government and Prof. Arun P. Kulshreshtha, Director, NAM S&T Centre. The Workshop was concluded with the adoption of the Kochi Declaration on Cleaner Production and Energy Conservation for Sustainability.

The present publication is based on the deliberations of the Cochin workshop and also includes papers contributed by eminent specialists in the field. It comprises 21 scientific articles and review papers classified under 5 chapters, namely, Energy Policies and Strategies, Cleaner Production, Renewable Energy, Energy Efficiency, and Case Studies.

I would also like to acknowledge the dynamic involvement and untiring efforts of Prof. (Dr.) Jurgis Staniškis, Director, Institute of Environmental Engineering, Kaunas, Lithuania, and Mr. C. Jayaraman from SEEM, India for technical editing of this publication, supplying the 'From the Editors' Desk' and for suggesting future course of action while reviewing the contents. I am indebted to Dr. Marianne Osterkorn, Director-General of the Renewable Energy and Energy Efficiency Partnership (REEEP) for sparing her valuable time in writing the 'Foreword'. Last, but not the least, the valuable services provided by the entire team of the NAM S&T Centre, particularly by Mr. M. Bandyopadhyay, Dr. V.P. Kharbanda, Mr. Gaurav Gaur and Mr. Pankaj Buttan in compiling the presented papers and giving a shape to this volume are deeply appreciated.

I hope that this publication will serve as a valuable reference material on energy efficient cleaner production for the entire scientist community, experts and policy makers in developing countries who may find it useful in working out appropriate plans and programmes in their countries, and will also create awareness among the masses to the increasing significance of the subject and its effect on their livelihoods.

**Arun P. Kulshreshtha**

*Director, NAM S&T Centre*

# Contents

# Part I

# *Energy Policies and Strategies*

# Chapter 1

# *Energy Efficiency Policies and Strategies for the Developing Asia to Cope with Rising Oil Prices*

**Brahmanand Mohanty**
*Regional Adviser for Asia*
*The French Environment and Energy Management Agency (ADEME),*
*13 Zamindar Garden, Puducherry – 605 001, India*
*E-mail: mohanty@vsnl.com*

## ABSTRACT

Around 85 to 90 per cent of the world's energy demand growth 15 years from now will come from infrastructures and facilities not yet developed and products not yet manufactured, majority of them in Asian developing countries like China and India which are more oil intensive and much less energy efficient than their industrial counterparts. There is therefore strong reason to adopt appropriate policies and implementation measures to lower the energy as well as oil intensity, through the adoption of structural changes, energy efficiency (regulations and standards, financial incentives and disincentives, institutional strengthening, capacity enhancement, awareness creation, training, etc.), and fuel diversification. Experiences of selected industrialized and developing countries have led to the conclusion that it may be easier to build an oil frugal economy from the start than to convert an oil economy into an efficient one. Asian countries that are late starters in industrialization have much to gain because of less energy needed to produce a unit of GDP today thanks to the technological progresses, improved knowledge and better energy management policies and strategies. This paper elaborates on the challenges to reduce the oil price vulnerability of Asian developing countries that are highly dependent on oil for their economic development. It then outlines strategies and the corresponding policies and tools that the countries can adopt in order to build an oil-frugal economy. The last part deals with the oil price vulnerability of the poor and recommends policies and strategies to achieve higher energy/oil productivity not only to protect the poor from oil price volatilities but also help in improving the HDI and achieving the MDGs.

*Keywords: Rising oil prices, Energy efficiency, Policies and Strategies, Asia.*

## Asia's High Dependance on Oil

Data available for Asian countries from 1995 to 2005 show that there is a linear correlation between the growth in oil consumption per capita and the growth in GDP (Gross Domestic Product) per capita, with the exception of a small number of Asian countries at the bottom and top of the graph (Figure 1.1).

This is the outcome of industrialization of Asian developing countries, relocation of industries from industrialized countries, rapid growth in personal and freight transport, and increased agricultural mechanization along with more land brought under cultivation. Also, countries achieving high level of economic growth have shifted from non-commercial energies and in some cases, from solid to liquid or gaseous fuels (China, India, Sri Lanka and Vietnam). With the increase in per capita income, there is a phenomenal growth of the personal modes of transport and other energy using appliances (Vietnam, for example). The fall in the per capita oil consumption in Pakistan and Philippines is associated with very little growth in per capita GDP, proving the fact that oil intensity may have to rise to ensure economic growth in countries with low per capita GDP. In the case of Myanmar, oil consumption per capita has gone up marginally in comparison with the growth of GDP per capita. This could be due to the State earning from the export of oil and gas as well as other natural resources without an appreciable rise in domestic oil demand to slow down growth of non-oil industries, transportation and other oil-consuming sectors.

The high rate of growth of per capita oil (expressed in ton of oil equivalent or toe per capita) in developing Asian countries with low to medium economic growth rates eclipses the fact that there is a large disparity in commercial energy consumption (the relatively few rich using more commercial energy such as electricity and oil) and a large population of the region does not have access to modern energy.

Figure 1.2 shows the changes in the share of oil in total primary energy supply between 1995 and 2005. Countries whose share of oil in the Total Primary Energy Sources (TPES) has dropped over the period are shown on the right. In economically more advanced countries like Malaysia and Thailand, the dependence on oil is reducing due to the substitution of oil by natural gas and coal for power generation and growth of the less energy/oil intensive service sector. In Pakistan and Philippines, the drop in the share of oil is associated with poor economic growth. All other countries have seen a rise in the share of oil, the increases being higher for Sri Lanka (more thermal power plants) and Vietnam (oil for transport and for industrial steam).

The evolution of the oil intensity of GDP for some Asian countries is shown in Figure 1.4 along with the per capita GDP, expressed in USD at 2000 prices and Purchase Power Parity (PPP). While Figure 1.1 showed an increase of per capita energy demand in many countries, here one can observe that the oil intensity has

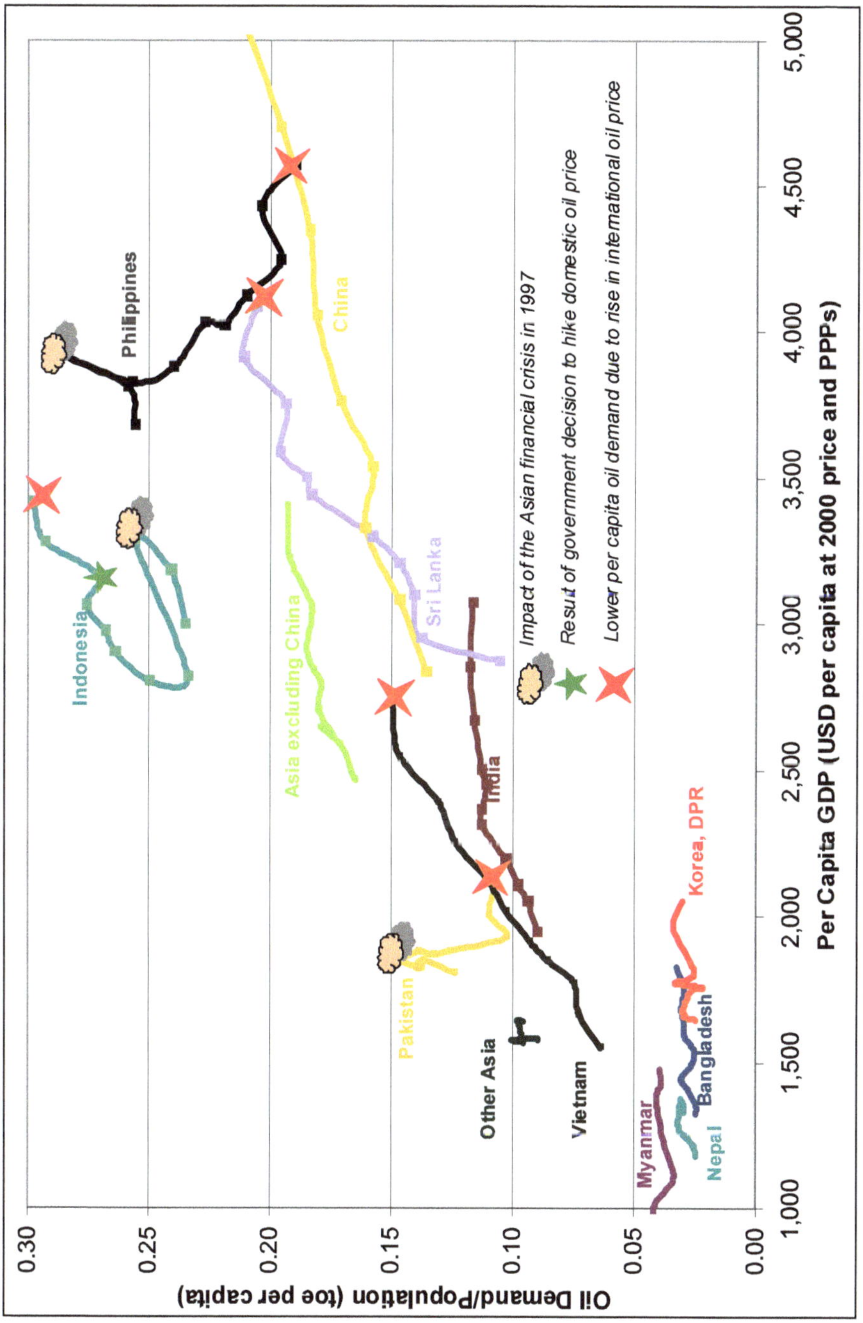


**Figure 1.1: Evolution of Per Capita Oil Consumption and Per Capita GDP of Selected Asian Countries During 1995–2005**
**(*Data Source*: IEA Energy Statistics, Energy Balance for Non-OECD Countries 2007)**

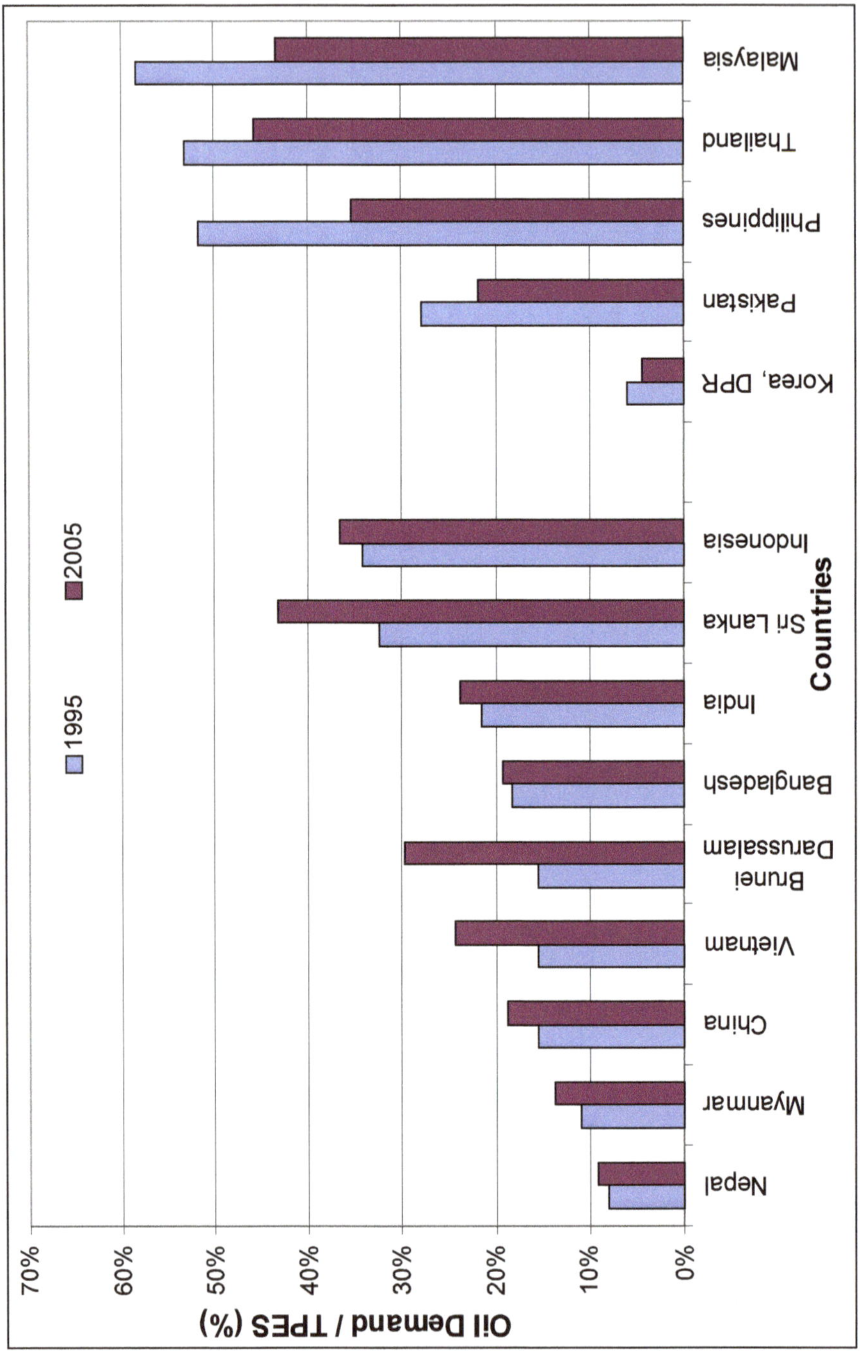


**Figure 1.2: Changes in the Share of Oil in Total Primary Energy Consumption Between 1995 and 2005**
**(*Data Source*: IEA Energy Statistics, Energy Balance for Non-OECD Countries 2007)**

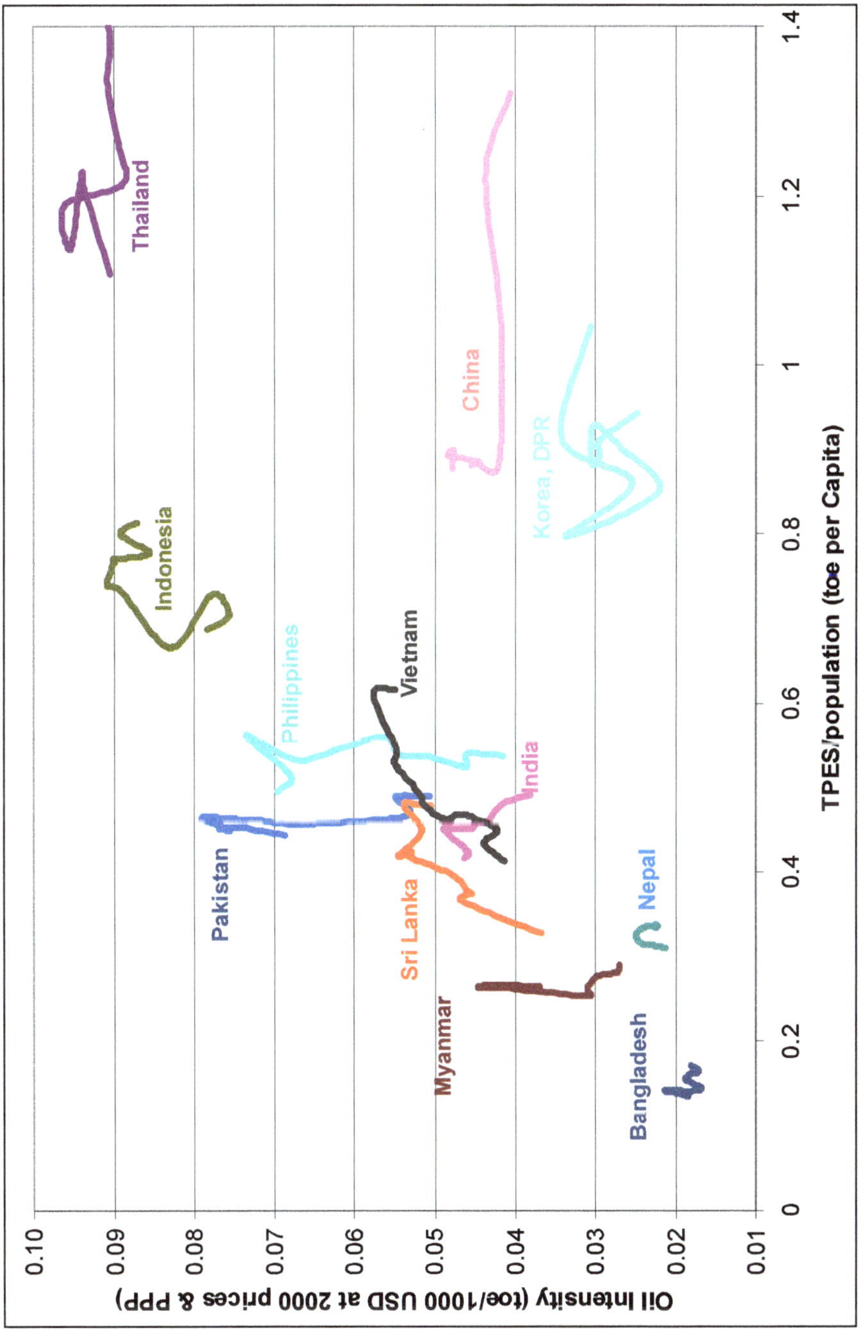


**Figure 1.3: Evolution of Oil Intensity Versus TPES Per Capita Between 1995 and 2005 (*Data Source:* IEA Energy Statistics, Energy Balance for Non-OECD Countries 2007)**

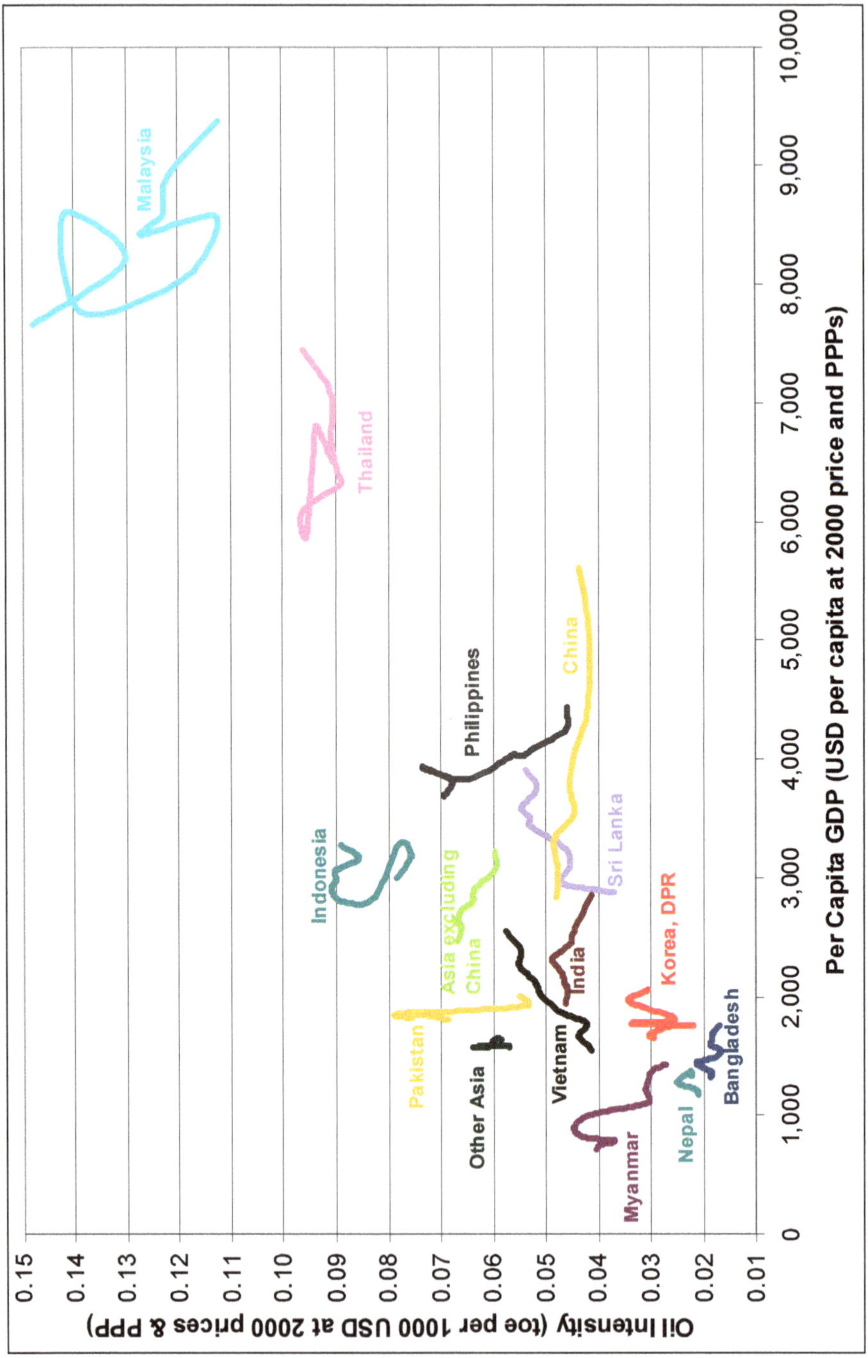


**Figure 1.4: Evolution of Oil Intensity and Per Capita GDP of Selected Asian Countries During 1995–2004 (*Data Source*: IEA Energy Statistics)**

come down or remained flat. This is partially due to the adoption of more efficient technologies in newer facilities, deliberate shift away from oil (*e.g.* industries and agricultural irrigation), and higher contribution of other forms of energy, such as coal, natural gas or hydro.

While OECD countries have the highest demand for oil, the growth of oil demand has been relatively high in the newly industrialized countries of Asia. Trends of the evolution of oil intensity show that China and India have improved their energy productivity to certain extent without sacrificing the per capita GDP. On the other hand, Pakistan and Philippines have not seen much increase in their per capita GDP with the lowering of their oil intensities.

The impact of the Asian economic crisis is reflected on Indonesia, Malaysia and Thailand which have seen their per capita GDP shrink in the late 1990's before rising again. In the case of Vietnam, the growth of per capita GDP is accompanied by a sharp increase in the oil intensity, due to the development process undergoing structural changes. Countries at the bottom left of the figure have a low per capita GDP and a very low oil intensity, because the majority of the population does not have access to or cannot afford commercial energy. "Other Asia" also falls in this category, representing all other smaller countries of the region which are not shown in the figure.

## Challenges of Reducing Oil Price Vulnerability

Energy has been and continues to be the engine for economic growth, especially for the developing countries that need to create the infrastructure necessary to cope with the rapid industrial growth, urbanization and higher standard of living.

Since the oil crisis of October 1973, the world has witnessed several oil supply disruptions, resulting in oil price volatilities and adversely affecting the national economies, particularly those with high oil intensity (oil needed to produce a unit of GDP) and greater reliance on oil import (indigenous oil production – consumption). Asian countries such as China, South Korea, India and Thailand are most at risk because of their high-energy intensities and significant oil import dependence. Also at risk are the small island nations that are highly dependent on oil to meet their energy requirements and a few smaller countries such as Cambodia and Nepal.

Many oil intensive Asian countries have adopted policies to reduce subsidies on petroleum products and let the users pay the additional costs, and this has led to price inflations and affected the poor adversely. Others have resorted to absorbing oil price shocks as a temporary measure to lower the economic burden on their population, and by adopting policies to reduce the dependence on oil products in the medium and long term. Some countries not in a position to absorb the additional economic burden have seriously contemplated rationing to tide over the shortage of fuel supplies, with a potential negative impact on the economic growth and the quality of life.[1]

---

1 Cases of fuel rationing at the petrol stations in the southern part of China, Indonesia and Thailand have been reported during 2005-2006; all government offices in the Philippines were directed to achieve a mandatory 10 per cent reduction in their fuel consumption in April 2005 (ESMAP Report 323/06, 2006).

During the first half of 2008, one has witnessed very rapid oil price hikes due to the worries of oil supply not being able to cope with the escalated growth in demand of the rapidly developing economies. Skyrocketing oil price during the last few months have adversely affected most countries in the Asia-Pacific region, creating uncertainties in food prices, financial markets and cost of living. Perhaps the most important impact of the oil price hike has been on the poor who are at the "bottom of the pyramid". Many of them who had barely managed to afford fossil fuel-based modern energy services have been forced to climb down the "energy ladder".

Small island countries of the Asia-Pacific region have also been adversely affected by the oil price hikes. Most of them are heavily dependent on oil products as the main energy source and have much less alternatives compared to most other countries. The landmass of small island countries is generally not suited for biomass production.

Asian developing countries can learn from the measures taken by the industrialized countries to reduce their dependence on oil. Following the oil shock of the early 1970s, industrialized countries took initiatives and appropriate measures to lower significantly their primary energy use per unit of GDP. In fact, major OECD countries consumed one-third less primary energy in 2000 to generate a unit of GDP compared to 1973, predominantly due to improved energy efficiency in the key economic activities as an outcome of implementing a wide range of policies and programmes (IEA 2004).

Energy efficiency improvements have resulted in multiple benefits at the micro and macro levels: improved bottom-line for the end-users and businesses, increased energy productivity and employment, enhanced comfort and public health, reduced energy cost burden on low-income households, enhanced national energy security. Moreover, EE has become a strategic element to slow down the depletion of conventional oil resources and for the fight against climate change over the last decade.

Since 1990, EE improvements have helped in reducing the global energy demand in 2002 by 20 per cent, which is equivalent to 2 Gtoe ($10^{12}$ toe) of energy (ENERDATA, 2006). Analysis of data available from 15 European countries over the period 1990 and 2002 shows that the EE of final consumers improved by 10 per cent on average, representing about 100 Mtoe ($10^6$ toe) of energy savings. While the primary and final energy consumption increased approximately 0.9 per cent per annum on an average between 1990 and 2003, the GDP grew twice faster during the same period, demonstrating a decoupling between energy use and economic activity. Structural changes in the economy had a marginal influence on the energy intensity reduction of final consumers, about 10 per cent of intensity decrease during 1990 and 2002.[2]

The above results are an outcome of energy efficiency policies and strategies adopted by countries around the world, in close partnership with industry and

---

2 This is due to the advanced state of development in these countries. Structural changes as a tool for energy intensity reduction is also valid for developing countries whose traditional agro-based economies are still evolving and/or whose industrial sectors are undergoing shifts towards services sectors.

consumers, typically as a combination of policy instruments and programmes, classified into the following main categories:

- Legislation and regulations to oblige market actors to lower their energy intensities
- Creation of institutions that are in-charge of implementing EE policies into action
- Assistance for research and development and for demonstration and dissemination of technological innovations
- Fiscal and financial instruments to enhance the adoption of EE measures
- Activities aimed at informing, communicating, creating awareness, education and training, assistance for decision making
- Support to facilitate EE investments, subsidies, fiscal incentives, development of financial mechanisms to enhance the adoption of EE measures through market transformation
- Strengthening the role of energy services, and introduction of voluntary approaches and public-private partnerships, etc.

Some of the above instruments are proven while others were found to be less cost-effective when the full costs and benefits of energy efficiency policies and programmes were taken into consideration. The shortcomings have been overcome by learning from the past experiences, particularly by integrating different policies into comprehensive strategies to transform the market for energy-efficient products and processes, and by taking into account non-energy benefits. What is important to keep in mind is the fact that energy efficiency is but a part of the solution to tackle high oil intensity. As economies prosper and efficient technologies and practices are adopted, and due to shift from high energy consuming industries to stronger tertiary activities, energy demand should decrease. But there is a potential rebound effect on energy demand due to the natural human tendency to opt for a more energy consumptive lifestyle as people's incomes increase, for instance, the tendency to acquire bigger houses, cars and appliances. Once the basic needs of human needs are satisfied, more attention needs to be devoted to influencing peoples' lifestyle and behaviour to pursue the challenging goals of securing energy supplies and ensuring environmental sustainability.

Based on the past experiences, following are some of the options that are available for the developing Asia to reduce the oil price vulnerability:

1. Managing oil price risks: creation of oil price stabilization fund, signing of long-term oil supply contracts, providing oil subsidies and making tax adjustments, designing specific compensation schemes targeting the poor, etc.
2. Enhancing oil supply: Increasing efforts for oil exploration, enhancing refining capacity, investing in foreign oil fields, engaging in barter of manufactured products against oil import, etc.

3. Restraining oil demand: Increasing energy use efficiency and imposing measures to reduce fuel wastage
4. Diversifying energy sources: Switching from oil to other alternatives, both fossil and renewable forms
5. Preparing for emergencies: Rationing oil use, and building strategic oil reserves.

## Strategies to Build an Oil-Frugal Economy

Though the oil prices have seen some downward trend in the 3rd quarter of 2008, it is widely recognized that the oil price is unlikely to go down to the low values we have seen in the past. Keeping this in mind, Asian developing countries should not limit their strategies to only adopting short-term policies of managing oil price risk or enhancing oil supply. Considering the various challenges of population growth and dwindling natural resources, the time is ripe now for these countries to look seriously at options to improve energy efficiency and switching to other alternatives, mainly renewable energies which serve the dual purpose of achieving an oil-frugal as well as a low-carbon economy.

Four broad strategies that are proposed to achieve an oil-frugal economy include:

1. Making structural changes by shifting to less oil intensive activities, from manufacturing to services, for example.
2. Reducing oil intensity by changing the patterns and practices of production and consumption
3. Upgrading the technology of production and conversion equipment, and making consumer devices more efficient
4. Displacing oil with other alternative forms of energies

With the exception of the small island nations where oil is used for meeting practically all the energy demands, the most significant determinant of oil dependence is transportation because the largest share of the increase in oil use is accounted for by the transport sector. In fact, the global oil consumption in transport increased four times faster than consumption in industry from 1971 to 2003 (IEA, World Energy Outlook 2005). Hence more emphasis needs to be given on addressing the rise in oil demand in the transport sector. Table 1.1 briefly outlines an inventory of measures for each of the four proposed categories that developing Asian countries could adopt to attain an oil-frugal economy.

Implementation of the above strategies would require that the governments play a decisive role by formulating and adopting coherent policies to steer in the direction by:

- Strengthening the existing institutions to enable them to develop and implement effective policies,
- Allocating adequate human and financial resources to achieve the goals,
- Expanding the choice and adoption options,

**Table 1.1: Strategies for the Transport Sector to Achieve an Oil-Frugal Economy**

| *Making Structural Changes* | *Reducing Oil Intensity* | *Adopting more Efficient technology* | *Substituting Oil* |
|---|---|---|---|
| **Transport Sector** | | | |
| • Urban mobility plan<br>• Intercity freight transport plan<br>• Transport plan for target groups<br>• Efficient intermode transport<br>• Improved land-use and smart growth<br>• Road use pricing<br>• Intermodal and mass transit options<br>• Parking fee | • Reduce travel<br>• Use smarter and efficient vehicles<br>• Use on-board fuel efficiency technology<br>• Focus on pedestrian neighbourhood<br>• Encourage car-pooling<br>• Control vehicle speed<br>• Improve mass transit service<br>• Encourage telecommuting<br>• Promote efficient driving style | • Vehicle efficiency<br>• Hybrid cars<br>• Use of advanced materials<br>• Intelligent highway systems<br>• Telematic system for freight transport<br>• Railways and marine transport | • Biomass liquid fuel<br>• Fuel blending<br>• Promote natural gas |
| **Industry Sector** | | | |
| • Low oil intensive process<br>• Recycle energy intensive products<br>• Outsource energy intensive products<br>• Demand side management | • Reduce peak electricity demand to avoid peaking power plant | • Improve efficiency of boilers and furnaces<br>• Adopt cutting-edge industrial technologies<br>• Ensure process optimization and process integration<br>• Explore thermal cascading<br>• Adopt district energy system<br>• Increase energy and material recovery | • Save natural gas and electricity to displace oil<br>• Substitute petrochemical feedstock by biomaterials<br>• Substitute oil by coal as feedstock<br>• Substitute oil as fuel for boiler to coal, natural gas and renewable energy |
| **Agriculture Sector** | | | |
| • Co-operative farming<br>• Intensive cultivation<br>• Marginal land cultivation | • Encourage efficient tractor and pump-set operation<br>• Promote drip and night irrigation | • Improve efficiency of diesel operated tractors<br>• Improve efficiencies of motors and pumps<br>• Develop hybrid systems to reduce dependence on oil | • Replace oil by biomass liquid fuel<br>• Accelerate rural electrification for pumping and irrigation<br>• Use hybrid systems to reduce dependence on oil for pumping |

*Contd...*

**Table 1.1–Contd...**

| *Making Structural Changes* | *Reducing Oil Intensity* | *Adopting more Efficient technology* | *Substituting Oil* |
|---|---|---|---|
| **Household Sector**<br>• Shift from home-based micro enterprises to community enterprises<br>• Shift from household cooking to community cooking | • Reduce heating for comfort in cold climates<br>• Create awareness for cooking fuel saving | • Design more efficient kerosene stoves<br>• Replace kerosene lighting by LED lighting | • Replace kerosene stove by other alternatives (LPG, biogas, coal/ briquette)<br>• Introduce more efficient biomass stove or solar cookers |

- Accelerating the development of advanced technologies to achieve greater business success and national energy security goals,
- Fostering alternative energy supplies,
- Recognizing and rewarding those who help phase out oil and contribute to non-energy benefits, especially targeted towards the poor,
- Enhancing competition between different modes of energy services as well as the ways to need them less.

## Energy Efficiency Policy: Tools and Instruments

For an effective implementation of policies and strategies, the following are essential: setting realistic targets and time-bound programmes, suitable implementing organizations, well-defined targets and actions, adequate human and financial resources, and monitoring of results achieved for developing better future programmes. For instance, in the transport sector, an inter-ministerial agency may be created, representing various ministries that are concerned with oil use in the transport sector: transport, public works, health, environment, etc. The different target groups could be big transport operators, public transport service providers, large enterprises, and the general public. The types of action could be linked to long distance, inter-city or urban travels.

The energy efficiency policy tools and instruments that can be employed to lower the oil intensity are broadly classified into 3 groups:

1. Legislation and regulation (laws, standards and specifications, mandatory quotas, targets and obligations)
2. Fiscal and financial measures (subsidies, taxes, duties, financial incentives/disincentives)
3. Direct public investment (investment in social infrastructure, market/trade infrastructure, technology R&D, demonstration and deployment, technology transfer and capacity building, institutional/human resource development)

Given the diverse circumstances in which Asian developing countries find themselves – different income levels, budgetary situations, amounts spent on price subsidies, availability of indigenous petroleum resources, sources of electricity, and the impact of weather on hydropower, to mention a few

- There is no simple or universal strategy for dealing with different oil prices. Moreover, no single government policy step can be expected to meet multiple government objectives but a package of measures needs to be formulated, whereby each measure may make only a small contribution but their combined impacts would be significant (ESMAP, 2006).

Table 1.2 recommends policies and strategies for energy conservation/efficiency improvement to reduce the oil intensity of regional economies.

**Table 1.2: Energy Efficiency Policy Tools and Instruments to Achieve an Oil-Frugal Economy**

| *Legislation and Regulation* | *Fiscal and Financial Measures* | *Direct Public Investment* |
|---|---|---|
| **Transport Sector** | | |
| • Mandate relevant authorities to develop transport master plans<br>• Adopt minimum fuel-efficiency standards and mandatory efficiency labelling<br>• Set target dates and penetration rates for biofuels and other cleaner fuels | • Remove price distortions and eliminate subsidies that mostly benefit the rich<br>• Introduce incentives for "scrap and replace" programmes aimed at low-income households<br>• Provide financial assistance for urban mobility plans to various stakeholders<br>• Provide financial assistance to purchase fuel-efficient public transport vehicles<br>• Impose on private vehicles higher fuel taxes that reflect the social cost of travelling<br>• Set annual registration taxes for cars on the basis of engine size, weight and efficiency<br>• Offer incentives for procurement of new technologies | • Strengthen existing institutions<br>• Set up observatories to gather relevant data and benchmarking<br>• Undertake "well-to-wheel" analysis<br>• Launch driver-training and mass awareness programmes<br>• Support R&D to develop enabling technologies<br>• Support R&D and help manufacturers to improve production facilities |
| **Industry Sector** | | |
| • Oblige power utilities to buy electricity from, and sell to cogenerators<br>• Allow cogenerators to sell electricity and thermal energy to third parties | • Give rebates for fuel-efficient appliances, by charging fees on inefficient ones<br>• Support energy audits for identifying oil saving measures<br>• Provide incentives and tax credits for investments in combined heat and power schemes<br>• Introduce time-of-use tariffs to avoid use of oil-fired power plants during utility peak periods | • Strengthen existing institutions<br>• Set up observatories to gather relevant data and to benchmark<br>• Support R&D to develop innovative processes, thermal integration and heat recovery |
| **Agriculture Sector** | | |
| • Mandate the creation of an oil saving fund to reduce oil demand in agriculture | • Allocate funds for technology demonstration and deployment<br>• Provide incentives to develop infrastructure for producing fuel derived from biomass | • Strengthen existing institutions<br>• Establish centres to train technicians who can assist farmers<br>• Support the development of rural energy service companies |

*Contd...*

**Table 1.2–Contd...**

| *Legislation and Regulation* | *Fiscal and Financial Measures* | *Direct Public Investment* |
|---|---|---|
| **Household Sector** | | |
| • Set minimum standard and develop labels for kerosene and LPG stoves and heaters<br>• Ask industry to produce fuel-efficient stoves and apply penalty clauses for non-compliance<br>• Introduce energy conservation codes for new buildings | • Remove price distortion and eliminate subsidies that benefit mostly the rich<br>• Allocate funds for technology demonstration and deployment | • Strengthen existing institutions<br>• Support training of rural youth to provide and maintain efficient lighting systems<br>• Use communication media to educate rural populations about the scope for oil saving<br>• Provide R&D support to develop high-efficiency cooking stoves and reliable and cost-effective, energy-efficient electric lamps |

Developing countries tend to focus more on laws and regulations because of the perceived low costs of their implementation. However, laws and regulations alone will not be adequate if there are no strong financial mechanisms in place to trigger the energy efficiency market or if the institutional, market and social infrastructures are not properly developed. Governments may do well to choose a set of policy measures to achieve the best result.

## Oil Price Vulnerability of the Poor and Safeguarding their Interest

While the strong correlation between energy/oil use and economic development was emphasized in the earlier part of the paper, energy is also considered as an essential ingredient of social development. The United Nation's Development Program (UNDP) has defined the Human Development Index (HDI) to measure how well a country is meeting not just the economic but also the social needs or the quality of life of its population. The HDI is calculated on the basis of a simple average of life expectancy, educational level, and real per capita GDP.

Figure 1.5 highlights the impact of primary energy per capita on the HDI for the Asian countries at different levels of development. At the top end are Malaysia and Thailand, with a higher HDI ranking associated with the primary energy per capita exceeding 1.5 toe/capita. At the bottom end are Bangladesh, Myanmar and Nepal with poor HDI ranking associated with primary energy per capita below 0.5 toe/capita. It is interesting to observe that for countries with very low per capita energy consumption, a small increment in the energy use results in a substantial HDI improvement, satisfying the basic needs of a vast lower income population. On the other hand, countries such as Malaysia and Thailand (and China, to some extent) have experienced quite important increase in per capita energy consumption due to structural changes and changes of lifestyle of the population benefiting from the higher economic growth (rise in transport demand and the use of household electrical appliances).

Energy in itself is not a basic human need but lack of access to modern forms of energy can result in poor health care, education, transport and communication facilities. Figure 1.6 shows the negative impact of low oil consumption on the HDI, especially for small island nations and in countries where a vast population lives in isolation from the electric grid and in the absence of traditional energies such as biomass, petroleum products remain the main source of energy for inter-island transport, power generation or even for lighting.

Energy poverty goes hand in hand with general poverty. Countries with low HDI generally have oil consumption below 1 barrel per capita. If one focuses much on reducing the oil intensity or oil dependence of such countries with very low per capita oil demand, there is a danger of denying the level of comfort, convenience and economic welfare demanded by the majority of consumers due to poor energy accessibility. Provision of greater quality of modern energy services to the poorest of the poor to meet their basic human needs will be a giant step towards achieving the Millennium Development Goals (MDGs) but have only a marginal impact on greenhouse gas emissions (UNDP, 2006).

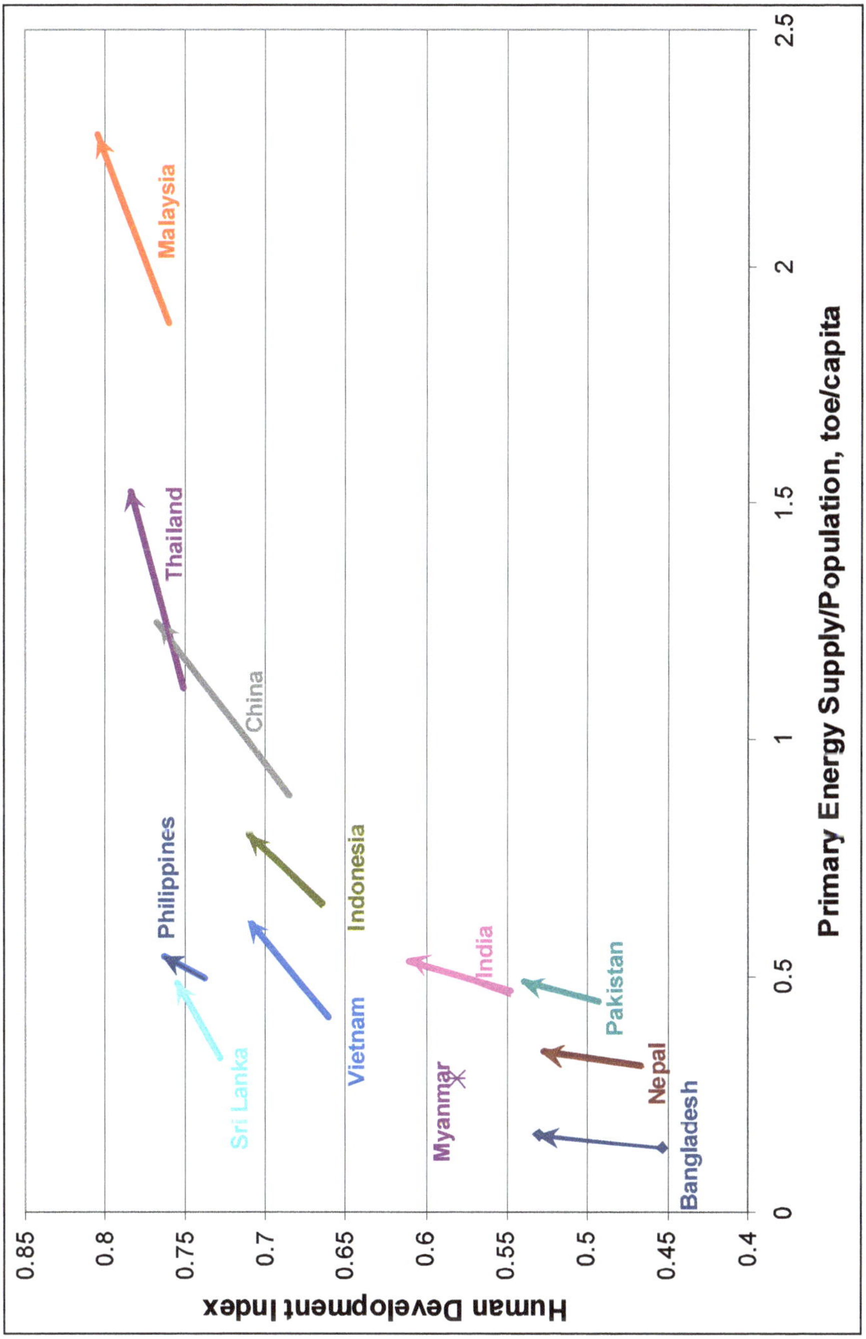


**Figure 1.5: Evolution of the Human Development Index and the Primary Energy Demand Per Capita Between 1995 and 2004**

**(*Data Sources*: IEA Energy Statistics, UNDP 2004)**

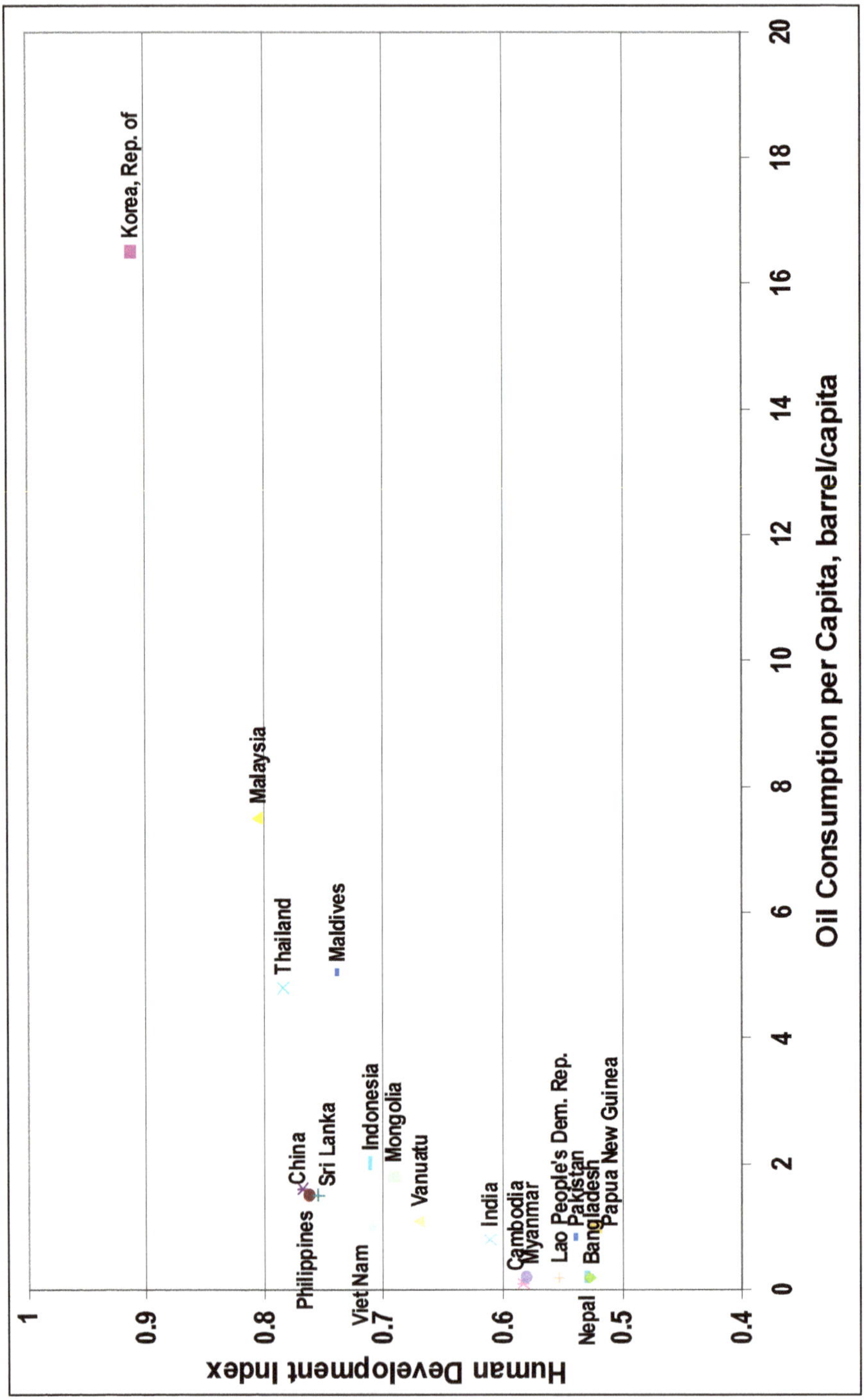


Figure 1.6: Correlation Between Oil Consumption Per Capita and the Human Development Index for 2003

(*Data Sources*: ADB 2005, UNDP 2004)

Energy security for the poor should go beyond providing energy for subsistence. There is increasing recognition that energy access to the poor can increase their livelihood opportunities, production capacities and income so that they can afford clean and convenient energy sources (Planning Commission of the Government of India, 2006). There are strong relationships between energy availability, accessibility and affordability for achieving the MDGs of eradicating extreme poverty.

Analysis of the final energy consumption of population with the share of population living below poverty line shows that kerosene and liquefied petroleum gas (LPG) are transition fuels and with the increase in income, these are replaced by natural gas or electricity (IEA World Energy Outlook 2002). Depending on geophysical characteristics of countries and the availability of alternatives to oil, several end-uses can be satisfied without growth in oil consumption while seeking higher HDI. However, oil would remain essential to satisfy many end-uses such as cooking in poor and intermediate groups of households, rural (including agricultural) transportation, irrigation (in countries with very low level of rural electrification rates), decentralized power generation (in island countries and others with low rural electrification rates), manufacture of chemical fertilizers and petrochemical products.

When the oil consumption increases to 2 barrels per capita, it can support a reasonable level of development with a substantial improvement in the HDI, provided it is efficiently used to improve the energy productivity, measured in terms of the output and quality of goods and services generated with a given set of energy inputs. Experience shows that the poorest are likely to benefit the most from the basic energy services extended to them. The demand for energy services is rising rapidly in developing countries; reducing energy investment requirements and making the best use of existing supplies to improve energy accessibility should rank high in their priorities (WEC 2006).

A sizeable population in Asia-Pacific countries faces poverty and the challenge of not having access to modern energy services that are essential for economic growth and human development. Policies and strategies other than those outlined in the above section would be necessary to protect these vulnerable poor people from the adverse impacts of the oil price volatilities. A recent study conducted by UNDP in China, India, Indonesia and Lao PDR showed that the fuel price hikes during the period 2002 to 2005 had very adverse impacts on the poor households as they experienced a very steep rise in the price of basic energy services. They had to pay 171 per cent more for cooking fuels, 120 per cent more for the transportation, 67 per cent more for electricity, 55 per cent more for lighting fuels and 33 per cent more for oil-based fertilizers and other agricultural inputs (UNDP 2007).

There is thus an urgent need to develop the necessary energy infrastructure, to make suitable energy carriers available and affordable, and to deploy efficient technologies for assisting the deserving poor to escape the vicious cycle of poverty perpetuated by the lack of access to modern energy sources. The latter have the potential to reduce poverty by raising agricultural and industrial productivity, extending the quality and range of products, and by increasing the earning of the deprived.

Energy related problems are quite different for the poor living in villages than those in the urban or peri-urban areas. With the exception of a handful of Asian developing countries, most rural population is dependent on traditional biomass for cooking, heating (using very inefficient conversion technologies and processes, and contributing to serious health impacts from indoor smoke pollution) and rural industries, kerosene for lighting, and diesel for water pumping, agriculture, rural transport, and even electricity in most island countries and remote rural areas. Kerosene and LPG for cooking, diesel for power generation, or electricity for lighting, pumping or refrigeration are either not available or not affordable. On the other hand, though the poor in urban and peri-urban areas may have access to modern forms of energy, such as kerosene and LPG for cooking, electricity for lighting, pumping and refrigeration, or diesel for transport, many of them may still resort to more inefficient alternatives such as fuel wood for cooking or kerosene for lighting because of the poor quality of energy infrastructure, fuel affordability or the high first-cost of equipment and appliances. Poor urban planning and renewal and lack of convenient public transport forces the poor to spend more money on commuting for work or studies. Some opt for alternative modes of transport, such as motorized two-or three-wheelers that are inefficient in design and lack regular maintenance.

Government policies to reduce the economic burden of the poor by providing subsidies on selected fossil fuels have often led to perverse effects: (1) subsidies for diesel and kerosene are subject to high leakages, with the more affluent benefiting more than the poor; and (2) subsidies in energy prices increase wasteful and nonessential use of energy and do not encourage market penetration of more energy-efficient technologies. At the macro level, across the board subsidies on petroleum products worsen the government's fiscal position and less government funding is available for undertaking other desirable activities aimed at poverty alleviation and meeting MDGs.

Biomass is projected to dominate energy demand in many Asian developing countries. When the wood and biomass become scarce due to the demand exceeding supply, the economically better-offs are likely to switch to kerosene and boost the demand for oil products. There is indeed great scope for improving the use of biomass for specific applications through the adoption of more modern and efficient technologies that in turn will alleviate poverty and the adverse impact on health. Moreover, it can help to boost the rural market and create more employment in rural areas. However, investment remains the key for the technological changes to occur and governments will have to play a crucial role by supporting the technology uptake by adopting comprehensive strategies and programmes.

There are several policies and strategies that governments can formulate to "kick-start" the adoption of technologies and fuel switching in order to assist the poor to go up in the ladder of development. To start with, let us analyze the options to deal with a common situation prevailing in many villages:

- An efficient electric lamp can have 100 times higher luminous efficacy than the kerosene-based lighting, and yet hundreds of millions of households depend on kerosene for lighting because of the lack of access to reliable and affordable electricity.

- On the other hand, energy required for cooking is 4 to 8 times less when wood is replaced by kerosene or LPG as fuel; but the poor continues to use wood or biomass for cooking because of the lack of capital to invest in an efficient kerosene/LPG stove and to pay for the high recurring cost of fuel.
- Biomass-based gasification technology for producing electricity at small scale has reached a level of maturity where it can produce cheaper power than the diesel-operated decentralised power generators.
- While the fuel needed to generate power would come in the form of wood and biomass that is presently used by the rural population for cooking, the kerosene saved from the lighting applications may be channelled to cooking.
- Such fuel cross-substitution (from wood to kerosene for cooking and from kerosene to wood-based power for lighting) accompanied by the adoption of highly efficient conversion technologies, would help to provide better and cleaner energy services to a larger population at a more affordable cost, using the existing energy sources.
- Both wood energy generation and electricity application in rural industries will help create rural employment and boost rural economy, lowering hardship for women and children, and reducing the pressure of migration from the thinning rural to over-populated urban areas.

Following are some of the possible government interventions to realize such options:

- Support rural entrepreneurs, with injection of capital, to develop proven and energy-efficient biomass-based power generation facilities in rural areas for lighting, irrigation and other electricity-based services.
- Provide training and education to rural unemployed youth for operating and maintaining decentralized power generation facilities, and providing reliable and cost-effective energy services to the local population on a fee-paying basis.
- Assist the rural population to acquire fuel-efficient kerosene stoves for cooking through support for R&D, testing, prototype and market development.
- Issue debit cards/smart cards to targeted households so that they get a given cash amount for the purchase of their monthly entitlements of kerosene.
- Design programme to engage the poor villagers in sustainable agro-forestry and wood plantation activities for the provision of indigenous energy to generate electricity (or even to run combined heat and power facility) in rural areas that are deprived of grid electricity; energy available would support rural agro-processing and small/micro enterprises.

Another such example is the development of a sustainable biofuel programme in rural areas with the following benefits:

- Reduced dependence on imported fossil fuel.

- Rural population that would eventually be engaged in growing the plant species for ethanol or bio-diesel production will have improved livelihood.
- Biofuels produced in poor quality uncultivable land can be used judiciously in rural development applications (rural transport, pump irrigation, decentralized power generation, etc.), fostering growth, reducing poverty and improving the quality of life of the poor.[3]

Possible government interventions to make this happen include the following:

- Introduction of land-use policy reforms, allocation of community land and provision of technical assistance and cash-subsidy to the rural poor willing to undertake energy plantation to produce biofuels
- Creation of village energy cooperatives to coordinate the sustainable supply of biofuel
- Supporting R&D to develop high-yield plant species and convert the plant produce into high-grade liquid fuel
- Assisting entrepreneurs involved in modifying rural transport vehicles, farm tractors, pumps and other motive devices that would work well with biofuel as energy source

A vast population in rural and peri-urban areas in Asia-Pacific region makes considerable efforts to gather animal manure, prepare it for use as fuel which burns inefficiently, creates smoke and indoor air pollution and adversely affects human health. Individually owned or collective/community biogas plants allow for anaerobic digestion of the animal manure and other organic wastes to produce biogas that will lead to multiple benefits:

- Provide clean energy for cooking, power generation or small-scale rural industries
- Spent slurry from the biogas digester with high carbon to nitrogen ratio can be used as an excellent plant nutrient, thus lowering the need for chemical fertilizer that is produced by using oil as feedstock
- Save the time and efforts needed to prepare the fuel
- Improve the combustion efficiency, thus allowing for more useful energy generation from the same amount of fuel, and even substituting diesel as fuel for decentralized power generation from community biogas plants
- Such schemes are already in operation in quite a few Asian countries but in the absence of sustained government policies and support, their numbers are quite limited. Government can provide support in various forms:

---

3 The Brazilian ethanol program cut 1975-2002 oil imports by a cumulative undiscounted total of US$50 billion (2000 USS) which is more than 10 times the total undiscounted 1975-89 real investment, and 50 times its cumulative 1978-88 subsidy; while ethanol has replaced about a quarter of Brazil's gasoline, the program has also provided nearly 700 000 jobs in 2003 (*Source*: Lovins *et al.*, 2005)

- Undertake R&D programmes to optimize the design of biogas digesters of various capacities to function with different qualities of organic wastes
- Provide technical and financial assistance to poor families and communities for developing individual or collective biogas plants
- Support R&D for developing efficient cookstove and power generators to operate with biogas as energy source, etc.

A package of policy measures is needed to protect the poor from energy price volatilities. Asian developing countries would do well to plan for the eventuality of future oil supply disruptions and higher oil prices and formulate policies that can alleviate the financial burden of the poor. Another key strategy is to support the development of energy service entrepreneurs who are able to aggregate demand across multiple social and income-generating needs within the rural community and rapidly scale up energy services. Thanks to the adoption of energy efficient techniques and practices, and innovative payment and recovery mechanisms, they can provide better service to their clients at lower cost while gaining higher benefits for themselves.

## Conclusions

Oil is a strategic commodity as all countries depend on it as a fuel of choice for the transport sector, even though there have been fuel switching to natural gas or LPG, and biofuel, as in the case of Brazil. The growth of oil demand is a function of several parameters: oil price, economic growth, changes in lifestyle, technologies of oil-using equipment and appliances, availabilities of substitutes at a competitive prices, etc.

Although there are high uncertainties regarding the duration and the magnitude of the increase in world oil prices, there is enough evidence to raise concern about their impacts on the fast-growing Asia-Pacific region, which produces about 11 per cent of the world's crude oil but consumes more than 20 per cent. Among the many policy stances that would be beneficial under all price scenarios, energy efficiency and energy diversification appear as important "no-regret" strategies to strengthen the economy's ability to cope with the higher oil prices or oil shocks. Regional energy policy makers may take note of some of the suggested strategies and policies to lower the oil intensity and pursue them with earnestness in order to avoid jeopardizing future growth.

A sizeable population of the region faces poverty and the challenges of not having access to modern energy services that are essential for economic growth and human development. Regional policy makers can better plan for the eventuality of future oil supply disruptions and higher oil prices by adopting some of the recommended policies and strategies to alleviate the financial burden of the poor and to achieve the MDGs.

Improved energy services can provide the low-income population with cheaper energy and higher-value energy products. A major challenge faced by the governments is to bring down the cost of providing clean and reliable energy for households and productive uses, and on finding ways and means to overcome the hurdle of high initial investment. There is considerable opportunity for developing innovative marketing and delivery approaches but access to finance remains a critical issue.

Industrial countries are striving to meet the emission targets set by the Kyoto Protocol. Asian developing countries can help them meet this challenge if the latter shared their policy experiences and best practices: a wide range of low-cost technological options that is already available in industrialized world could be effectively transferred, helping to reduce the risks associated with specific technologies and facilitating sound investment decisions. Finally, global climate change-related funding mechanisms offer help for agriculture, energy savings and efficiency, renewable energy and better technologies for transport and industries. Rapidly developing Asia-Pacific countries should develop clear roadmaps for exploiting such global financing mechanisms to reduce their oil dependence and achieve the MDGs.

## Acknowledgements

The contents of this paper are mainly derived from a more comprehensive unpublished background paper that was prepared by the author in 2007 as input to the UNDP publication entitled "Overcoming vulnerability to rising oil prices: options for Asia and the Pacific".

## References

1. ADB [Asian Development Bank]. 2005. The Challenges of Higher Oil Prices, Manila, Asian Development Bank Outlook 2005 Update.
2. EIU [Economist Intelligence Unit]. 2006. World Energy: EIU's 2006 Oil Intensity Index.
3. ESMAP [Energy Sector Management Assistance Programmes]. 2006. Coping with Higher Oil Prices, ESMAP Report 323/06.
4. Geller, H. and Attali, S. 2005. The Experience with Energy Efficiency Policies and Programmes in IEA countries: Learning from the Critics, IEA Information Paper.
5. Gitonga, S. and Clemens, E. (eds.). 2006. Expanding Access to Modern Energy Services: Replicating, Scaling Up and Mainstreaming at the Local Level, New York, United Nations Development Programme.
6. IEA [International Energy Agency]. 2002. Chapter 13: Energy and poverty, World Energy Outlook, Paris, France: OECD/IEA.
7. IEA [International Energy Agency]. 2004. Oil Crises and Climate Changes: 30 years of Energy Use in IEA Countries, Paris, France: OECD/IEA.
8. IEA [International Energy Agency]. 2005. Saving Oil in a Hurry, Paris, France: OECD/IEA.
9. IEA [International Energy Agency]. 2005. World Energy Outlook, Paris, France: OECD/IEA.
10. Lovins, A., Datta, E. Kyles, Bustnes, O.E., Koomey, J.G. and Glasgow, N.J. 2005. Winning the Oil Endgame: Innovation for Profits, Jobs and Security, Rocky Mountain Institute.

11. Modi, V., McDade, S., Lallement, D. and Saghir, J. 2005. Energy Services for the Millennium Development Goals, New York, Energy Sector Management Assistance Programme, United Nations Development Programme, UN Millennium Project, and World Bank.

12. Planning Commission of the Government of India. 2006. Integrated Energy Policy: Report of the Expert Committee, New Delhi.

13. UNDP [United Nations Development Programme]. 2004. The State of Human Development, Human Development Report 2004.

14. UNDP [United Nations' Development Programme]. 2007. Overcoming vulnerability to rising oil prices: options for Asia and the Pacific, UNDP Regional Centre, Bangkok.

15. WEC [World Energy Council]. 2006. Energy Efficiencies: Pipe dream or Reality? London, WEC Statement 2006 for Asia and the Pacific, UNDP Regional Centre, Bangkok.

# Chapter 2

# *Government's Initiatives on Energy Conservation and Efficiency in Tanzania*

**Hosea A. Mbise**
*Ministry of Energy and Minerals, Dar-es-Salaam, Tanzania*
*E-mail: mbise@mem.go.tz*

## ABSTRACT

Energy conservation and efficiency activities in Tanzania started about 25 years ago and the Ministry of Energy and Minerals through its Energy Department has actively been promoting related activities during the last two decades. More emphasis has mainly been the industrial, household and commerce sectors although potential exists in the others sector including transport, commerce, mining and agriculture. Barriers to energy conservation and efficiency have been lack of awareness, financial constraints and lack of legal enforcement. Efforts are underway to find means and ways of encouraging utilization of energy efficient appliances and equipment.

This paper focuses on energy efficiency in electricity supply and demand sides. The total annual power consumed by about 800, 000 customers in Tanzania is about 4,000 GWh, while the generated amount is close to 5,000 GWh. Industrial, commercial and residential consume about 93 per cent of the total electricity delivered.

Most of the electrical equipment and appliances imported to Tanzania are inefficient because there are no energy efficiency standards and label. Major end-users of electrical energy include motors, air conditioning, lighting, fridges, freezers, electric cookers, water heaters, air compressors, pumps, ICT systems, dishwashers, washing machines, etc.

It has been established that a saving of more than 25 per cent through introduction of energy efficiency standards and labels for all the appliances in the market, which translates to up to 200 MW saving! In monetary terms, this could be more than US$ 350 million.

On the other hand, replacement of for instance, incandescent lights in all Tanzanian residential households connected to electricity supply system can result into a saving of more than 40 MW, which is more than the installed total capacity of all isolated diesel power stations in the whole country.

Improving energy efficiency in electricity use will have three benefits:

- Supply more consumers with the same electricity production capacity;
- Slow down the electricity demand growth, and reduce the investment needed for the expansion of the electricity demand.
- Environment protection

The paper concludes by urging for the need to promote Efficiency Standards and Labels to enhance energy security. Efficient equipment and appliances help energy and money saving as well as protection of the environment. The initiative is also considered to be a very sustainable route towards poverty reduction.

***Keywords:*** *Energy conservation and efficiency, The Ministry of Energy and Minerals, National Cleaner Production Centre and Renewable energy.*

## Introduction

The roles of the Tanzanian Government in the energy sector include facilitation of development, provision of conducive environment for private investment and initiatives; promotion of effective regulation; as well as monitoring and co-ordination of sectoral activities and programmes.

The Ministry of Energy and Minerals (MEM) which is responsible for implementing Government's policies, oversees the activities of the different institutions operating in the energy sector. MEM supervises implementation of the national energy policy, which is backed by legislation and regulations.

Specific roles of the MEM as far as the energy sector is concerned include:

1. Provision of guidance on sectoral development;
2. Policy formulation and facilitation towards implementation; and
3. Advising on legislation, regulation and fiscal matters related to the sector;

Major actors or institutions in the energy sector include Tanzania Electric Company (TANESCO), Tanzania Petroleum Development Corporation (TPDC), Energy and Water Utilities Regulatory Authority (EWURA), Rural Energy Agency (REA), Oil marketing Companies (OMC), and Independent Power Producers (IPPs) as well as Petroleum exploration companies

The Energy Sector's Mission is to create conditions for the provision, of safe, reliable efficient, affordable and environmentally appropriate energy services to all sectors on sustainable basis.

Energy Conservation and Efficiency (ECE) activities, projects and programmes are promoted, coordinated, monitored and supervised within the Department of

Energy. The revised National Energy Policy document of 2003 document provides very clear policy statements towards energy conservation and efficiency.

## Importance of Energy Conservation and Efficiency

Energy Efficiency and Conservation (ECE) efforts in Tanzania were necessitated by the following factors:

1. Oil crisis (as with other countries) in terms of availability and prices; Tanzania imports about 1.8 million tonnes of petroleum products, spending more than US$1.4 billion, under the present price of US$140/barrel!
2. Growth in electricity use, which stands at 10 per cent ; as well as growth in petroleum use in industries, transport, household, commercial, mining and agricultural sectors;
3. ECE is regarded as a quick win towards poverty alleviation; and
4. Concern for environmental protection.

It is now clear that the main options to improve energy supply situation in order to meet the current demand in Tanzania are:

1. Energy imports (electricity and oil)
2. Further petroleum (liquid) exploration and development of coal and natural gas utilization;
3. Development and utilization of renewable sources of energy particularly solar, wind, hydro, hydro geothermal, etc.
4. Energy conservation and efficiency
5. The need to import oil, poverty threats, as well as the existing constraints on energy supply have certainly enhanced prioritization of energy conservation and efficiency initiatives in our country.

Energy conservation and efficiency in Tanzania started about 25 years ago. The Ministry through its Energy Department has been promoting related activities during the last two decades. More emphasis has mainly been the industrial sector although large potential exists in the other sector like transport, mining, agriculture, etc.

Improving energy efficiency, for instance in electricity use will have two benefits:

1. Supply more consumers with the same electricity production capacity;
2. Slow down the electricity demand growth, and reduce the investment needed for the expansion of the electricity demand; and
3. Protection of the environment through reduced green house gas emissions

## Selected Experiences of MEM in Energy Conservation and Efficiency in the Past Ten Years

The Ministry (MEM) has executed a number of projects and programmes on energy conservation and efficiency in close collaboration with local institutions and Development Partners during the past two decades. Such institutions include

Universities, research organizations as well as the Power utility company. Facilitation has particularly been made through various Development Partners, including CIDA/SADC, Sida, UNIDO, UNDP/World Bank, GTZ, and USAID.

Specific experiences by the Ministry are shown on the matrix below:

| | |
|---|---|
| 1996 | Commencement of CIDA funded (US$10 million) project, that is SADC industrial Energy Management (SIEMP) training workshops in SADC countries including Tanzania |
| 1999 | a) Two partnership arrangements between two tertiary institutions (DIT and UDSM) and SIEMP were finalised in Tanzania through SIEMP); |
| | b) Revision of the National Energy Policy 1992, to include issues on energy conservation and efficiency more clearly. |
| 2002 | Launching a project on energy audit of Dar es Salaam Water Supply Systems covering three water pumping stations through MEM's financial support. The audit findings indicated a saving of US$ 2. 4 million (out of US$. 6 million) in annual electricity bill giving a pay-back period of less than a year, with a modest investment |
| 2003 | Promotional seminars on energy audit findings on Dar es Salaam Water Supply System, as well as Southern African Development Community (SADC) Industrial Energy Management project |
| 2003-5 | Energy audit of Government and other public buildings |
| 2006 | a) National workshop to deliberate on energy audit reports in public buildings in some of the Ministries |
| | b) Energy Conservation and Efficiency (ECE) promotional activities |
| | c) National workshop on Energy Efficiency Standards and labels. |
| 2007 | Facilitation of projects to reduce power transmission losses from the present 23 to 15 per cent by year 2011. |
| 2008 | Preparation Master Plan on energy conservation and efficiency for the 5 East African Community Countries |

## Tanzania's Power Sub-Sector: Supply and Demand

The Tanzanian electric power supply system is comprised of both hydro and thermal generation units. The total electricity installed generation capacity at present is 1252 MW out of which 561 MW (545 MW available capacity) is hydro. The installed capacity of the thermal power generating sets is 690 MW (or 605 MW available capacities). There are about 15 generating units serving isolated locations, having a total generation capacity of 36 MW. Presently, import from Uganda and Zambia and small IPPs amounts to 10 MW. The maximum (peak) demand is about 800 MW, serving about 800,000 customers countrywide.

The annual electricity consumption at present is about 4,000 GWh out of about 5,000 GWh generated. The Peak Demand is expected to increase significantly in the next six years (Table 2.1). Residential, commerce and industries (light and heavy) altogether account for about 93 per cent of the total power consumption. Public lighting and bulk sales/export to Zanzibar account for 6.7 per cent, while the balance caters for other sectors (Table 2.2).

**Table 2.1: Estimated and Projected Power Generation and Consumption**

| *Year* | *2006* | *2010* | *2031* |
|---|---|---|---|
| Net Generation (GWh) | 3,300 | 5,030 | 21,100 |
| Power demand (MW) | 800 | 1452 | 4,800 |
| Peak (MW) | 733 | 1011 | 4,328 |

*Source*: Power System Master Plan, 2008.

**Table 2.2: Electricity Consumption by Major End-users**

| *Consumption Sector* | *Share to the Total Energy Consumption (per cent)* |
|---|---|
| Residential, small commercial and light industries | 59.20 |
| Industrial | 34.00 |
| Transport | 0.00 |
| Agriculture | 0.00[1] |
| Public lighting | 0.10 |
| Zanzibar bulk supply | 6.70 |

*Source*: Tanzania Electric Supply Company.

About 10 per cent of the population has access to electricity supply while the situation is worse in rural areas where 2 per cent of the population is connected to electricity grid system. The annual per capita consumption of electricity in Tanzania is about 100 kWh, which is quite small compared to a country like India having close to 500 kWh.

The total number of customers in the country is close to 800,000 and almost all of them are connected to the national grid system. While electrification level in urban areas is less than 37 per cent, the level of access to electricity in rural areas (where 75 per cent of the population lives) is about 2 per cent.

## Potential for Energy Efficiency Standards and Labels in Tanzania

### Experiences and Good Results

Demand side management could mean good housekeeping measures; energy efficient appliance, load shedding, load shifting, as well as with energy efficiency in appliances.

In 1993, UNDP/World Bank through the Energy sector Management assistance programme (ESMAP) assisted implementation of the Demand Side Management (DSM) with financing from Sida where about 30 large industrial consumers were

1 Insignificant for irrigation at Sugar, rice and other small scale farms and in agro processing *e.g.* cotton ginneries.

audited based on electricity consumption; it was found that savings of up to 10 MW or energy savings of 35,642,440 MWh was possible[2] with a simple payback period of 2.5 years!

In 2002, MEM and TIRDO demonstrated through energy audit on Dar es Salaam Water Supply Systems that about 40 per cent of the total expenditure on electricity costs for water pumping could be saved easily through energy conservation and efficiency (Ref. Matrix on page 5).

It has also been established that there could a saving of more than 25 per cent through introduction of energy efficiency standards and labels for all the appliances in the market, which translates to up to 200 MW saving! In monetary terms this is about US$ 350 million, [Table 2.3 indicates installed capacity cost in US$/kW for various energy sources] apart from the environmental benefits. On the other hand, experiences have shown that replacement of incandescent lights (tungsten filament lamps) in all Tanzanian residential households connected to electricity can result into a saving of about 40 MW!

**Table 2.3: Summary of Installed Capacity Costs for Various Energy Sources**

| *Source* | *Installed Capital Cost (US$/kW)* |
|---|---|
| Small hydro | 1,500–3500 |
| Large hydro | 1,100–3000 |
| Wind-off shore | 900–1200 |
| Solar (PV) | 22,000–35,000 |
| Coal | 1,000–2,000 |
| Natural gas | 400–900 |
| Diesel | 400–1000 |
| Nuclear | 1000–1500 |
| Biomass | 500–2500 |

*Source*: World Energy Council-Renewable Energy Projects Handbook-February 2004.

## Market

Wide range of appliances and equipment in the Tanzanian market, include motors, air conditioners, lighting equipment, refrigerators, freezers, electric cookers,

---

2 The survey revealed cost saving opportunities from the following:

- Reduction of Maximum Demand
- Good housekeeping practices and maintenance programs
- Retrofitting refrigeration and air conditioning systems
- Improvement of water pumping system
- More efficient operation of boiler
- Improved insulation of walls and hot water system
- Reduced number of lights and replace some with efficient lights.

water heaters air compressors, pumps, ICT systems, dishwashers, washing machines, etc.

High energy costs cause a rise in the price of goods and services and therefore there is an urgent need to reduce energy costs in all sectors of the economy so as to enhance competitiveness particularly in an industrial production setting. Most of the electrical goods imported into Tanzania are inefficient because there are no energy efficiency standards. This situation results into energy wastage and consequently high costs to the economies of the country.

Efforts are underway to strategise ways of reducing importation of energy inefficient appliances and equipment.

### Relevant Stakeholders in Tanzania

Major stakeholders towards promotion of energy efficiency and conservation include:

*MEM*, Ministry of Trade, Industry and Marketing; Tanzania Bureau of Standards (TBS); *National Cleaner Production Centre* (NCPC); TANESCO; Tanzania Petroleum Development Corporation, Tanzania Revenue Authority; Ministry of Finance and Economic affairs, Energy and Water utilities Regulatory Authority (EWURA); Confederation of Tanzania Industries (CTI); Research and Tertiary learning Institutions; and the National Environment Management Council (NEMC).

## Conclusion/Recommendation

Experiences elsewhere have shown that energy conservation and efficiency can enhance energy security since efficient equipment and appliances help energy and money saving as well as protection of the environment.

Tanzania sees Energy Efficiency Standards and labels as a very powerful weapon towards promoting Energy Conservation and efficiency in three sectors–industrial, residential and commercial where electricity is most consumed.

Certainly there is a need now to come up with legal interventions in order to enforce efficiency standards and labels.

## References

1. Mbise, and Dr. Mwakahesya, L. U., 2004. "Energy security for Households" a paper presented in a Workshop on Country Water Partnership (CWP), Integrated Water Resources Management, Dar es Salaam, 25th–27th February 2004.
2. Mrindoko, B. J., and Mbise, H. A., 1998. Energy conservation and Efficiency and Renewable Energy in Tanzania; unpublished report, Dar es Salaam.
3. Mrindoko, B., Mwihava, N. and Mbise, H., 1999. Government Policy Framework on Renewable Energy; a paper presented at an international seminar on renewable energy, Arusha, May.
4. Lutengano Mwakahesya (PhD) and H.A. Mbise, 2005. "Performance of the Energy Sector from 1995-2005; report submitted to ESAURP, Dar es Salaam, October.

5. Ministry of Energy and Minerals, 2003. The National Energy Policy; Dar es Salaam February.

6. World Energy Council, 2002-2004. The potential for Regionally Integrated Energy Development in Africa: a Discussion Document; WEC Work Programme Document.

7. TANESCO Master Plan, 2001. Power Sector Master Plan, Dar es Salaam.

   a. B. J, Mrindoko, 2002. Assessment of Energy sector in Tanzania, A paper prepared for a meeting on Economic Infrastructures-Nairobi, Kenya, April.

   b. Vice President's Office, United Republic of Tanzania, 2005. National Strategy for Growth and Reduction of Poverty, final draft, 15th January 2005.

# Chapter 3

# *Energy Sources, Utilization and Energy Access in Tanzania*

**Rwaichi J.A. Minja**
*Chemical and Process Engineering Department, University of Dar es Salaam, Dar es Salaam, Tanzania*
*E-mail: rminja@udsm.ac.tz or rwaichi@hotmail.com*

## ABSTRACT

The energy sources in Tanzania include: bio energy (mainly biomass), electricity, petroleum oil (imported), natural gas, coal, solar and wind. Among these sources, biomass, hydro, wind and solar are renewable sources of energy. As a result of high capital costs and technologies associated with harnessing solar and wind energies, these have been utilized to a very small scale in Tanzania. Some of the potential hydropower sources have been developed for production of main grid electricity, but due to climate change, which has lead to drought in some years, electricity production from the installed hydropower plants sometimes goes below the production capacity. Other electricity production sources are from natural gas, petroleum oil and coal. Bio energy resource (mainly fuel wood and charcoal from both natural forest and plantations) accounts for 93 per cent of total energy consumption. The challenge which lies ahead is how to increase clean energy access. The paper explores the current energy status in Tanzania, energy access and problems associated with energy access and initiatives in place.

***Keywords:*** *Indigenous energy resources, Electricity, TANESCO, Bio energy, CAMARTEC and Energy resources.*

## Introduction

Tanzania has large reserves of indigenous energy resources including natural gas, coal and water to meet the demands of the power sector for many years. There is also undetermined potential of geothermal energy[1]. Currently petroleum, hydropower, natural gas and coal are the major source of commercial energy in the country. The biomass energy resource, which comprises fuel-wood and charcoal from both natural forest and plantations, accounts for 93 per cent of total energy consumption. Petroleum is imported. The transport sector is the main consumer of petroleum products. Presently there are a number of companies exploring for oil, but so far no sound results have been reported. Wind and solar energy is another source of energy. Very little attempt has been made to utilize this source of energy which could be a viable alternative source to reduce use of wood and oil for heating purposes. Tanzania has an area of 954 000 square kilometres and has a 220kV transmission network covering 2248 km. It has 1400 km of 132kV transmission lines. 55 isolated and mini hydroelectric generators having a capacity of 23MW supply 135 km of areas not connected to the grid.

## Electricity Sources and Supply

Electricity generation, transmission and distribution in Tanzania are through the Tanzania Electric Supply Company (TANESCO). The company is 100 per cent government owned and is responsible for 98 per cent of the country's electricity supply[2]. TANESCO's generation system consists of hydro, and thermal. Hydro contributes the lion's share of TANESCO's power generation. The total generation from TANESCO own sources in 2003 were 2,662,027,682 kWh out of which 2,551,416,842 was from hydro power stations. It is reported that Tanzania has an estimated 3800 MW of economic Hydro electric potential capacity. Energy from water (hydro) is a very important source of electric power in Tanzania currently accounting for about 67 per cent of total available installed capacity[3]. The Hydropower is prone to draught effects-so some thermal power stations have been installed. This includes electricity form natural gas which started to be produced from 2006.

## Electricity Generation

TANESCO has been operating an interconnected grid system comprising of hydropower generating stations. The installed hydro generation capacity are: Kidatu 204 MW; Kihansi 180 MW; Mtera 80 MW; Pangani 68 MW; Hale 21 MW; and Nyumba ya Mungu 8 MW totaling to 561 MW. Several diesel generating stations with capacity of 80 MW are connected to the national grid in Dar es Salaam, Mwanza, Tabora, Dodoma, Musoma and Mbeya. Because of running problems their contribution is about 35 MW. Some regions, districts and townships are dependent on isolated diesel–run generators (Kigoma, Lindi, Njombe, Mafia, Mpanda, Tunduru, Songea, Liwale, Ikwiriri, Masasi and Kilwa Masoko). These have installed capacity of 31 MW but they effectively contribute about 15 MW due to aged machinery and lack of spare parts. Two private independent power projects (IPP's) which are connected to TANESCO grid are IPTL (Independent Power Tanzania Ltd) with 100 MW installed

---

1 http://www.sovereign-publications.com/tanzania.htm (accessed April 21, 2008).

2 http://www.areed.org/country/tanzania/energy.pdf (accessed April 21, 2008).

3 http://www.sovereign-publications.com/tanzania.htm (accessed April 21, 2008).

capacity and SONGAS (Songo Songo gas – to electricity project) which by the end of 2006 was producing 182 MW. TANESCO also imports 10 MW of electric power for Kagera Region from Masaka substation in Uganda while Sumbawanga, Tunduma and Mbozi districts receive about 3 MW from neighboring Zambia. Bulk supply of electricity is made to Zanzibar from Ras Kilomoni substation at the Indian Ocean coast in Dar es Salaam[4].

## Bottlenecks of Mainly Depending on Hydropower

Tanzania experienced poor hydrology conditions throughout the years 2003 up to 2006 with significant drop of water inflows to hydropower dams, which resulted into significant reduction of hydro electricity generation. Hydropower was up to year 2001 contributing about 97.5 per cent of the energy needs in the national grid system, but in year 2005, the contribution of hydropower had dropped to about 50 per cent and by mid – year 2006 to 30 per cent of the national grid energy needs because of drought. To subsidize this deficit, TANESCO resorted to importing more thermal energy from IPTL and SONGAS. The cost of this imported thermal energy is so high that the Government has had to give substantial amount of subsidies to TANESCO. However, as the hydrology situation worsened in 2006, TANESCO had no option but to institute load shedding in order to stretch the little available water in the hydropower dams for a longer period. On the other hand, the Government approved TANESCO's plans to implement short term generation projects to avoid severe energy shortages which had serious negative consequences in the national economy[5].

**Table 3.1: Short Term TANESCO Power Generation Plans**

| *Power Generating Company* | *Location* | *Ownership* | *Expected Capacity* | *Financier* | *Status* |
|---|---|---|---|---|---|
| Rental Gas Based Aggreko Plant | Ubungo DSM | Rental contract for 2 years | 40 MW | The Government of Tanzania | Fully operational |
| Rental Gas Based Dowans Holdings Plant | Ubungo DSM | Rental contract for 2 years | 100 MW | The Government of Tanzania | Fully operational |
| Rental Diesel Engines Mw Alstom Plant | Mwanza | | 40 MW | The Government of Tanzania | Logistical arrangements going on |
| Permanent Gas Based Tanesco Plant | | TANESCO | 100 MW | The Government of Tanzania | Completed end 2008 |
| Permanent Plant by Warsila | Tegeta DSM | TANESCO | 45 MW | The Government of the Kingdom of the Netherlands | Construction not yet |
| Coal Fired Kiwira Power Plant | Kiwira Mbeya | Kiwira Coal and Power Limited | 200MW | | Expected to increase from 6 MW to 50 MW |
| Conversion of IPTL diesel units to gas | | | | | Still under negotiation |

*Source*: Information in this table was compiled from http://www.tanesco.com/

4-5 http://www.tanesco.com/ (accessed April 20, 2008).

It was expected that, when all these rental and permanent upcoming generation projects become fully generational before the end of year 2007, they would added up a total of 375 MW into the national grid. This was apart from 282 MW from IPTL and SONGAS, and 561 MW from the hydro power plants.

## Long-Term TANESCO Power Generation Plans

Some of the projects recommended in the TANESCO Power System Master Plan (PSMP) of 1999 which is updated yearly are as shown in Table 3.2.

**Table 3.2: TANESCO Proposed Power System Master Plan**

| *Project* | *Year* | *Cost USD (Million)* |
|---|---|---|
| Zambia–Tanzania (200 MW) interconnector | 2010 | 168.10 |
| Ruhuji hydropower (358 MW) | 2012 | 400.63 |
| Mchuchuma Coal fired plant (200 MW) | 2018 | 201.72 |
| Mchuchuma Coal fired plant (200 MW) | 2023 | 176.50 |
| Rumakali hydropower (222 MW) | 2024 | 338.16 |

*Source*: Information in this table was compiled from http://www.tanesco.com/

The objective of PSMP is to determine the least cost programme for development of Tanzania's generation and transmission expansion plans, and to enable TANESCO understand the existing future level of demand for supply to customers.

## Electrical Energy Accessibility Expectations

Tanzania has per capita electricity consumption of 46/KWh per annum, which is growing at the rate of 11–13 per cent. Hence the government is encouraging investment to expand generating capacity, distribution system and developing indigenous sources of energy[6]. It is intended that the rest of the country, including an estimated 8,200 villages should be supplied with electricity to curb deforestation. In addition there are plans to supply power to Kenya and Malawi from Tanzania.

By the end of year 2006, Tanzania Mainland had 21 regions and 113 districts. All the Regional Headquarters so far have electricity. Of the 113 districts, 95 district headquarters had electricity by the end of 2006, and 7 were expected to have electricity by the end of 2007[7].

Tanzania cannot be transformed into a modern economy without a dramatic improvement in the access of electricity to the rural areas. The major challenges facing rural electrification in Tanzania are to ensure that 89 per cent of the population has access to electricity (as only 11 per cent have access to electricity and majority of TANESCO customers are in urban areas). Majority of rural electrification projects are expensive, this requires huge investment, hence the need for local and internal investors to participate in rural projects so as to expand the country's power sector[7].

---

6 http://www.areed.org/country/tanzania/energy.pdf (accessed April 21, 2008)

7 http://www.tanesco.com/ (accessed April 20, 2008).

Thus, there are two major objectives of Tanzania's rural energy strategy. Firstly, modern energy including electricity, modern biomass technology and other alternative modern energy sources, must be harnessed to promote rural productive uses such as job creation, stimulation of investment and revenue earnings. Secondly, access to modern energy must be availed to key rural services, particularly in the health, education, security and water sectors. While it is not possible for many years for every rural household to have "a light bulb in every home", it is possible for all Tanzanians to have access to education, health, water and security if access to use of modern energy is emphasized. The Ministry of Energy and Minerals has recently come up with the proposal of providing institutional systems that can facilitate faster implementation of rural energy projects. Rural Energy Agency (REA) and Rural Energy Fund (REF) will be instruments to promote access to rural energy projects. REA is the institution with the task of facilitating implementation of the rural power projects, through the private sector, non-governmental organization, community based organization and other ownership. REF is the financial instrument that the REA will be able to use to raise capital for its projects[8].

## Other Alternative Energy Resources in Tanzania

### Biomass and Forest Energy Resources

About 90 per cent of the total energy used in Tanzania is generated from wood. The amount of wood fuel consumed will continue to increase with the increasing population. Wood fuel consumption estimates for 1987 was 42.4 million $m^3$ and the projected amount for year 2000 was 64.2 million $m^3$. The demand was observed to exceed the capacity of forests to regenerate by about 12 million $m^3$ per year. In rural areas, about 90 per cent of the households use traditional tree stoves with low thermal efficiency while about 62 per cent of the urban households use inefficient charcoal stoves with an efficiency of 15 per cent. Although reliable figures are not available, estimates suggest that there is no nationwide fuel wood crisis. However, in many areas surrounding main urban centres, densely populated settlements, and those around rural industrial centres and in drier regions, fuel wood shortages are being experienced. Due to high prices, non-wood energy sources are hardly accessible for the major part of the population[9]. Improving wood burning devices, charcoal stoves, charcoal production techniques and planned tree planting for energy source are necessary measures which should be employed to ensure efficient and sustainable utilization of forest resources.

Tanzania generates huge amount of biomass resource each year in the form of agricultural (about 15 million tonnes) and forest residues (about 1 million tonne). The percentage of crop residue left in the field after harvesting varies with the crop. However, depending on their heat value, bulk density, and distance from the village to the farms the field residues or crop remainder that are left in the field after harvesting

---

8 http://www.tanesco.com/ (accessed April 20, 2008).

9 http://www.fao.org/DOCREP/003/AB575E/AB575E03.htm.

are not very attractive for fuel application due to economic and technical reasons. Most small-particle biomass residues are characterized by low bulk density, which makes handling and transportation for disposal costly. For economic reasons, crop and wood processing SMEs dispose most excess residues that have no alternative use. The disposal method commonly used is to dump them nearby and sometimes burn them. This disposal practice creates air, soil and water pollution within the locality, and downstream rivers and streams.

## Coal Resources

Tanzania has 1,200 million metric tonnes of coal, which could be used for both industrial and power generation[10]. The coal reserves are found in Mchuchuma in south western Tanzania near the northern tip of Lake Nyasa. Recent studies indicate that Mchuchuma coal deposits can provide fuel for 400MW generation capacity for up to 35-40 years. There is an existing small coal mine at Songwe-Kiwira which started production in 1988 with initial production planned at 150,000 tonnes and reaching a maximum of 300,000 tonnes at maturity of the project[11].

## Natural Gas Resources

The proven natural gas reserves are located in the SongoSongo Island offshore on the Indian Ocean and Mnazi Bay, South East of Mtwara town. The Songo Songo gas reserves are estimated to be 726 billion cubic feet (20.5 billion $m^3$). Songas is an internationally-owned company that generates electricity using gas from Songo Songo. The business consists of two different operating streams, Gas Processing and Transportation and Power Generation. Processed gas is transported through a 225 kilometre pipeline to Dar es Salaam where it is used in Songas' Ubungo power plant. This electricity is fed into the national electricity grid from where it is distributed to end users by TANESCO. Gas is also delivered to other large energy users in Dar es Salaam[12].

Mnazi Bay gas was discovered in 1982 at 27km south east of Mtwara town in Mtwara region (Figure 3.1). The gas reserve has a potential of about 0.5 to 1.0 trillion cubic feet (14.1 to 28.3 billion $m^3$). Feasibility study for the exploitation of the gas was done 1994. A production sharing agreement for Independent Power Production (IPP) was signed between the government of Tanzania and Artumas Contractors in May 2004 for electricity generation, transmission and distribution in the regions of Mtwara and Lindi. The initial electricity production capacity was 16.5 Mw. Commercial operations were to commerce by the end of 2005. The Mnazi bay project was operational by 2007. Mnazi bay natural gas is a source of reliable, cheap and affordable power which will save foreign money used to buy fuel. There is also a potential of other investment opportunities as more gas is expected[13].

---

10 http://www.areed.org/ country/tanzania/energy.pdf (accessed April 21, 2008).

11 http://www.sovereign-publications.com/tanzania.htm (accessed April 21, 2008).

12 http://www.songas.com/ (accessed March 23, 2008).

13 http://www.artumas.com/html/mtwara_energy_project.aspx (accessed April 21, 2008).

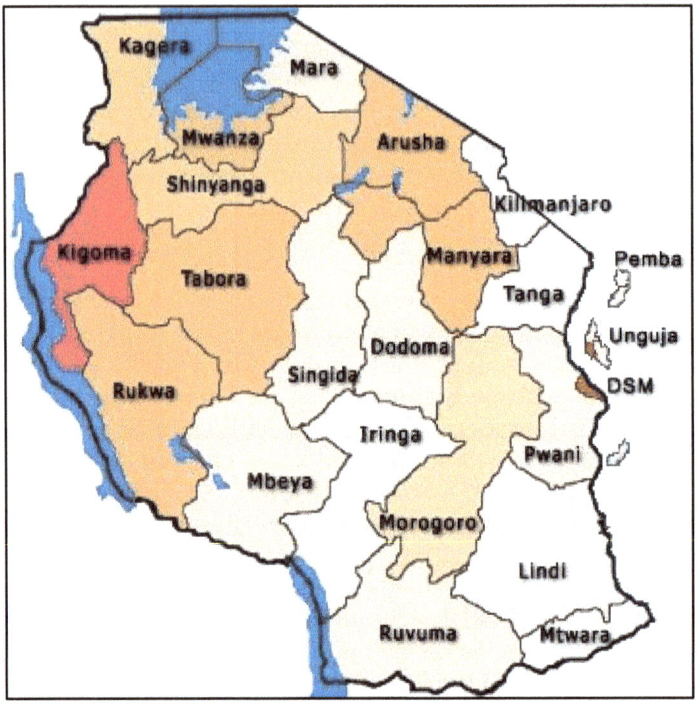


**Figure 3.1: Map of Tanzania Showing the Regions**

## Biogas Resources

In Tanzania, biogas digesters were first introduced in 1975. To date the Centre for Agricultural Mechanization and Rural Technologies (CAMARTEC) disseminates the so called Fixed Dome digesters (the Chinese adopted fixed dome type) attached to cowsheds, pigsties or toilets and effluents go out through an outlet man hole to an overflow point and out as liquid fertilizer. Biogas digesters have been installed in various places including health clinics and mission hospitals. However, the household biogas digester is by far the most widely attempted model using domestic animal excreta. The biogas produced from these household systems has been used mostly for cooking, with some use for lighting. The advantages associated with the Fixed Dome digesters are: Maintenance free, all masonry work with less metal, easy to free gas leakage, gas storage at the same structure, very durable with a life span of 20–30 years and it is also sustainable. The disadvantages of the design include: high initial investment cost and replacement of appliances[14].

The Tanzanian CAMARTEC program targeted farmers in the "Coffee and Banana Belt". This is a fertile region with highly productive small scale farmers. The

14 Ngowi, H.Z., 2006. Biogas Technology Development in Tanzania: Design, Construction, Management and Utilization of Biogas, COSTECH, Energy Resources in Tanzania, Vol. 1.

programme adopted a focused cluster approach in which participants were selected from a concentrated area. Standardized biogas plant designs and administrative procedures were adopted. The farmers also practice extensive animal husbandry, keeping their animals in pens. The farmers are relatively well off and could pay for the biogas plants. CAMARTEC constructs biogas plants either on a turnkey basis or with a contribution of the farmer in form of material supply. They give a two year guarantee on the plants. There was a complementary program by the church organization Evangelical Lutheran Church in Tanzania (ELCT) which targeted lower income earners but involve them in actual construction work in order to reduce the costs. This approach led to 150 biogas plants build in the country.

Logistics for training programs as well monitoring of projects were simplified. Advice to farmers was readily given cost effectively by a resident biogas officer. Farmers in the same area could also learn from each other and exchange ideas. The biogas plant design was integrated into the day-to-day activities of the farmer. In collaboration with the livestock unit, cattle pens were modified to facilitate collection of dung into the digester.

Low-cost plastic digesters were introduced in Tanzania in 1993 through a FAO technology transfer project and through a local NGO, SURUDE. By 1995, SURUDE installed almost 100 units of the polyethylene tubular film digesters in three areas, about 60 of them were installed in Turiani, Morogoro region.

Socio-cultural constraints include the lack of knowledge about biogas, the social stigma associated with handling any type of waste, fear and dangers involved in using any type of gas, beliefs that fuel wood imparts a better flavor to cooked food, and the low level of literacy which makes information dissemination about biogas more difficult. Biogas, which is free of smoke, has the potential to greatly reduce this health problem. Some potential users are reluctant to try biogas digesters out of concern about sanitation. If biogas is to compete and find its rightful place as an important technology for rural and peri-urban energy and sanitation then social marketing and branding it as a symbol of social progression is imperative.

The production of a high value fertilizer is a value added proposition in selling biogas technology. A major difficulty is in utilizing manure sources properly. Liquid manure is preferred for most technology, but households were/are not accustomed to storing and handling it. Promotion of liquid manure digesters requires additional education and training to maximize benefits and ensure sustainability. Operation costs increase upon consideration of acquiring water, but these would be mitigated by rainwater collection methods. Similarly, operation of biogas technology such as kitchen burners improves energy and labor efficiency.

## Wave Power Resource

There has been an initiative where by Zanzibar was considering the possibility of turning Indian Ocean currents and waves into electric power to make the utmost of its geological position as an archipelago off east Africa. If studies prove viable, the Zanzibar Utilities Company will build a power plant on the Pemba Island, Ocean energy constitutes to a large unexploited source of renewable energy and wave power therefore commands a good economical potential[15].

## Solar Energy Application

Tanzania is situated near the equator, is one of the "solar belt" countries of the world, *i.e.* countries with 2800–3500 hours of sunshine per year and a global radiation of 4–7 kW per square meter. This enormous amount of solar energy makes solar energy preferable to any other energy source, from the sustainable energy supply point of view.

The application of solar energy in Tanzania using modern conversion techniques began in the early 1980's. The growth in the number of applications and systems installed in Tanzania was very slow but now it is a lot faster. By the mid 1980's, there were only a handful of organizations that were dealing with solar energy. Most of the systems that were installed at that time were solar photovoltaic (PV) installations for powering repeater stations of the Tanzania Posts and Telecommunication Corporation (TP&TC), a solar PV pumping system for irrigation and some solar water heaters. There is no complete data on solar energy systems that are installed in the country. Applications cover both solar thermal and photovoltaic applications. The applications range from centralized community applications for villages, missions, hospitals, schools, hotels etc. to individual household applications. The capacities of the systems range from 10kWp to less than 100Wp. Companies that are using solar photovoltaic energy in large scale in Tanzania include the Tanzania Telecommunication Company Limited (TTCL) for powering repeater stations, Tanzania Railways Corporation (TRC) and Tanzania Zambia Railway Authority (TAZARA) for communication signals, Tanzania Zambia Pipeline Company for cathodic corrosion protection, Tanzania Parking Systems for powering car packing meters in Dar es Salaam city, and Tanzania Habours Authority (THA) for lighting signs for navigation in the sea.

Solar thermal application is regarded as being less complicated and less capital intensive than photovoltaic application. The common applications are for water heating, drying, cooking, cooling, and thermal applications on industrial and institutional level. Some of the common applications of PV in Tanzania are refrigeration, water pumping system, and power supply systems for lighting, powering equipment and other uses. In all of these systems, all the components are imported from abroad, mainly Europe, North America, East Asia (including Japan) and South Africa. Experience with other countries in Africa including neighboring countries, has revealed that many of the components for the solar PV systems can be manufactured locally.

The Government of the United Republic of Tanzania in collaboration and with support from various international and national stakeholders is implementing various pilot projects on removing barriers to the transformation of rural solar photovoltaic market in Tanzania[16].

---

15 http://alt-e.blogspot.com/2005/01/alternative-energy-tanzania-wave-power.html (accessed April 21, 2008).

16 Kimambo, C.Z.M., 2006. Solar Energy: Utilization Status and Constraints in Tanzania, COSTECH, Energy Resources in Tanzania, Vol. 1.

The barriers that hinder widespread dissemination and adoption of solar PV technology in Tanzania are as follows:

1. Limited awareness of, and experience with solar energy technology
2. Inadequate business knowledge and capacity for distribution, aggressive marketing and sales of solar energy appliances and systems
3. Limited technical knowledge of proper sizing, installation, operation and maintenance
4. Lack of solar energy professionals
5. Policy implementation
6. Difficult access to finance for end users
7. High cost of doing business
8. High cost of solar energy systems, initial capital investment and operation and maintenance
9. Low purchasing power of the rural people
10. Lack of established dealer network
11. Lack of solar energy standards and poor/inappropriate installations

## Wind Energy Resources

Wind can be defined as moving air and therefore contains kinetic energy that can be converted into mechanical energy. Wind energy being a renewable source if developed could make a contribution in addressing the energy access to the rural communities. Wind technology is well known and wind resources are also known to be available in abundance and have no effect on the environment. Wind energy has not been effectively utilized in Tanzania although the resource is available. The factors which contribute to failure of exploiting this particular type of natural resource include lack of the long term wind data[17].

Wind energy applications in the majority of cases fall into two basic categories namely electricity generation and water pumping using either a horizontal axis or a vertical axis wind machines. Performance of electricity generating wind machines generally require high wind speeds greater than 5 m/s as opposed to water pumping wind machines for which lower wind speed say 2.5 m/s are sufficient. The speed range of at least above 5 m/s is suitable electricity generating.

In Tanzania wind energy has been used in isolated cases for water pumping in Dodoma, Singida and Tabora regions for livestock, irrigation and domestic use. Majority of water pumping wind turbines were installed between 1975 and 1980 by the Government of Tanzania through Australian Government support, missionaries, other private organizations and individuals. Other parts include Mara, Shinyanga, Arusha, Kilimanjaro, Tanga and Mwanza regions. These windmills used machinery

17 Nzali, A.H. and Mushi, S.J.S., 2006. Wind Energy Application in Tanzania, COSTECH, Energy Resources in Tanzania, Vol. 1.

manufactured by Tozi Baldi in Italy, Southern Cross in Australia, Bob Harries Engineering Ltd in Kenya and Ujuzi Leo Industries in Arusha – Tanzania. By 1996 a total of 129 water-pumping windmills were installed in Tanzania. Electricity generating wind turbines have been installed at Goba in Dar es Salaam (for experimental purposes), Mang'ola in Karatu, Itigi, Chunya, Sikonge and Kili at Bukene.

## Geothermal Energy Resources

Geothermal energy refers to energy from the earth's interior, where igneous rocks are in molten or partly molten state, conducted to the surface. Geothermal energy is considered to be renewable because the earth's interior is and will continue to be in the process of cooling for indefinite future. Geothermal energy has practically no intermittency and has high energy density compared with other renewable energy sources[18]. Tanzania has a significant potential of geothermal energy but the utilization of this resource is not yet in practice. Current estimates show that about 250 MW is available around the country[19].

Heat energy from the earth's interior reaching the earth surface through conduction and magmatic activities. Hot magmatic bodies from the earth's interior are pushed near the surface of the earth and the energy manifests as hot springs, geysers, volcanoes and fumaroles at earth surface. These features occur along specific areas controlled by plate boundaries and in Africa such boundaries are represented by the East African Rift System (Figure 3.2), which passes through Tanzania.

Most of the geothermal anomalies in Tanzania occur as a hot spring along a small creek or river. Logically, all these geothermal anomalies are within a drillable depth with land rig and are located on fairly rough to smooth terrains, which are easily accessible during the dry spell. There is great potential for existence of geothermal resources in hot springs at Mbeya to the south, Arusha (Lake Manyara, Natron Eldonyo Lengai, Ngorongoro Crater, Lake Eyasi) and Musoma to the north, and Rufiji (Luhoi and Utete) to the east (Figure 3.2). Geothermal exploration was carried out between 1976-79 by SWECO, a Swedish consulting group, in collaboration with Virkir-Orkint, with the financial support of the Swedish International Development Agency (Sida). Reconnaissance missions and surface exploration were carried out in the north (near Arusha, Lake Natron, Lake Manyara and Maji Moto) and in the south (Mbeya region). Approximately 50 hot springs were mapped which are associated with block faulting and recent volcanicity[20].

---

18 Gwang'ombe, F. and Mwihava, N., 2005. Renewables in Tanzania: Status and Prospects of Biomass-based Cogeneration and Geothermal Technologies, Publication of AFREPREN/HBF – HA Initiative on Renewables in Eastern and Horn of Africa: Status and Prospects, Nairobi.

19 Mwihava N.C.X., Mbise, H. A. Mnzava L., Kibakaya G., 2004. Draft Country Study Research Report, Ministry of Energy and Minerals, Dar es Salaam.

20 Marobhe, I. M., 2006. Geothermal Energy as Alternative Energy Source in Tanzania, Workshop on Sustainable Energy Options for Tanzania, Commission of Science and Technology, Dar es Salaam.

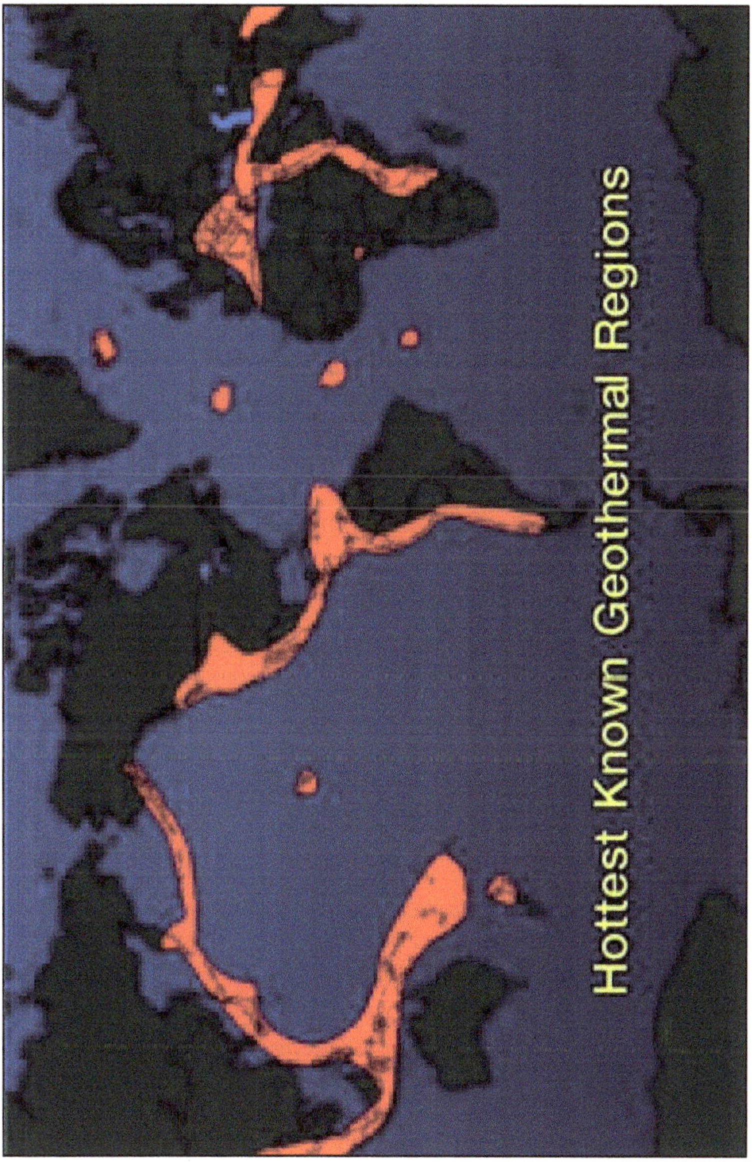


**Figure 3.2: Areas with Potential Geothermal Resources in Global Perspective (Marobhe, 2006)**

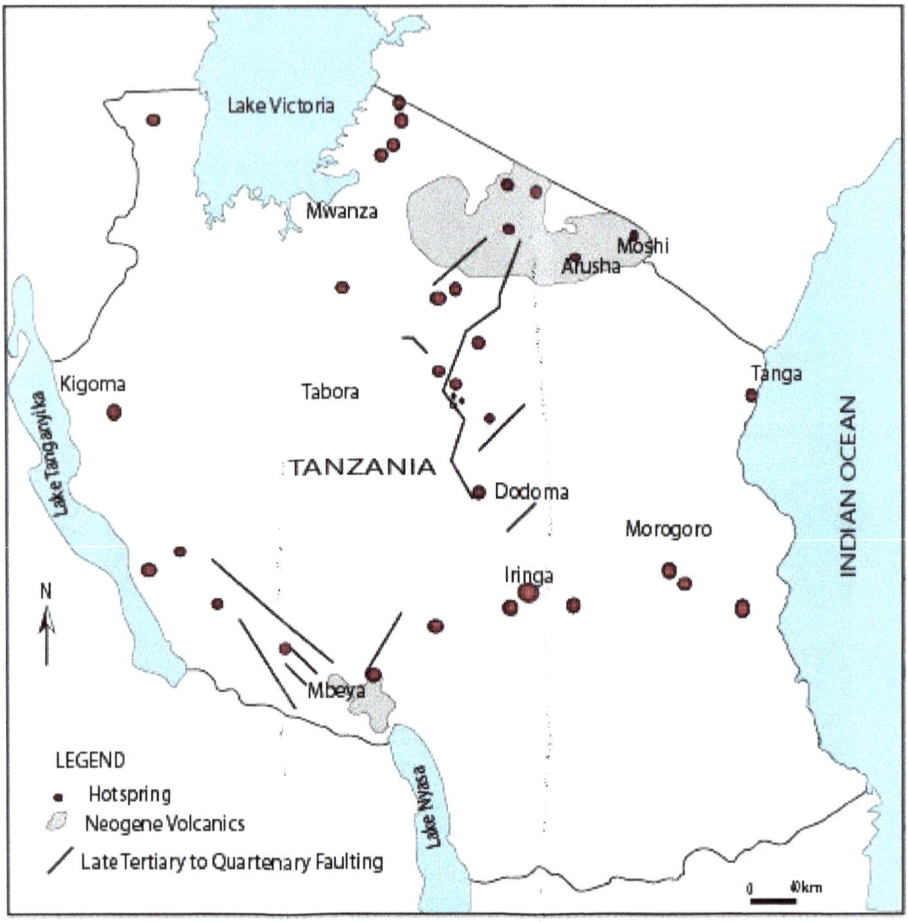


**Figure 3.3: Areas with Geothermal Potential in Tanzania (Marobhe, 2006)**

## Bio Energy Initiatives

The interest in bioenergy processing and use in Tanzania is increasing although the level of exploitation is still low and at small scale. Most of the initiatives are focusing on energy access services to the urban and peri-urban leading to the majority of rural poor continuing to be marginalized in their access to energy services[21].

For economic reasons, crop and wood processing SMEs have disposed of excess residues that have no alternative use. The disposal method commonly used is to dump them nearby and sometimes burn them. This disposal practice creates air, soil and water pollution within the locality, and in downstream rivers and streams.

As fuel wood scarcity increases due to population growth, increasing farmlands, and human settlements, rural villagers are beginning to rely more heavily on

21 Shuma, J., 2006. Rural Energy Agency: Opportunities and Challenges in Tanzania, TATEDO at http://www.ease-web.org/pdf/2006_access3_shuma.pdf (on 24th October, 2007, 1pm).

agricultural residues for fuel. "Large-particle" crop residues (such as maize cobs and coconut shells) can readily be burned directly in conventional household and institution stoves, and in furnaces and kilns used in small industries and are widely used (though with low thermal efficiency and higher pollution).

"Small-particle" crop residues and sawdust cannot be burned directly in conventional stoves or grate-fired furnaces because the small particles choke the primary air supply, thus extinguishing the fire. For the case where forced primary air supply is employed, the small particles are entrained with the air, causing increase in fly ash and very poor efficiency, since the particles leaves the combustion chamber unburned or partially burned (incomplete combustion).

Gasifier-stoves for rice husk have been introduced successfully in some countries (India, Thailand, Philippines, Peru, China, and Gambia). In Dar es Salaam and other places where sawmills exist, sawdust gasifier-stoves are commonly used by food vendors and restaurants. The sawdust gasifier-stove has not been successful in domestic application. This is mainly because of the fragility of the compacted sawdust in the stove when the stove is moved or shaken slightly (like when cooking *ugali*), the sawdust column crumbles down, blocking the air draft and extinguishing the fire.

In an effort to solve the disposal problem of jatropha seed cakes generated after pressing the oil, *Kampuni ya Kuendeleza Tecknologia* (KAKUTE) of Arusha has developed gasifier-stove to utilize seed cakes. The stove burns with intense heat (due to oil left in the cake), but at a price of Tsh. 35,000/= [In 2007], the intended users finds the cost of stove too high. For these reasons, adoption of low-cost gasifier-stoves to suit available feedstock, cooking habits and fire management skills is of great consequence to most areas with crop processing activities.

For the case of biodiesel engines, despite the high interest shown by the government in bioliquid fuels production and use in Tanzania, the level of exploitation is still at an infant stage. Most of the initiatives are small scale, and the large scale activities are aiming at exporting the fuels. However, all these energy initiatives have recognized the link between energy and economic development although most of the initiatives are focusing on urban and well to do communities, often in the production of biodiesel for transport purposes. Very few initiatives, if any, are focusing on empowering the rural poor to access modern energy services.

There is therefore a need to increase awareness and to demonstrate how diesel engine powered by the fuel that can be processed locally can increase productivity in farming, crop processing and electricity generation at the village level. The idea is to introduce a technology that could be owned and operated by the village or an entrepreneur in the village, using local labor and resources[22].

## Initiatives on Biofuels Utilization

Recently, Tanzania has experienced growing interest from both industries and NGO's in sugarcane and oil seed crop cultivation for biofuels production. Most of the

---

22 PISCES Inception phase Report for DfID 2007. http://www.research4development.info/SearchResearchDatabase.asp?OutputID=176614.

stakeholders have already established individual contacts with Government Ministries, such as the Ministry of Energy and Minerals and the Ministry of Agriculture and Food Security in order to secure institutional support for their activities. Roughly, the current biofuels activities and opportunities in Tanzania can be divided into large-scale and small-scale approaches.

The large-scale biofuels production, such as the production of ethanol from sugarcane promoted by the sugar industries will have a prime focus on biofuels for transportation and will require supportive policies and regulations in place for start-up in order to secure the rather large investment. The sugar producers have already conducted internal feasibility studies for ethanol production, which may serve to quickly proceed to the ethanol production implementation stage.

On the other hand, smaller-scale activities by organizations such as KAKUTE are mainly concerned with the creation of rural income and revenue opportunities from oil seed crops, either through the production of plant oils (for food and/or fuel) or other commodities such as soap production from jatropha oil. Also, the Department of Energy Engineering (University of Dar es Salaam) has two on-going projects with KAKUTE on development of cooking stoves and lamps to use jatropha oil.

In 2006, the Government through the Ministry of Energy and Minerals established a Biofuels Task Force. The job of the task force is to develop biofuels guidelines and strategy, legislation and regulations. The task force shall also ensure close co-operation between the different Government Ministries involved in the development of biofuels policies, as well as to provide an information channel between Government and biofuels stakeholders from industry, farmers associations, NGOs and civil society.

## PISCES Bioenergy Project

The Policy Innovation Systems for Clean Energy Security (PISCES) Consortium is a project which aims at researching on improving access to energy and livelihoods for the poor in developing countries. PISCES Consortium considers the interdependence of energy, water and food security–and that Bioenergy is the pivotal issue at the intersection of these three factors. Thus, concerns have been rising at equally high political levels about the potential competition of biofuels production with food production in particular, and about the real carbon reduction contribution of biofuels. A framework for Bioenergy has thus been developed (Figure 3.4) to provide the foundation of research, policy approaches and decision-making with relation to bioenergy. The PISCES framework on bioenergy conceptualizes Bioenergy as deriving from one of three sources: Bioresources (natural wood growth), Biowaste (by-products from human Agricultural and Industrial Processes) and Biofuel (crops purpose grown for energy). To turn these resources into the Energy Services needed by end users, it is necessary to apply Production/Extraction Technologies, Processing Technologies and Appliance Technologies which are specific to the various types of bioenergy resource and end-use intended. The PISCES project is carried out in four target countries of Kenya, Tanzania, Sri Lanka and India. The project will last for 5 years commencing in January 2008 and is funded by DFID. The PISCES research design and activities have been developed in order to balance two expected outcomes, that is, demonstrable policy change leading to energy access impacts in each country of

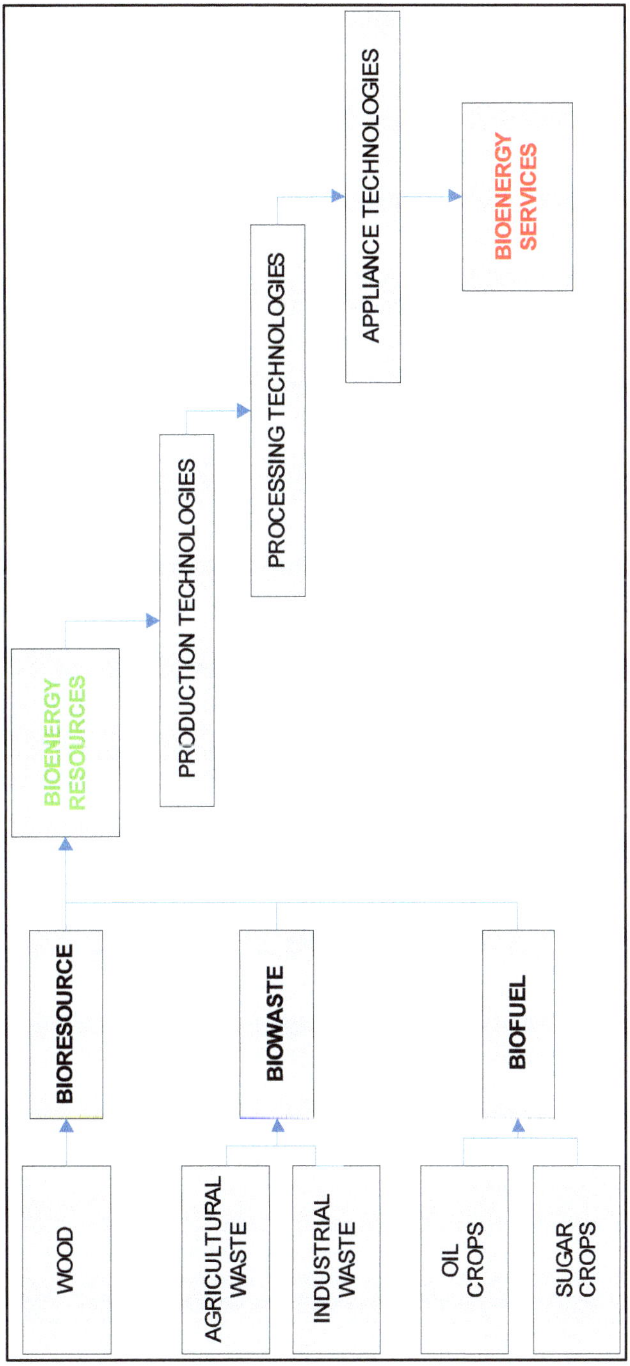


**Figure 3.4: PISCES Bioenergy Framework**

action and contribution to global knowledge regarding bioenergy and development. The key partners in this research consortium are: University of Edinburgh in UK, MS Swaminathan Research Foundation (MSSRF) in India, University of Dar es Salaam in Tanzania, and Practical Action in Kenya and Sri Lanka[23].

## Conclusions

According to the Tanzanian Delegation at an EU Energy Initiative Workshop in Maputo[24], the country's strategy and direction on Priority Energy Initiatives include:

1. Promoting affordable and reliable energy supplies in the whole country
2. Reforming the market for energy services and establish adequate institutional framework–to facilitate investment, services expansion, efficient pricing mechanisms and other financial incentives
3. Enhancing development and utilization of indigenous and RE sources and technologies
4. Taking account of environmental concerns in all energy activities
5. Promoting energy efficiency and conservation
6. Increasing energy education and building gender-balanced capacity in energy planning, implementation and monitoring
7. To achieve that, the following 3 programmes were proposed:

### Programme 1

Provision of financial and technical support to the Rural Energy Fund with the objective of facilitating increased access to modern energy services for the rural areas or rural poor. In this 5 years programme the following were to be expected:

1. Provide financial mechanisms to rural energy service delivery companies as part of a strategy to make energy affordable (work with local financial institutions and NGOs)
2. Provide technical and financial support to energy authorities incl. rural based credit institutions
3. Provide finance, technical assistance and training to technological development of Programmes to promote rural energy
4. Design support instruments (and capital) to assist commercial enterprises dealing with clean and RE technologies

---

23 PISCES Inception phase Report for DfID 2007, http://www.research4development.info/ SearchResearchDatabase.asp?OutputID=176614

24 Uisso, J., N. Mwihava, D. Mbekenga, R. Mtengule, E. Sawe, D. Biswalo, 2005. Priority Energy Initiatives for Tanzania Developed by the Tanzanian Delegation at an EU Energy Initiative Workshop in Maputo, 12-14 April 2005 at http://www.e4d.net/euei/Maputo/presentations/Tanzania working/ 20group.ppt#271, 2, Members of the Delegation (assessed March, 2008).

## Programme 2

Establishment of a programme to strengthen institutions' capacities to deliver better energy services with the objective to ensure that institutions are delivering better and sustainable energy services to the poor by 2010. In this 5 years programme the following were to be expected:

1. Introduction of energy programmes in schools and college curricula (this incl. extension training institutes)
2. Establish energy and development institutions (in order to improve insights on energy-poverty links, as well as energy R&D)
3. Initiate an awareness raising and information dissemination programme on modern energy services as well as local participation in Programme development and implementation
4. Build capacity on energy sector reform management, *e.g.* contract negotiations

## Programme 3

Design and implement a service delivery model to facilitate large scale uptake and use of sustainable energy (clean fossil fuel and renewable energy technologies) in rural areas with the objective to increase the large scale uptake and use of sustainable energy including clean fossil fuels and RE technologies in rural areas. In this 5 years programme the following were to be expected:

1. Design and implement a model that will promote the use of renewable energies, such as biogas technology, RE power water schemes, energy crops and plants (Jatropha and oil seeds), biofuels, cogeneration, etc
2. Design a model that will improve the efficient use of energy, *e.g.* improved stoves, improved charcoal techniques and management
3. Develop and implement a model that facilitate a transition to modern fuels (LPG, compressed natural gas) as well as the dissemination of efficient cooking appliances
4. Support and/or upscale existing biofuel Programmes

## References

1. German Technical Cooperation (GTZ), 2005. Liquid Biofuels for Transportation in Tanzania Potential and Implications for Sustainable Agriculture and Energy in the 21st Century at http://www.gtz.de/de/dokumente/en-biofuels-for-transportation-in-tanzania-2005.pdf.
2. Kearney, A.T., 2006. Biofuel Projects at http://www.unido-aaitpc.org/unido-aaitpc/new1/tanzania/BIOFUEL%20PROJECTS1.pdf.
3. Manyanga, L., 2006. Jatropha Sub-Sector Evolution in Tanzania at http://www.jatropha.de/Journal/Jatropha%20Developpement%20in%20Tanzania.pdf.

4. Manyanga, L., 2006. Jatropha Sub-Sector Evolution in Tanzania at http://www.jatropha.de/Journal/Jatropha%20Developpement%20in%20Tanzania.pdf.
5. Tanzania Energy Policy, 2000. United Republic of Tanzania. Ministry of Energy and Minerals.
6. http://www.ndctz.com/power.htm, TATeDO, 2006. Energy Sector in Tanzania.
7. http://www.tatedoease.or.tz/energy.html.
8. http://www.unep.org/GC/GCSS-IX/DOCUMENTS/Tanzania-1G-EnergyGen.doc.

# Chapter 4

# *Energy Efficiency Strategy of the Republic of South Africa*

**M. Bipath**
*Department of Minerals and Energy,*
*South Africa*

## ABSTRACT

This is the first Energy Efficiency Strategy for South Africa. The Strategy takes its mandate from the *White Paper on Energy Policy*, published in 1998, and links energy sector development with national socio-economic development plans as well as being in line with other Government departmental initiatives. In addition, it provides clear and practical guidelines for the implementation of efficient practices within our economy, including the setting of governance structures for activity development, promotion and coordination. This Strategy allows for the immediate implementation of low-cost and no-cost interventions, as well as those higher-cost measures with short payback periods. These will be followed by medium-term and longer-term investment opportunities in energy efficiency. The Strategy acknowledges that there exists significant potential for energy efficiency improvements across all sectors of our national economy. The Strategy covers all energy-using sectors and will be implemented through Sectoral Implementation Plans as outlined within. Systems will be put into place in order to periodically monitor progress against the target that will be reviewed at the end of each phase.

***Keywords:*** *South Africa, Energy efficiency strategy, Sectoral implementation plans, socio-economic development, Efficient practices, Energy consumption.*

## 1.0 Introduction

### 1.1 Why do we Need an Energy Efficiency Strategy?

Worldwide, nations are beginning to face up to the challenge of sustainable energy – in other words to alter the way that energy is utilised so that social, environmental and economic aims of sustainable development are supported.

South Africa is a developing nation with significant heavy industry, which is by its nature energy intensive. This energy intensive economy largely relies on indigenous coal reserves for its driving force. At first sight there would appear to be an apparent paradox between using less energy and developing a healthy and prosperous nation based on energy intensive activities. This is not the case. In recent years energy efficiency has significantly gained in stature and has become recognised as one of the most cost-effective ways of meeting the demands of sustainable development.

The benefits of energy efficiency upon the environment are self-evident. These benefits are of particular relevance, as South Africa remains one of the highest emitters of the Greenhouse gas $CO_2$ per capita in the world. At a local level the problems of $SO_2$ and smoke emissions have been the focus of concern for many communities living adjacent to heavily industrialised areas. Energy efficiency can address both the macroscopic and microscopic aspects of atmospheric pollution.

The economic benefits of improving energy efficiency have been well documented since the first Oil Crisis in the early 1970's. Many forward-thinking industrial and commercial concerns have already adopted energy efficiency as a key policy towards maximising profits. The national electricity utility, Eskom, has itself embarked upon a Demand Side Management programme to help reduce the requirements for investment in new power generation capacity.

Such positive contributions to both our physical and economic environments will inevitably benefit our social well being also; the alleviation of fuel poverty, job creation, improvements to human health, better working conditions–the list goes on. All of these factors will significantly contribute towards the aims of sustainable social development.

It is for these reasons that South Africa needs an Energy Efficiency Strategy.

## 1.2 The Status Quo

The existing energy policy of South Africa is captured within the *White Paper on Energy Policy* (1998). The policy aims to provide the nation with wider access to energy services, by various means, whilst ensuring that the environmental impacts of energy conversion and use are minimised as far as possible. This is of relevance to Africa as a whole, as South Africa uses some 40 per cent of the total electricity consumed within the continent.

South Africa is endowed with rich deposits of minerals and fossil fuel in the form of coal. It is no surprise, therefore, that the economic development of our country has historically been focused upon the extraction and processing of these resources. This has led to the development of a national economy heavily dependent upon energy as its driving-force, and has resulted in the core of our industries being those concerned with energy-intensive activities, such as iron and steel production and other raw materials processing. Furthermore, coal has inevitably emerged as the major source of primary energy to meet the demands of industry and the country as a whole. In 2000 the total primary energy supply to the nation was nearly 4,300 Petajoules, of which 79 per cent was attributable to coal.

Our abundant coal reserves have partially contributed towards an economic environment wherein the unit price of electricity is amongst the cheapest in the world. One of the undesirable side effects of this has been that energy efficiency has been demoted to make way for "priority" considerations, such as plant expansions and increases in production throughput. Indeed, by international standards the South African economy uses a relatively high amount of energy per unit of national economic output, or GDP (8,1 PJ per billion US$ in 2000).

In recent years the issue of energy efficiency has attracted more interest in South Africa, and a number of initiatives and projects have proven the merits of enhanced energy performance. The 2002 World Summit on Sustainable Development, held in Johannesburg, recognised energy efficiency as a key tool to enhance clean energy development and to mitigate the negative effects of energy use upon the environment. A further development is Eskom's Demand Side Management programme, which is aimed at reducing the national peak power demand, thereby deferring the immediate need for additional power generation capacity.

In short, energy efficiency is fast gaining ground as a cost-effective means to approach all aspects of sustainability. It is generally accepted that South Africa holds numerous opportunities for energy savings, together with pollution mitigation measures of international significance. This Strategy offers a consolidated approach in order to capture these opportunities in the best interests of our nation.

## 1.3 The Strategic Process

*The White Paper on Energy Policy*, published in 1998 states:

"Significant potential exists for energy efficiency improvements in South Africa. In developing policies to achieve greater efficiency of energy use, government is mindful of the need to overcome shortcomings in energy markets. Government would create energy efficiency consciousness and would encourage energy efficiency in commerce and industry, will establish energy efficiency norms and standards for commercial buildings and industrial equipment and voluntary guidelines for the thermal performance of housing. A domestic appliance-labeling program will be introduced and publicity campaigns will be undertaken to ensure that appliance purchasers are aware of the purpose of the labels. Targets for industrial and commercial energy efficiency improvements will be set and monitored."

It also states that:

"Vehicle purchasers do not generally consider the vehicle's fuel consumption as a major criterion. This is due in part to a lack of accurate information on vehicle fuel efficiency. The Department of Minerals and Energy will provide information on the fuel use characteristics of new vehicles. Energy consumption information should be included in all advertising, vehicle test reports and vehicle specifications."

The *White Paper* gives a mandate to the DME to promote Energy Efficiency through various means. Although Government's present capacity to undertake energy efficiency programmes is limited, the DME will finalise and consolidate considerations to ensure appropriate leadership in the sector.

The Strategy sets a national target for energy savings, of at least 12 per cent, to be achieved by 2015. This target is expressed in relation to the forecast national energy demand at that time, based on the 'business as usual' baseline scenario for South Africa modelled as part of the National Integrated Energy Plan (2003), which uses energy consumption data for the year 2000. The target also assumes that the Energy Efficiency interventions outlined in this Strategy are undertaken, these measures being primarily focussed on low cost interventions that can be achieved with minimal investments.

Assuming the target is achieved in 2015 the following table gives a very high level indication of the possible monetary savings that could be achieved by implementing the strategy.

**Energy Efficiency Savings for the Next 5 Years in Monetary Value**

| | **Electricity Target 12 per cent by 2015** | | |
|---|---|---|---|
| *Saving in Peta Joules* | *Million kWh* | *Average cost* | *Total* |
| 29 | 8.0 | 0.2 | R1,611,111.11 |
| 45 | 12.5 | 0.2 | R2,500,000.00 |
| 63 | 17.5 | 0.3 | R5,250,000.00 |
| 81 | 22.5 | 0.3 | R6,750,000.00 |
| 101 | 28.05 | 0.4 | R11,222,222.22 |
| Total for 5 years | | | R27,333,333.33 |
| | **Petrol Target 9 per cent by 2015** | | |
| *Saving in Peta Joules* | *k litres* | *Cost per litre* | *Total* |
| 6 | 184352.9679 | R4.00 | R737,411,871.73 |
| 7 | 215078.4626 | R4.50 | R967,853,081.65 |
| 8 | 245803.9572 | R4.70 | R1,155,278,599.04 |
| 9 | 276529.4519 | R4.90 | R1,354,994,314.31 |
| 10 | 307254.9466 | R5.00 | R1,536,274,732.77 |
| Total for 5 years | | | R5,751,812,599.50 |

In cost-benefit terms the best measurement stick is the payback period. In Phases 1 and 2 the majority of interventions will involve no cost or low cost. This means that the South African economy will make low cost gains in efficiency. In the case of low cost measures the payback period will be less than 3 years during which period the investment in equipment will be off-set by the savings.

In excess of 50 per cent of all EE measures fall within these no-cost/low-cost categories. The DME will monitor the realised cost benefits as part of the strategy implementation.

Energy efficiency improvements will be achieved through enabling instruments and interventions including economic and legislative means, information activities, energy labels, energy performance standards, energy audits, energy management and the promotion of efficient technologies.

The Strategy will cover all economic sectors, and be implemented through Sector Programmes. Systems will be put in place to monitor and evaluate progress in energy efficiency improvements and a periodic strategic review of the implementation will be undertaken.

Various public events will be organised in order to raise the profile of energy efficiency in the minds of the key stakeholders and the public at large.

## 2.0 Visions and Goals

### 2.1 Vision

| |
|---|
| *To encourage sustainable energy sector development and energy use through efficient practices*<br>*thereby*<br>*Minimising the undesirable impacts of energy usage upon health and the environment,*<br>*and*<br>*Contributing towards secure and affordable energy for all.* |

### 2.2 Goals

The Strategy's eight goals are outlined below. The order in which the goals appear does not relate to any particular priority, although they are grouped in terms of social, environmental and economic sustainability.

**Social Sustainability**

***Goal 1: Improve the Health of the Nation***

Energy efficiency reduces the atmospheric emission of harmful substances such as oxides of Sulphur, oxides of Nitrogen, and smoke. Such substances are known to have an adverse effect on health and are frequently a primary cause of common respiratory ailments.

***Goal 2: Job Creation***

Studies show that jobs will be created by the spin-off effects of energy efficiency implementation. Improvements in commercial economic performance, and uplifting the energy efficiency sector itself, will inevitably lead to nationwide employment opportunities.

***Goal 3: Alleviate Energy Poverty***

Energy efficient homes not only improve occupant health and wellbeing, but also enable the adequate provision of energy services to the community at an affordable cost.

**Environmental Sustainability**

***Goal 4: Reduce Environmental Pollution***

Energy efficiency will reduce the local environmental impacts of its production and use. These impacts include the atmospheric emission of harmful and odorous gases.

***Goal 5: Reduce $CO_2$ Emissions***

Energy efficiency is one of the most cost-effective methods of reducing Greenhouse Gas emissions, and thereby combating Climate Change. Addressing Climate Change opens the door to utilising novel financing mechanisms, such as the CDM, to reduce $CO_2$ emissions.

**Economic Sustainability**

***Goal 6: Improve Industrial Competitiveness***

It has been demonstrated that one of the most cost-effective ways of maximizing commercial profitability is the adoption of appropriate energy efficiency measures. Nationwide, this will improve South Africa's export performance and improve the value that her economy derives from indigenous energy resources.

***Goal 7: Enhance Energy Security***

Energy conservation will reduce the necessary volume of imported primary energy sources, crude oil in particular. This will enhance the robustness of South Africa's energy security and will increase the country's resilience against external energy supply disruptions and price fluctuations.

***Goal 8: Reduce the Necessity for Additional Power Generation Capacity***

It is estimated that the country's existing power generation capacity will be insufficient to meet the rising national maximum demand by 2007-2012. Energy efficiency is integral to Eskom's Demand Side Management programme, which is intended to reduce the level of load growth by a cumulative value of 4255 MW by 2025, equivalent to a saving of a six unit coal-fired power station. Efforts will be made to give Eskom responsibility for meeting a portion of the target set out in this strategy through its annual shareholder compact.

## 3.0 Targets

### 3.1 Baseline Statistics

The energy usage data illustrated in this section has been taken from the *Digest of South African Energy Statistics, 2000* published by the Department of Minerals and Energy. Data concerning energy intensity has been taken from the International Energy Agency's study *Energy Balances of OECD countries, published in 2002.*

**Primary Energy Supply**

The total *primary* energy supply to South Africa increased from 3,924PJ in 1993 to 4,295PJ in 2000, an increase of 9.5 per cent. In 2000 coal contributed 79 per cent of the total national primary energy supply, as illustrated in Figure 4.1.

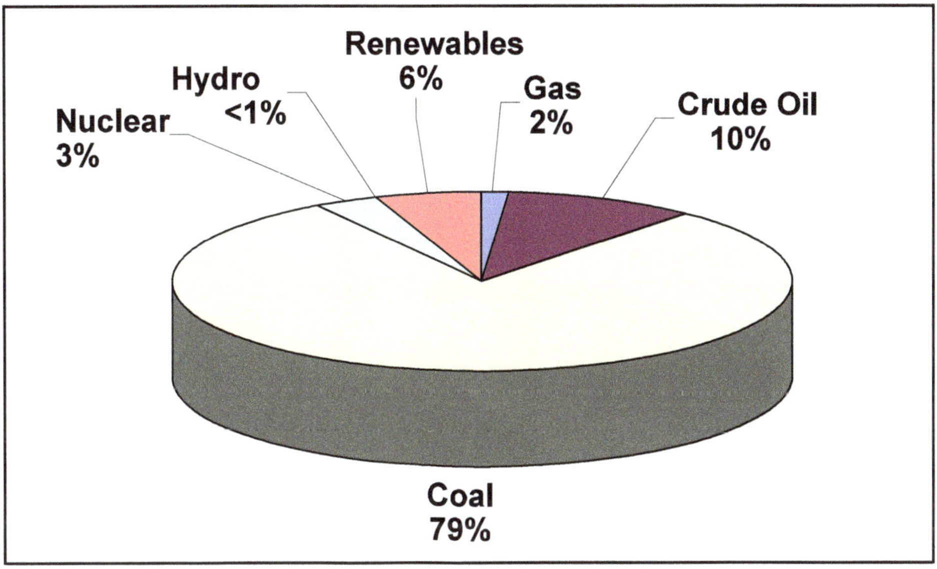


**Figure 4.1: 2000 Primary Energy Supply**

**Sectoral Usage**

The final *end-user* energy usage in 2000 was 2,193PJ. The three largest energy-consuming sectors were industry, residential and transport. The remaining sectors accounted for less than 10 per cent of final energy demand in 2000. Figure 4.2 depicts the sectoral split of final energy use, and excludes "non-energy use" carriers, such as solvents and lubricants.

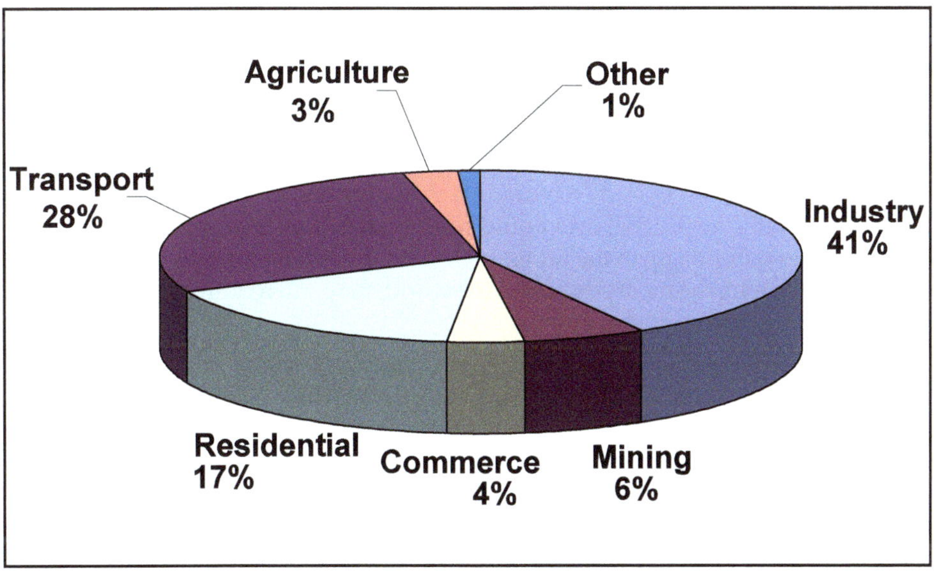


**Figure 4.2: 2000 Final Energy Use by Sector**

Figure 4.3, below, also refers to 2000 and illustrates the split of final energy use by each individual energy carrier. It is of significance to the national Balance of Payments that the largest of these, in energy content terms, is petroleum products. Crude oil is South Africa's single largest import, and the vast majority of the downstream products are utilised by the Transport Sector.

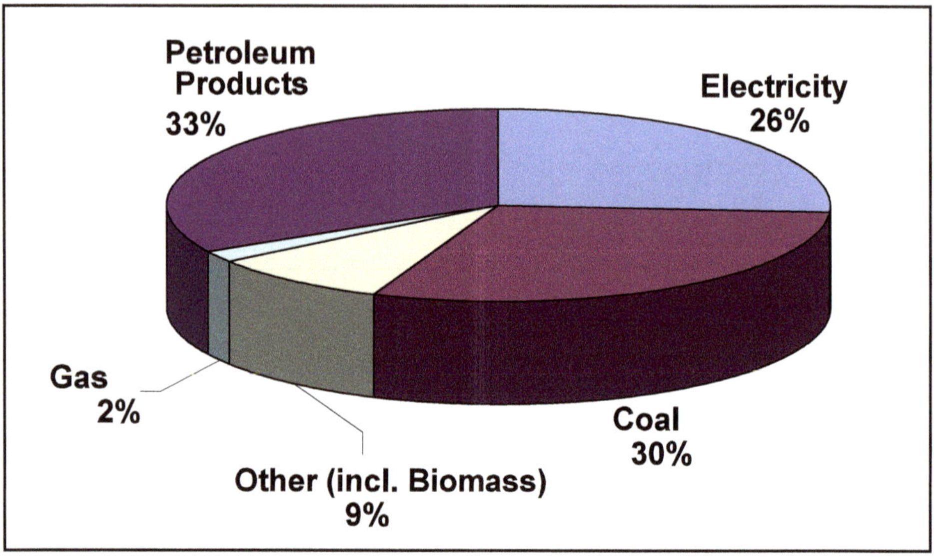


**Figure 4.3: 2000 Final Energy Use by Carrier**

**Energy Intensity**

Compared with developed countries, the South African economy uses a lot of energy for every Rand of value added. In 2001 South Africa had the 26th biggest GDP in the world but was the world's 16th largest consumer of energy. There are two reasons for this. The first is the nature of the activities, which dominate the economy. Mining, minerals processing, metal smelting and synfuel production are inherently intensive users of energy. South African gold mines are very deep with low ore concentrations, so it necessarily requires much energy per ounce of gold. The process used by Sasol to convert coal into liquid fuels is such that only about a third of the energy in the coal ends up in the liquid fuel. Even though South Africa's aluminium smelters are among the most efficient in the world, they still require large amounts of electricity to produce one ton of aluminium.

This will change as South Africa moves into high value manufacturing and service industries, which is already happening. It will also change with changes to processes, for example, when Sasol swaps from coal to natural gas as a feedstock for chemical production at Sasolburg, making production more efficient. However, it must be noted that the low prices of electricity gives South Africa a comparative advantage in high energy intensive industries such as aluminium smelting. If the aluminium could be beneficiated into products of higher value, South Africa would gain even more.

The second reason for the high-energy intensity is that South Africa is sometimes wasteful in the use of energy. Low energy costs have not encouraged industry, commerce, transport and households to adopt energy efficiency measures. There is a lack of awareness of energy efficiency. However, with the prospects of higher energy prices in future and with growing environmental awareness, especially about the emission of greenhouse gases, there is a growing concern in South Africa to promote energy efficiency.

The Figure 4.4 gives an indication of the energy consumption in Petajoules per capita of some selected developed and developing countries. It is clear that South Africa is closer to the developed countries in terms of energy intensity than to the developing countries.

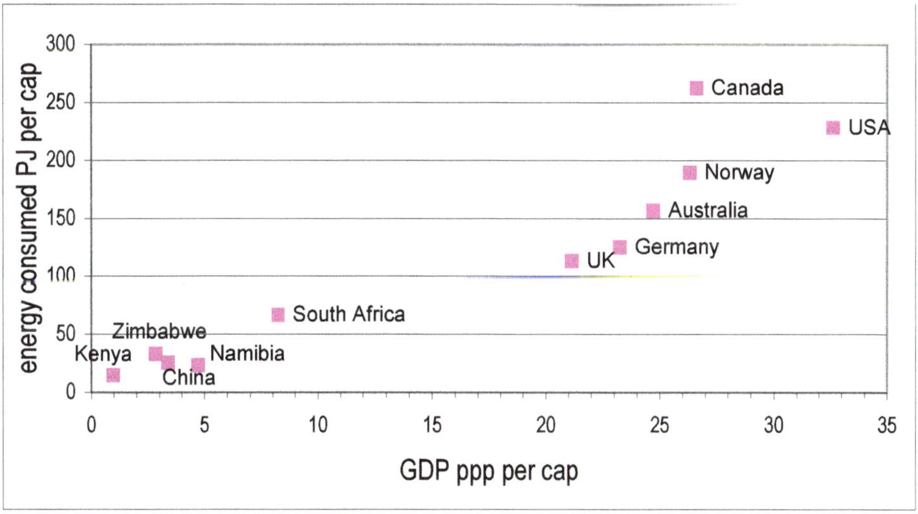


**Figure 4.4: Final Energy Demand per Capita (2000)**

The practice of 'benchmarking', comparing the energy intensities of different sectors in other countries, aids in estimating where reductions in energy use may be achieved in South Africa, and thus determining those interventions which have worked internationally will be likely to be successful in South Africa.

The trend over recent years has been for energy consumption in both the industrial and mining sectors to increase at a steady rate generally concomitant with growth in that sector (exceptions exist such as gold mining where increased mining depth and decreased ore quality have resulted in increased energy consumption per unit output), whereas the trend for other sectors has been to remain level or even decrease where new, more efficient, technology has been introduced to replace old technology *e.g.* Private motorcars.

## Climate Change

At the Rio de Janeiro Earth Summit of 1992 the United Nations Framework Convention on Climate Change stated that its fundamental objective was to achieve stabilisation of the concentrations of Greenhouse gases (GHGs) in the atmosphere at a level that would prevent dangerous anthropogenic interference with the climate system. South Africa ratified the Convention in 1997, which enables South Africa to apply for financial assistance for climate change related activities from the Global Environmental Facility.

The Kyoto Protocol (1997) is an agreement under which industrialised countries (Annex 1 countries) will reduce their combined greenhouse gas emissions by at least 5 per cent compared to 1990 levels by the period 2008 to 2012. Following recent ratification by Russia the United Nations Protocol has become legally binding on 16 February 2005, thereby committing the Annex 1 parties accounting for 61,6 per cent of the total 1990 carbon dioxide emissions to achieve the 5 per cent reduction by 2012.

South Africa acceded to the Kyoto Protocol in March 2002. Although the Kyoto Protocol does not commit the non-Annex 1 (developing) countries, like South Africa, to any quantified emission targets in the first commitment period (2008 to 2012), there is potential for low cost emission reduction options in these countries. The Clean Development Mechanism provides for trade in certified emission reductions between non-Annex 1 countries and Annex 1 countries and thus supports sustainable development with respect to greenhouse gas emissions in developing countries while helping Annex 1 countries to comply with their commitments under the Kyoto Protocol.

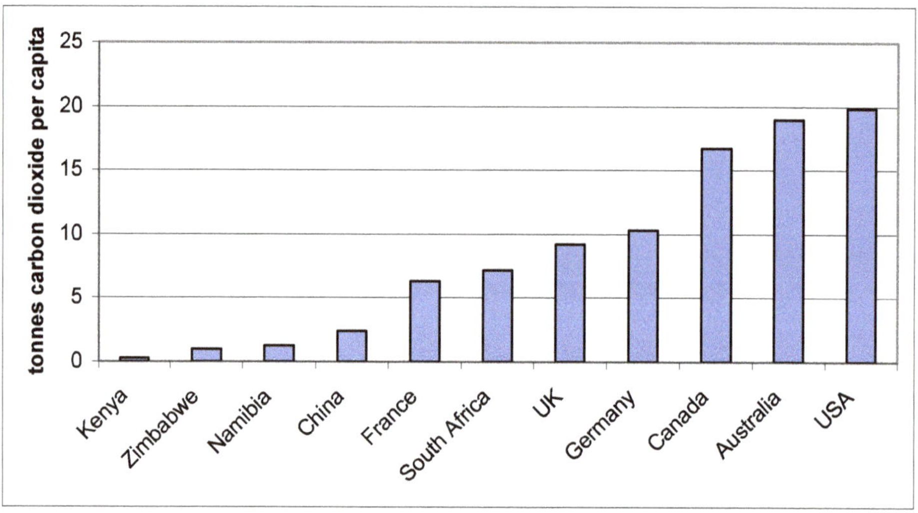


**Figure 4.5: Carbon Dioxide Emissions per Capita (IEA, 2001)**

South Africa is by far the largest emitter of GHGs in Africa and one of the most carbon emission-intensive countries in the world, annually emitting some 7 tonnes of carbon dioxide per capita, as shown in Figure 4.6, due to the energy intensive economy and high dependence on coal for primary energy. Certain energy efficiency initiatives have potential for financial support under the Clean Development Mechanism. Nevertheless over the past few years with the restructuring of the economy there has been a decline in several energy intensive activities. The DME has also

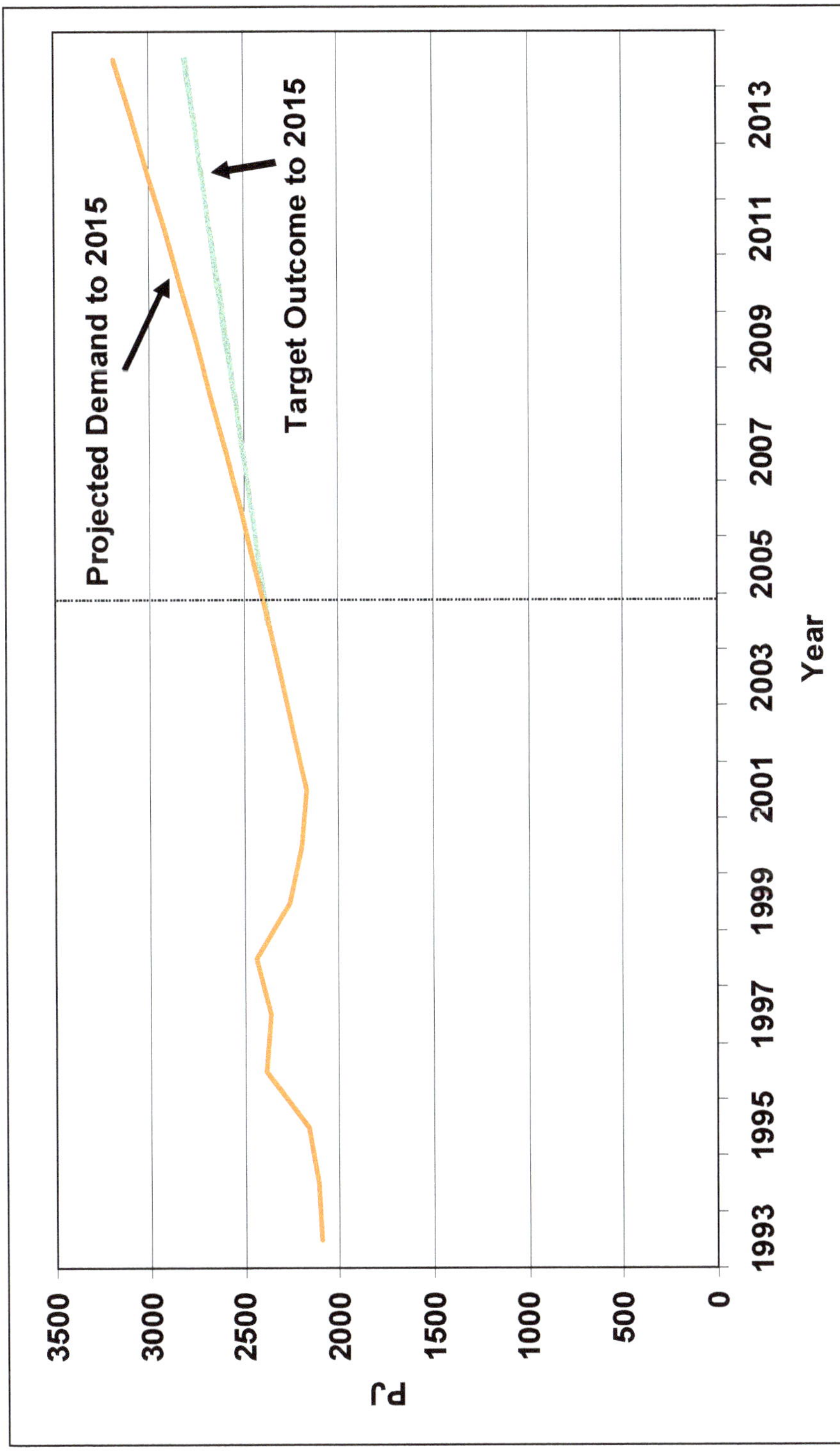


Figure 4.6: Final Energy Demand–Target Outcome to 2015

established the Designated National Authority to process CDM projects and certain other clean development initiatives have taken place such as the publication of a draft "Cleaner Fuels Strategy" (*Joint Implementation Strategy for the Control of Exhaust Emissions from Road-going Vehicles in the Republic of South Africa, 2003*) for public comment. It is expected to be finalised in 2005.

## 3.2 Barriers

One of the fundamental steps necessary to enable successful implementation of any Strategy is the need to understand the barriers confronting it, and how to overcome those.

Several of the more traditional barriers are self-evident, and are described briefly below. In addition to these, however, is a barrier relating to the state of the country itself; the argument being that energy efficiency should be a relatively low priority when compared with other pressing national issues such as quality of life and education. It is important to bear in mind that energy plays an integral part in the solution of these problems, and that without clean and affordable energy such issues will be difficult to resolve.

### Energy Pricing

This is a perceived barrier that stems from South Africa's historically low unit price of coal and electricity, although there has been a gradual and incremental rise in energy prices over recent years. This barrier still holds strong amongst the mind-set of many commercial and industrial organisations that argue that medium and high-cost interventions cannot be justified due to the lengthy paybacks involved.

Energy efficiency makes sound economic sense. Although the unit price of energy may be low, for the time being, the overall *cost* per unit to many industries is high because of the energy-intensive nature of their operations. If energy efficiency is approached correctly and with the right emphasis, payback on investment is frequently less than three years. Education and awareness programmes are some of the first steps to take towards overcoming this barrier.

### Lack of Knowledge and Understanding of Energy Efficiency

Energy efficiency opportunities are frequently overlooked due to the simple fact that industry and other consumers are unaware that they exist. It is the intention of the Strategy to enhance awareness in such matters and to bring knowledge and understanding into the various sectors. This will be achieved through awareness campaigns, demonstration programmes, audits and education, and publicising corporate commitment programmes, and public building sector energy efficiency implementation initiatives. Use of the mass media and electronic options such as websites will be fully explored to publicise energy-saving tips, energy management tools and best practice methods. Where possible joint resources for Demand-Side Management and Energy Efficiency will be capitalised upon for the purposes of promotion, since the cost of awareness campaigns and related measures is too high to be sustained continually if executed individually.

**Institutional Barriers and Resistance to Change**

Institutional barriers often stem from a fear that outsiders will identify previously overlooked opportunities, thereby uncovering apparent incompetence. There is also a frequently encountered misconception, particularly within industry, that energy efficiency will disrupt production processes and that changes should not be made unless absolutely necessary. Typically energy audits are conducted at a plant level in order to determine the costs and benefits of various energy efficiency options that present themselves. Energy service companies do this and advise their clients on the optimal path to follow.

It is important to understand that to a large extent these are emotional barriers. An approach is required, therefore, that is not only professional and technically competent, but also sensitive to such issues.

**Lack of Investment Confidence**

Achieving optimum energy performance sometimes involves the installation of costly plant and equipment, and investors may be reluctant to tie-up financial resources in long-term projects. Recent history has seen a degree of uncertainty, both nationally and internationally, due to the fluctuations in the strength of our currency. This is an ongoing problem, and investors as well as local stakeholders and institutions should be encouraged to cost all externalities when considering energy efficiency investment opportunities. Furthermore, appropriate risk-weightings should be attributed to fossil fuel prices when considering plant lifetime running costs. The notion of introducing incentives on energy efficient appliances and equipment will be considered during the lifetime of this Strategy.

**The Practice of "Bounded Rationality"**

Decision making with limited management resources requires the use of imperfect, or incomplete, information and less than fully rational procedures. This is significant as the majority of energy consumers currently have imperfect information regarding the range and performance of energy efficient products. This fact inevitably results in poor decision-making when purchasing goods or specifying equipment.

It is an intention of the Strategy to enhance the decision-makers' awareness of issues such as running costs, environmental costs, etc. This will be achieved via the adoption of appropriate standards, awareness and education, and by the use of instruments such as appliance labeling.

## 3.3 Targets

To date, only a handful of countries worldwide have set comprehensive targets for energy efficiency improvements. These countries include Slovenia, Japan, The Netherlands and New Zealand. The World Energy Assessment, published by the UN and the World Energy Council, suggests that specific energy usage in industrialised countries could be cost-effectively reduced by 35 per cent over a period of 20 years, if accompanied by effective policies. In the United States, the Electric Power Research Institute (EPRI) has proposed an energy efficiency improvement target of 2 per cent per annum.

This Strategy provides for the implementation of sector programmes in a three-phase approach, timed as follows:

- Phase 1: March 2005 to February 2008;
- Phase 2: March 2008 to February 2011;
- Phase 3: March 2011 to February 2015.

The broad principle of this phased approach is to initiate actions with rapid returns during the early phases. However, it is likely that interventions such as technical standards will also be addressed at an early stage to enable the long-term benefits to be maximised.

This Strategy proposes the following energy efficiency target for South Africa:

| *A Final Energy Demand Reduction of 12 per cent by 2015* |
|---|

The target stated above is expressed as a percentage reduction against the projected national energy usage in 2015. The target will be monitored continuously for progress, using a monitoring system, and an annual report will be issued. The projected usage is forecast at the present increase in economic development over the period and without any additional efficiency interventions. The forecast is derived from the Long range Energy Alternatives Planning tool (LEAP) utilised for developing the National Integrated Energy Plan for South Africa. The baseline scenario is similar to the base case scenario of the IEP ('business-as-usual') in which the following assumptions are made:

- Population growth: 2000=44 million, 2015=53,3 million (1,3 per cent per annum)
- GDP growth: 2,8 per cent average per annum growth over period
- Economic growth: 2,8 per cent over period
- Fuel switching limited apart from general increase in electricity consumption in residential sector

The target is voluntary at present but sub-sectoral targets may become mandatory in the course of time.

The national target is illustrated further in Figure 4.6, where final energy demand is shown as a total of all sectors and is expressed in Petajoules. The *Projected Demand to 2015* is as forecast at an annual growth rate of 2,8 per cent per annum. The *Target Outcome to 201* is shown assuming that the national target is achieved, and that savings are implemented uniformly across the three phases of the Strategy. In actuality it is likely that savings will begin to materialise towards the latter stages of Phase 1 and into Phase 2.

The national target is calculated using the individual targets for each economic sector, and by weighting these according the sectoral contribution to final demand, as previously shown in Figure 4.2. A great deal of research has been undertaken prior to deriving the specific targets for each economic sector. The DME commissioned detailed research projects (referenced in Section 7) to assess the baseline scenario of energy usage in South Africa, together with modelling the outcomes of technical efficiency interventions across the full range of sub-sectors. The emphasis of the DME research projects has been upon technical interventions alone, and the assumptions made in arriving at sectoral targets are considered to be conservative. Additionally, and of equal importance, are the non-technical opportunities for energy savings which exist within most sectors, in particular the buildings, industry and mining sectors. Such opportunities can be broadly defined as *Energy Management Best Practice*, and by inference tend to revolve around "soft" issues such as behavioural change arising from increased awareness, training, accountability and information systems. The importance of effective Energy Management has been demonstrated time-and-again, both in South Africa and abroad, and numerous case studies bear testament to this fact. This Strategy recognizes that Energy Management Best Practice will play a vital role in achieving the national target. DME has commenced an initiative to develop and rollout an Energy Management training and awareness programme to be implemented within the industry and mining sectors.

**Review of Targets**

A review of the national and sectoral targets will be undertaken at the end of each phase. This review will be carried out by the DME with the objective of assessing progress towards targeted outcomes and to address any areas where additional input may be required from stakeholders. The monitoring and measurement of targets is discussed in more detail in Section 3.5.

It is important that the targets are seen to be both *challenging* and *achievable*. In most cases the sectoral targets comprise a conservative estimate of the likely impact of technical interventions, coupled with the additional impact of Energy Management initiatives and behavioral changes described above.

**Industry and Mining Sector**

*A Target Final Energy Demand Reduction of 15 per cent by 2015*

The industrial and mining sectors combined are the largest users of energy in South Africa. A relatively high theoretical potential for energy saving exists, in the magnitude of 50 per cent of current consumption in comparison with international best practice, on a sector-by-sector basis. Notwithstanding this, research has shown that a savings potential of at least 11 per cent is readily achievable using low-cost to medium-cost technical interventions. Furthermore, an additional 5 per cent–15 per cent energy saving would be achievable via proven no-cost and low-cost techniques of energy management and good housekeeping. It is considered, therefore, that the prescribed target of 15 per cent is realistic and achievable.

**Power Generation**

*An interim Target of 15 per cent reduction in "parasitic" electrical usage by 2015*

The issue of targeting energy efficiency improvements within the power generation sector has been the subject of much dialogue between the DME and Eskom throughout the development of this Strategy. When addressing energy efficiency in fossil fuel power stations, there are two generic areas within which savings may lie:

***Within the Core Business Activity***

This constitutes the central thermal plant and steam turbines in existing power stations, together with their immediate ancillary equipment. The efficiency of energy conversion in the core activity is largely dictated by the technology, size and age of plant, together with its prevailing loading characteristics. In the majority of cases these aspects are largely outside the control of the generator and efficiency improvements can usually only be expected upon major capital refurbishments. Notwithstanding this, Eskom is developing its own detailed strategy to address long-term Supply Side issues and to establish targets for its existing fossil fuel stations. Eskom's Supply Side Management strategy and associated target are expected to be released during 2005.

***Within Non-Core Ancillary Equipment***

This is plant and equipment peripheral to the core business, such as minor pumps and fans, compressed air, lighting and air conditioning. It is reasonable to expect that energy savings opportunities for non-core users would be in-line with those expected within industry in general and, where appropriate, within the buildings sector.

An interim target of 15 per cent is therefore set for this sector based upon *savings measures applicable to parasitic electricity losses only*. Subsequent to issuing its Supply Side Management Strategy during 2005, Eskom will incorporate the longer-term savings associated with its core business activity into the overall target for this sector. It is expected to review the target for this sector, therefore, early during Phase 1.

**Commercial and Public Building Sector**

| *A Target Final Energy Demand Reduction of 15 per cent by 2015* |
|---|

Although this sector contributes a relatively minor percentage to national energy usage, savings here are known to be significant. Research and experience have shown that savings through low-cost and medium-cost technical interventions can exceed 25 per cent. In order to allow for partial sectoral penetration of technical measures, however, a figure of 11 per cent has been calculated using detailed modeling analysis. Additionally, a further 4 per cent is thought to be realistically achievable through managerial intervention and behavioral changes.

**Residential Sector**

| *A Target Final Energy Demand Reduction of 10 per cent by 2015* |
|---|

If a target of 10 per cent by 2015 were anticipated that would mean 1 per cent to be achieved per year and the bi-annual target would then be half a per cent every 6 months.

This sector used 360.5 PJ in 2000 in the form of coal (11 per cent), petroleum products (7 per cent), biomass (53 per cent) and electricity (29 per cent). The transition towards the use of higher calorific value fuels and a reduction in the use of thermal energy consumption will be driven by energy efficiency standards in housing, generally higher standards of living accompanying economic growth of 2.8 per cent pa, and the electrification programme. The measures outlined in the interventions section (see 5.4 Residential Sector Programme); mandatory standards, appliance labeling, efficient lighting and standards for non-electric appliances such as energy efficient coal stoves, wood stoves and liquid fuel stoves, should be comprehensive enough to achieve the goals proposed. However, the sector is very diverse, and many energy saving decisions lie at the individual household level, requiring an initial investment. Therefore an ongoing public awareness drive will be necessary to achieve a saving of 10 per cent by the year 2015, based on a projection from the present consumption. An easy to follow guide for households will be developed, such as energy savings tips, taking into consideration that changing people's lifestyle is by no means straightforward.

**Transport Sector**

*A Target Final Energy Demand Reduction of 9 per cent by 2015*

Transport is the second largest sectoral consumer of energy and is expected to grow considerably in the medium-term. Measures to address energy efficiency will not necessarily be easy to implement, as has been the experience internationally where motor vehicles have become the main means of transport. The projected savings will, therefore, only begin to impact during Phases 2 and 3 when planned interventions have begun to be implemented. Several measures will be regulatory in nature, but in order to be effective will rely heavily upon behavioral changes. Such changes are, by experience, the most intractable elements to influence. Another major challenge is the rapid swing in demand towards low efficiency 4x4s and sports utility vehicles.

A target of 9 per cent has, therefore, been established as a realistic but challenging objective for 2014. This target has assumed the introduction of a labeling system for vehicle energy consumption accompanied by other measures (legislative and otherwise) to promote vehicle energy efficiency on South Africa's roads, technology upgrade leading to more efficient vehicles/turnover in the vehicle park, but has excluded taxi-recapitalisation. As the impacts of taxi-recapitalisation may only be evident in the later phases of this Strategy the target will be reviewed in Phase 3. The impact of measures such as public transport systems, moving road to rail and spatial planning are also difficult to assess at this stage and remain interventions with impacts in the long-term.

## 3.4 Outcomes

Table 4.1 summarises outcomes by the eight goal areas of the Strategy, assuming that all targets are met. The goals are largely an expression of the objectives of the *White Paper on Energy Policy*, which represent Government's commitment on a number of socio-economic aims. It should be noted that not all goal outcomes are quantifiable at this stage, so qualitative commentary is provided against some outcomes. In addition outcomes such as job creation, energy poverty alleviation and improved industrial competitiveness are factually substantiated by international experience and studies, although no local investigations have been done in South Africa yet.

**Table 4.1: Projected Outcomes by 2015**

| *Goal Areas* | *Outcomes* |
|---|---|
| **Goal 1 Improve the health of the nation** | • Health benefits realised through reduced atmospheric pollution and improved living conditions, in particular a reduction in respiratory-related illnesses. |
| **Goal 2 Creation of Jobs** | • Long-term employment opportunities increased by reducing costs in commerce and industry;<br>• Employment opportunities increased within the energy efficiency sector and related activities. |
| **Goal 3 Energy Poverty Alleviation** | • Access to affordable energy services improved by promoting low energy alternatives in the marketplace;<br>• Lower energy costs for households by improving domestic energy efficiency. |
| **Goal 4 Reduce local Pollution** | • Atmospheric pollutant levels reduced by a reduction in fossil fuel combustion at power stations;<br>• Local atmospheric pollutant levels reduced by a reduction in fossil fuel combustion within industry and commerce;<br>• Transport-related atmospheric pollutant levels reduced by a reduction in combustion of petroleum products in motor vehicles. |
| **Goal 5 Reduce $CO_2$ Emissions** | • National $CO_2$ emissions reduced by improving energy efficiency across all economic sectors; |
| **Goal 6 Improve Industrial Competitiveness** | • Improved industrial and commercial profitability by controlling and minimising energy overheads;<br>• Improved international acceptability of South African products by minimising the environmental impacts of their manufacture. |
| **Goal 7 Increase Energy Security** | • Increased national resilience against oil price fluctuations by reducing the country's dependence upon imported crude oil supplies;<br>• Increased resilience against internal supply disruptions by reducing load demands placed upon power distribution systems. |
| **Goal 8 Defer Additional Generation Capacity** | • Construction of additional power generation plant deferred as far as practicable by contributing towards Eskom's peak load reduction target. |

## 3.5 Monitoring and Measurement

The *White Paper on Energy Policy* states that:

*"Government will ensure that the necessary resources are made available to establish structures, systems and legislation to facilitate the specification, collection, storage, maintenance and supply of energy data, and energy-related data, according to the requirements of integrated energy planning and international standards."*

Government must be able to measure the effectiveness and efficiency of measures implemented and initiatives put in place for the promotion of energy efficiency, as prescribed by Energy Policy, and must sustain this capacity.

There is a need to establish a system for continuous updating and registration of figures related to energy efficiency, in particular, indicators for efficiency measurement. A formalised system for collecting, managing data and calculating indicators are necessary for monitoring the implementation and success of activities initiated as part of the strategies for energy efficiency. It is DME's responsibility to establish such a monitoring and verification system and ensure that it is implemented. This is also a compliance measure that fulfils requirements by the Draft Energy Bill (section 17 (1) (2)) on the collection of energy information. The DME will take responsibility to ensure that progress towards targets is reviewed and monitored on a continual basis.

The success of such a system will inevitably depend upon using a multi-stakeholder approach, including consultation with representative bodies within each sector. The relevant players are outlined in Section 4.4. The DME has embarked upon a programme to develop detailed methodologies for the monitoring and tracking of sectoral targets. The monitoring plan will take into account all sectoral variables, which are likely to impinge upon the targets prescribed for each sector. Such variables will include factors such as *actual* economic and sectoral growth per sector, thereby enabling energy usage data to be normalised against representative data, which describe sectoral activity over time.

Independent external parties may be used to verify the Department's methodologies and findings.

## 4.0 Implementing Instruments

The Strategy makes use of a range of generic implementing instruments, which are applied as appropriate to meet specific needs within each Sector Programme. The relative maturity of the marketplace will determine which instrument is used where, and how. For example, where interventions are seen as novel and may require some development, the appropriate instruments may include trials and incentives. By contrast, where interventions are already somewhat developed awareness may be the driving need.

### 4.1 Support Mechanisms

The following paragraphs illustrate and describe the main supporting mechanisms and instruments used with each sectoral programme. Such mechanisms are intended to be independent of financial and policy instruments.

**Efficiency Standards**

Efficiency standards have been successfully applied overseas and have brought about significant improvements in efficiencies. South Africa has a well-developed system of standards and codes of practice that, in some cases, may be amended to include efficiency aspects without the need to establish completely new standards. The Draft Energy Bill gives the Minister of Minerals and Energy substantial authority to make standards compulsory.

Mandatory energy efficiency standards will be an important and integral part of the Strategy.

**Appliance Labeling**

Energy labeling of appliances is an internationally tried and tested tool to build awareness and raising capacity about energy consumption.

A number of studies, the latest completed as recently as 2003, have looked into the feasibility of introducing labeling for some household appliances in South Africa. Potential savings with labeling or higher efficiency standards are estimated at 3 PJ in 2012.

The adoption of European Union standards for labeling is considered, as this has already been tested and is widely approved. Furthermore, the success of the EU standards in conveying the message of efficiency to a diverse target-group, comprising a variety of cultural backgrounds, would be of particular benefit to South Africa.

In addition, energy efficiency labeling of motor vehicles will be introduced.

Mandatory appliance labeling for household appliances forms an important element of the Strategy and will be promoted and implemented.

**Certification and Accreditation**

The Strategy makes use of several instruments where inspectors or auditors will be expected to carry out certain technical functions, or studies. These functions will necessarily require a minimum level of technical competence on behalf of the party concerned. Examples include the certification of energy auditors for buildings, industrial plant, and the accreditation of inspectors for Efficiency Standards. The outline requirements of relevant accreditation procedures will be specified by the DME, professional associations and the certification made by SABS.

It is the intention of the Strategy to help develop such accreditation procedures and to enable appropriate certification to be awarded to the relevant aspirants.

**Education, Information and Awareness**

Information and generic awareness are key elements to achieve success in terms of changing South Africa into a more energy efficient society. Once laws and regulations are established, architects will need guidance (from standards, codes of practice, etc.) on how to design houses according to the new regulations, and plumbers should also have be informed about the need to insulate geysers.

Awareness-raising starts with pre-schooling education and runs through all learning fields into the adult education system, under the auspices of the National Qualification Framework (NQF) up to level 8. The DME will engage with the institutions responsible for education and support, and facilitate the inclusion of appropriate education on energy efficiency in the curriculum.

The DME will strive to ensure that:

- Energy Efficiency is taught and examined at all levels in all appropriate subjects, in particular engineering and architecture;
- Energy Efficiency is a competence requirement under the National Qualifications Framework training programmes for skilled workers in the relevant construction and buildings services trades

**Research and Technology**

Technological options represent significant potential for energy efficiency improvements and, in many instances, are well researched and already developed.

However, the majority of these technologies are not manufactured locally and require importation. The latter point will represent a challenge for the Government, particularly as the drive to promote energy efficiency gains momentum.

The National Energy Research Institute will be funded to carry out a dedicated programme of research and development for energy efficiency.

The Strategy will support appropriate research and the possible adaptation of internationally available technologies and processes.

**Regulation**

The historically low unit price of energy, coupled with limited awareness on energy savings potential, may result in only modest success arising from voluntary measures and other non-legislative instruments. For this reason, regulatory means will be applied to achieve further improvements where necessary. Efficiency Standards will have limited impact unless made mandatory, and energy audits should be accompanied by an obligation to implement, for example, all no-cost recommendations identified. The NER will contribute to or develop regulatory measures for guiding reporting and compliance.

**Energy Audits**

Energy audits have internationally been used across all sectors to identify efficiency measures that can be implemented in a cost-effective manner. However, to be effective it has often required both the audits as well as the implementation of measures to be compulsory and to be paid for by the client.

The Strategy will promote energy audits as a means of improving efficiency. Studies will be undertaken to design ways in which audits will achieve the greatest impact.

**Energy Management Systems**

Energy management enables the formalisation of monitoring, evaluating and targeting energy consumption as well as providing sector-specific benchmarking information. Within industrial and commercial applications, the concept of energy management must also embody other key areas, including Training, Motivation and Awareness, Green Accounts (where companies audit the environmental performance of their operation, as well as its economic performance), Energy Policy and formalised Monitoring and Targeting (M&T). The importance of effective M&T cannot be over-emphasised, as it provides the yardstick against which savings are targeted and improvements are measured. Without the key information that M&T provides, attempts to save energy within an organisation can be frustrating, futile and de-motivating.

The implementation plan for the industrial sector will execute a special project designed to develop a monitoring and targeting system in collaboration with the database function at the DME.

The Strategy will support the proliferation of energy management and the establishment of necessary information, including the introduction of Monitoring and Targeting and "Green Accounts".

## 4.2 Policy, Mandate and Governance

The mandate to govern and undertake energy efficiency initiatives is derived from the following documents:

- The South African Constitution;
- The White Paper on Energy Policy, 1998;
- The Municipal Systems Act No. 32 of 2000;
- The Electricity Act No. 41 of 1987 (as amended);
- The Draft Energy Bill;
- The Standards Act;
- The Draft Electricity Regulation Bill

The DME will prepare appropriate legislation and regulations for the Governance and Implementation of this Energy Efficiency Strategy. The regulations will take into account the capacity of the Government to enforce implementation.

The mandate given by the *White Paper on Energy Policy* says that the DME should promote Energy Efficiency through various means as well as consider the establishment of an agency to be instrumental for the Coordination, Leadership and sector capacity development for the implementation of Efficiency. Government's present capacity to undertake energy efficiency programmes is limited. Other countries in similar circumstances have found the establishment of an agency to be an effective means of providing the necessary leadership and capacity to implement programmes.

The DME will, during the early stages of implementation, finalise and consolidate considerations to ensure appropriate leadership in the sector.

## 4.3 Finance Instruments

The formulation and implementation of this Strategy is geared towards self-finance, positive savings and job creation. The majority of Energy Efficiency improvements will materialise through the implementation of standards, regulation and management tools, which at the end of the day lead to short pay back periods for the individual enterprises, house owners and government.

Costs related to investments in equipment or refurbishing of production flows, houses, etc. are to be born by the direct beneficiaries, which is reasonable to expect, due to the short payback periods. Where payback periods are less favourable, the option to partake in other enabling mechanisms should be encouraged. Such mechanisms include the Clean Development Mechanism (CDM), developed under the 1997 Kyoto Protocol.

For Government the real costs to implement this Strategy relate to the massive information and coordination requirements – such costs are to some degree flexible and therefore, kept at the implementation level rather than the Strategy level.

As an example, the successful implementation of energy efficiency in the electricity sector leads to reduction in costs to end users due to deferred generation investments. Therefore subsidising energy efficiency to accelerate implementation will prove beneficial to all users of electricity. This is the justification for the current subsidising of DSM and energy efficiency projects through the Eskom DSM Fund. The costs of intervention to the end consumer are at this stage best quantified by comparison with the costs of a 'business-as-usual' alternative – increasing electricity tariffs to pay for new generating plant. The justification for an energy efficiency strategy is identical to Eskom's DSM programme vis-à-vis deferred capacity.

**Incentives**

At this stage of South Africa's development it is difficult to justify government subsidies for Energy Efficiency when there are so many other pressing needs nationwide.

However the continuous process of fiscal reform does present opportunities to promote energy efficiency as part and parcel of the reform process. For example National Treasury has announced that it is reviewing the vehicle allowances in personal income tax that favour larger (and therefore less efficient) engines.

**Fee Bates**

Fee bates are under consideration for vehicles. The basic principle of fee bates is that levies are imposed upon less efficient vehicles and the funds collected used to cross-subsidise more efficient vehicles. National Treasury is simultaneously developing *"A Framework for Considering Market-Based Instruments to Support Environmental Fiscal Reform on South Africa"*. These two developments are being aligned to achieve the desired impact.

The Government's Motor Industry Development Plan has already enjoyed some success in shifting demand towards small more efficient cars. It presents an alternative instrument that could be used.

**Financing the Public Sector Implementation Plan**

Presently it appears that some cost effective capital measures in the Public Building Sector remain unimplemented because there are no specific budgets to cater for them. However, National Treasury have approved that within an existing Medium-Term Expenditure Framework it is possible to transfer operating cost budget items to capital cost budgets on the proviso that this is accomplished within the normal three year budget cycle. This three-year cycle constraint will probably require energy efficiency investments to be divided up and staggered over several years. The incentive for National Departments will be that they will be able to retain the savings that arise from the energy efficiency measures that they implement, once they have paid off the costs thereof.

The same budgeting and delivery mode will be extended to all Provincial and Local Government authorities and state owned entities that are funded by Parliamentary appropriations.

This presents another enabling instrument for cost effective capital measures with short to medium term payback periods.

**Energy Service Companies**

The business of an Energy Service Company (ESCO) is to sell energy services. Energy management being one of the most common activities with saved energy as the main product. The ESCO meets client needs to reduce costs, improve energy efficiency, manage risk and enhance a competitive edge. The ESCO will typically offer this through a package, which includes a comprehensive energy audit service, financing mechanism, equipment procurement, and installation and commissioning, operation monitoring and performance guarantees. This addresses those situations where companies do not have the expertise and resources to devote to energy management activities themselves.

Saved energy cost is typically used to meet the implementation cost of saving energy. There are several payments options that include lump sum payment or once off payment and performance contracting paid through shared savings where a percentage of the cost savings are split between owner and the ESCO. This approach can be used to implement resource management measures, such as energy management in industrial and commercial and even in the residential sectors. The strength of the approach is in the delivery of results i.e. actual energy cost savings by the ESCO enabling the customer to focus on their core business.

Although the ESCO concept is not new, it is not yet fully developed in South Africa. Players do exist in the marketplace, but they are relatively few and many of them have limited experience. As a result a confidence gap exists in the market where customers are sceptical and reluctant to use this service delivery mode. This is exacerbated by an absence of a well-established performance standard and service provider certification and accreditation. DME's role in respect of ESCOs will be primarily devoted to addressing this deficit by creating a more formal framework within which they can operate. This will consist of accredited performance standards and approved methodologies for energy efficiency audits as well as skills training accreditation.

Despite this, it is accepted that the ESCO model does have an important role to play in the delivery of energy management services in South Africa. Indeed Eskom's own Demand Side Management programme uses the ESCO concept as its key delivery mechanism. There is, therefore, merit in supporting and strengthening this service delivery mode given the potential to achieve energy savings at minimum cost to the Fiscus. Regulation of ESCOs is probably unnecessary as the risk of non-delivery is carried by the ESCO. However, the type of contract is quite crucial.

**Clean Development Mechanism**

The Kyoto Protocol was adopted at the third Conference of the Parties in 1997. The Protocol provides that developed nations accept commitments to limit, or reduce, the emission of greenhouse gases according to differentiated targets. For the signatory parties, termed Annex 1 countries, this represents an overall reduction of 5 per cent by the period 2008 to 2012 in relation to their combined emissions of greenhouse gases in 1990. South Africa ratified the UNFCCC in August 1997 and acceded to the Kyoto Protocol in March 2002 as a non-Annex 1 signatory.

Achieving such goals will result in significant costs to the economies of each Annex 1 country, and a number of mechanisms were developed to assist these countries to comply with their respective targets. One of these mechanisms is the Clean Development Mechanism, or CDM. The basic principle of the CDM is simple: developed countries can invest in low-cost abatement opportunities in developing countries and receive credits for the resulting emissions reduction. Such credits would then count towards their own abatement targets.

The CDM can positively contribute towards the sustainability objectives of a developing nation by:

- Transferring technology and financial resources;
- Developing sustainable methods of energy production;
- Increasing awareness of energy efficiency and environmental issues;
- Alleviating poverty through income and employment generation;
- Helping define investment priorities in projects that meet sustainability goals.

The CDM encourages developing countries to participate by promising that their own development priorities will be addressed as part of the package. This recognises that only through long-term development will all countries be able to play a role in protecting the environment.

**Demand Side Management**

At current energy usage levels and national economic development projections, it is estimated that without any interventions to reduce peak electricity demand, there will be a need to invest in new power generating capacity by around 2007. As the construction of additional generation plant is an extremely costly and lengthy process, Eskom has embarked upon a Demand Side Management (DSM) programme in order to reduce the capacity and costs of such an investment, by using a combination of energy efficiency measures, load management and negotiated interruptible supplies.

The DSM provision in Eskom's current Integrated Energy Plan is for a peak load reduction target of 1.37GW by 2015. In addition, there is an interruptible supply agreement target of 1.511GW by 2007 and 0.409GW by 2015. This target is likely to change dynamically over time in response to the actual requirement for DSM in South Africa. It is also likely to change in response to the effectiveness of interruptible supply agreements. Presently, however, the load reduction target is comprised of the following elements:

- 0.81GW from load shifting;
- 0.56GW from energy efficiency measures.

Eskom has adopted the ESCO methodology as the key tool to implement the DSM process. Energy efficiency and load management projects are implemented and managed via a third-party (the ESCO) and savings resulting from the intervention are shared between the host site, Eskom and the ESCO in approximately equal proportions. Capital funds are currently gathered through supply tariffs and administered by Eskom. The NER is responsible for ensuring that Eskom meets its Energy Efficiency and DSM targets. (The current funding mechanism is under review and discussion between Eskom and NER).

It is acknowledged that this methodology is inherently subject to the constraints and barriers previously outlined for ESCOs in the section above. Despite this there are 80 ESCOs already registered and projects have been implemented yielding a demand reduction of 187,2MW in 2004. The aim is to achieve an annual target of 153MW over the period 2003-2010.. In addition, approximately 500MW worth of potential projects are in the pipeline (as at end of October 2004).

**Energy Pricing**

Following from the specific policy objectives of the *White Paper on Energy Policy* (1998), energy pricing will be based on an assessment of the full economic, social and environmental costs and benefits of policies, plans, programmes and activities of energy production and utilisation. That is to say, a process of moving away from cross-subsidies towards cost-reflective prices will generally be adopted.

### 4.4 Stakeholders

The South African energy arena is characterised by a number of diverse role players each with a mandate within the fields of energy supply, conversion, efficiency and regulation. Only through well co-ordinated initiatives and promotion to activate the different role players will South Africa be able to effectively promote energy efficiency.

The DME will prompt the different stakeholders to take a leading role in their areas of responsibility on a sector-by-sector basis. The means will be information, regulation, promotion, and facilitation of an enabling capacity development framework, as well as the coordination of knowledge and actions where necessary, as well as publicised public comments.

Figure 4.8 shows how the key stakeholders will be involved in the strategic processes described in this document. Stakeholder relationships are shown against each implementing instrument, or focal area, in terms of primary stakeholders and secondary stakeholders. Primary stakeholders may be broadly defined as those whose main functionality deals directly with the associated focus area. Secondary stakeholders may be described as those whose responsibilities partly overlap with a particular focus area, or where their involvement would be of an *ad hoc* nature.

## 5.0 Sector Programmes

This section outlines the planned programme of interventions relating to each economic sector.

In order to meet the objectives of this Strategy, it is intended that energy efficiency interventions will be implemented through a phased approach. The timing of the three Phases is as follows:

- Phase 1: March 2005 to February 2008;
- Phase 2: March 2008 to February 2011;
- Phase 3: March 2011 to February 2015.

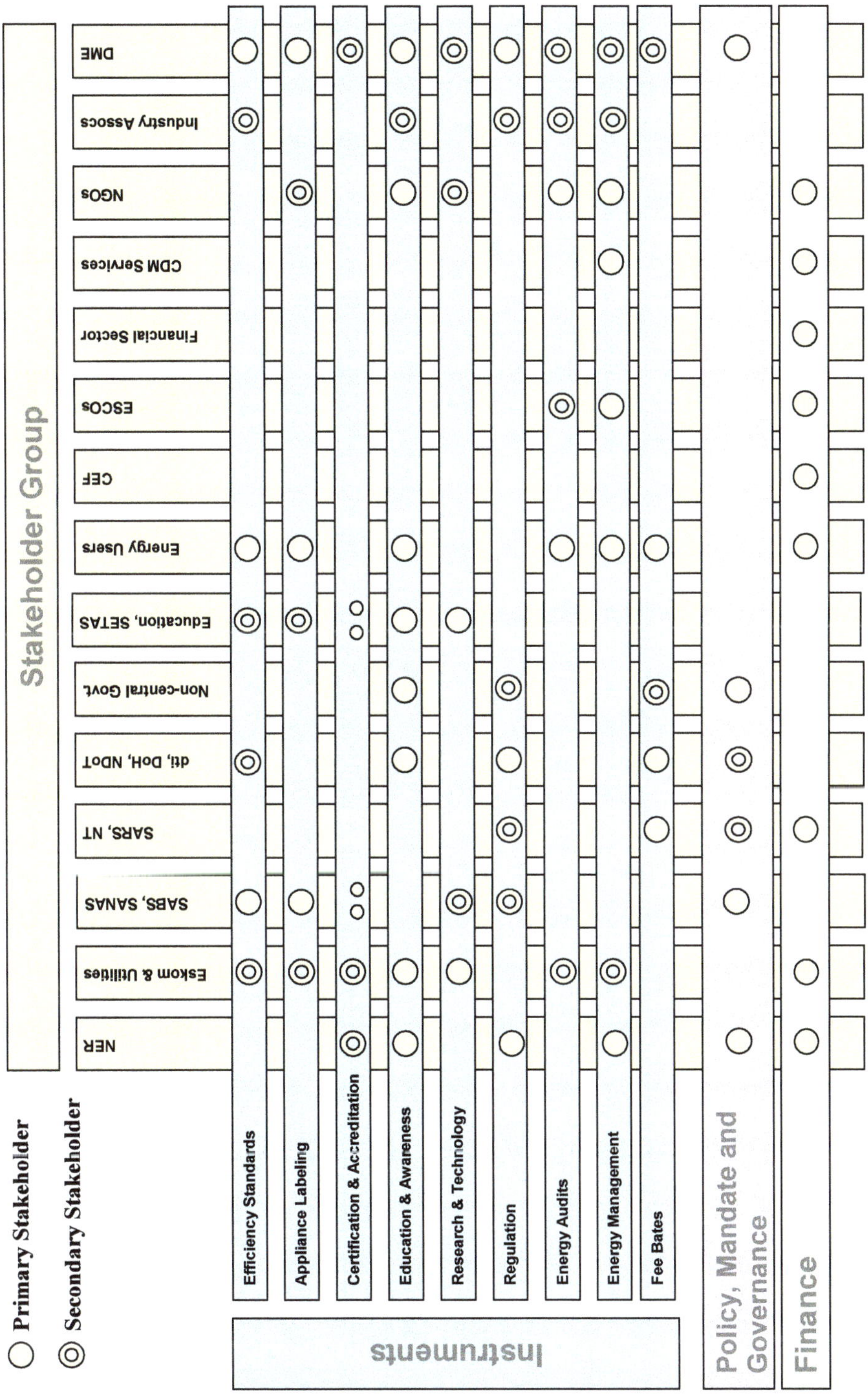


**Figure 4.7: Stakeholder Responsibilities**

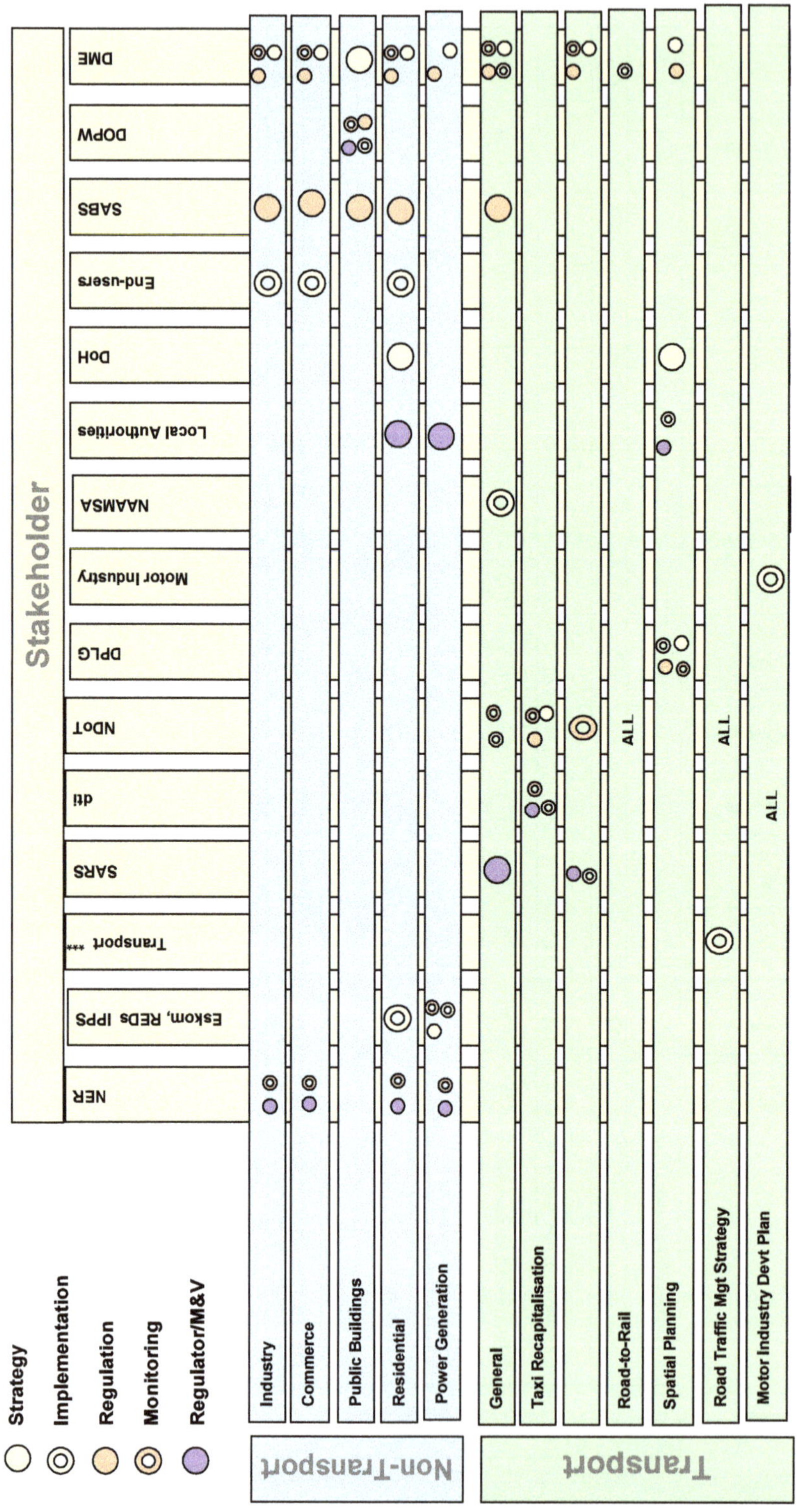


**Figure 4.8: Sectoral Stakeholder Responsibilities**

**** Transport includes the Department of Public Enterprises, Transnet, Spoornet, the SAA, the Ports Authority and Port Operations.

It is the broad intention of the Strategy that these phases will be linked to the economic characteristics of each activity. For example, it is probable that a low-cost intervention with a rapid payback would be implemented during the early stages of Phase 1. Conversely, a high-cost or complex intervention may only be initiated during the latter phases of the Strategy.

Whilst the economic criteria are important in determining the phasing of interventions, issues such as technical standards will be addressed at an early stage so as to maximise the long-term benefits. The Strategy will be reviewed at the end of each Phase. It should be noted that the different Phases are not mutually exclusive and that some interventions may transverse more than one Phase.

The interventions for each sector are described in outline on the following pages. These sections will form the basis for the elaboration of detailed implementation plans that will be developed for each sector. Implementation plans for the Industrial Sector and Commercial and Public Building Sector are currently being finalised for action following approval of the Strategy. Plans for the Residential and Transport Sectors will be developed during 2005.

## 5.1 Industry and Mining Sector Programme

### Key Facts

- Industry and Mining account for 47 per cent of total end-user energy demand in South Africa;
- Industrial energy usage is dominated by a small number of energy-intensive industries. These include ferrous and non-ferrous metals processing, mining, pulp and paper, and the petrochemical industry;
- The energy intensities in these industrial sectors are typically higher than those of other first world countries;

### Core Objectives

- To decouple the rate of growth of industrial energy consumption from the rate of growth in industrial output;
- To bring the energy intensities of major industrial sectors into-line with international standards and best practice.

### Approach

- A suite of "leadership by demonstration" programmes will be implemented. The DME will drive this activity, which, it is intended, will be actioned by voluntary initiatives within industry itself. There may be opportunity to utilise donor funding for certain specific projects. These programmes are intended to build industrial capacity in the area of energy management and best practice, and to give incentive for replication nationwide;
- A series of mandatory standards will be introduced in phases. The intention is to ensure that life-cycle costs are considered where the purchase of "horizontal technologies" is concerned. The term horizontal technology

refers to any specific technology which is commonly used across many industrial sectors;

- An obligation to carry out energy audits within the energy-intensive industries will be introduced in phases. This will assist in awareness raising and in ongoing improvements in energy efficiency.

## 5.2 Commercial and Public Buildings Sector Programme

### Key Facts

- Commercial and Public Buildings account for 3.5 per cent of final energy demand. The Commercial sector alone contributes 45 per cent towards total national GDP;
- The majority of energy is used in the form of electricity, the main end-uses being HVAC systems, lighting and office equipment;
- The Commercial sector is undergoing significant growth, which presents the opportunity to capture energy efficiency at the design stage of new stock.

### Core Objectives

- To demonstrate the Government's commitment to sustainable energy development within its own building stock;
- To progressively upgrade the energy performance of existing public and commercial building stock;
- To achieve best practice energy performance in new public and commercial building stock.

### Approach

- The Government will lead by example through raising energy efficiency awareness and by implementing specific measures within its own estate;
- Energy efficiency standards for buildings will be introduced and made mandatory, together with a building Energy Audit programme;
- Emphasis will be placed on incorporating energy efficiency into building design and energy efficient technologies will be introduced in existing buildings;
- Energy management systems for buildings will be tested, demonstrated and promoted;
- In conjunction with the implementation of SANS 204, energy labels will be developed to assist with compliance rating.
- The standard for office buildings (SANS 204) will be made mandatory by its incorporation into the National Building Regulations.

**Table 4.2: Industry and Mining Sector Programme**

| *Output Activity* | *Measures* | *Timeframe* | *Responsibilities* |
|---|---|---|---|
| **Norms and standards for horizontal technologies** Technical standards to be developed for Industrial boiler efficiency, electric motors and thermal insulation. These horizontal technologies are common to many industries and such standards will encourage lifetime cost assessments to be carried out when purchasing new plant. | • Research into existing national and international standards<br>• Formulation and adoption of standards<br>• Training of inspectors and operators<br>• Implementation of standards<br>• Monitoring and results dissemination | Phase 1<br>Phase 1<br>Phase 1–2<br>Phase 2<br>Phase 3 | DME, dti, SABS, Education Providers, Energy Services Industry engaging with Industry Associations and wider industry |
| **Energy Audit Scheme** This will encourage capacity building within audit providers and will promote energy efficiency best practice within industry itself. The scheme will be tested initially via a series of trial audits in industry. It is intended that regular energy auditing will be made mandatory for high-energy users in due course. | • Develop certification process for auditors<br>• Certification of auditors<br>• Trial audit scheme through specific sectors<br>• Ongoing mandatory scheme (non-subsidised) in specific sectors<br>• Monitoring and results dissemination | Phase 1<br>Phase 1<br>Phase 1–2<br>Phase 2–3<br>Phase 1–3 | DME, NER, SANAS, ESETA and Education Providers engaging with Industry Associations and Industry generally |
| **Energy Management Best Practice** A solid Energy Management foundation is essential in any firm in order to optimize energy efficiency best practice. The key tenets of good Energy Management are information (Monitoring & Targeting), Training & Awareness (Motivation) and corporate commitment (company policy). The importance of these will be demonstrated and promoted. | • Undertake Monitoring and Targeting demonstration projects<br>• Investigate suitable industrial training schemes and develop a programme of industry training support<br>• Establish a Corporate Commitment programme to encourage adoption of energy efficiency policy into existing corporate policies.<br>• Monitoring and results dissemination | Phase 1<br>Phase 1<br>Phase 1–2<br>Phase 2–3 | DME, NER, Eskom, Education Providers, Funding Agencies engaging with industry |

*Contd...*

**Table 4.2–Contd...**

| *Output Activity* | *Measures* | *Timeframe* | *Responsibilities* |
|---|---|---|---|
| **Technology Information and Research**<br>It is important to have a sound technology information base to industry in general. This will ensure that appropriate efficient technology is adopted and will potentially lead to further research-based projects. | • International benchmarking study to enable more meaningful benchmarking by industrial sector<br>• Establish a technology and information base, using existing literature<br>• Promote awareness of available information | Phase 1<br>Phase 2<br>Phase 2–3 | DME, Education Providers, International EE Agencies, Industrial Associations |
| **Promotion of Energy Service Companies (ESCOs)**<br>ESCOs are already playing a vital role in Eskom's Demand Side Management (DSM) programme. There is potential to further develop the energy service provision market to encourage greater energy efficiency within the industrial sector. | • Investigate the status of the existing ESCO market<br>• Establish means of enhancing the promotion of ESCOs within industry | Phase 1<br>Phase 1, ongoing | DME, Eskom DSM and NER engaging with ESCOs and wider industry |
| **Maximise the Value of Energy Efficiency Investments**<br>This activity will encourage and facilitate the conversion of eligible Carbon Credits into real cash benefits for industry, via the Clean Development Mechanism (CDM). | • Establish DNA function and procedures<br>• Improve CDM awareness and publicise successful CDM projects | Phase 1<br>Phase 1, ongoing | DME, dti and CDM service providers engaging with industry |

**Table 4.3: Commercial and Public Buildings Sector Programme**

| *Output Activity* | *Measures* | *Timeframe* | *Responsibilities* |
|---|---|---|---|
| **Energy Efficiency Standards for Commercial and Public Buildings** | • Develop Energy Efficiency Standard for Office Buildings (SANS 204) | Phase 1 | DME, SABS, CSIR, Architects, Building Industry |
| | • Research knowledge gaps | | |
| | • Stakeholder consultation | | |
| | • Finalise SANS 204 | | Government Departments, dti, DME, National Treasury, Legislature, SABS |
| | • Incorporate SANS 204 in National Building Regulations | Phase 2 | |
| | • Energy Labels (Compliance Rating–re SANS 204) | Phase 3 | |
| | • Implementation of standards | | DME, SABS, Building Industry |
| **Mandatory Energy Audits for Commercial Buildings** | • Prepare audit standard and framework | Phase 1 | DME, Training Certification Authorities – SETAs, Public Works |
| | • Prepare monitoring and evaluation protocol | Phase 2 | |
| | • Identify and prioritise categories of buildings to be audited | | DME |
| | • Identify/train/certify both trainer and trainee auditors (BEE) | Phase 3 | Treasury/DPW/DME |
| | • Determine/clarify financing mechanisms | | DME |
| | • Prepare and commission audit scheme | | DME |
| | • Monitor quality of audits as well as effect on overall consumption | | |
| **Energy Management Systems** | • Test and showcase energy management systems | Phase 1 | SABS, Manufacturers, DME, NER, DPW, Building Industry |
| | • Promote energy management systems | | |
| | • Initiate Green Accounts systems | Phase 1 | |
| | • Implement Green Accounts | | DPW, DME, NER |
| **Technologies** | • Efficient Lights Programme | Phase 1 | DPW, DME, NER |
| | • Part of SANS 204 | | |
| **Thermal Measures (HVAC)** | • Monitor Programme | Phase 2 | DPW, DME, NER |

## 5.3 Residential Sector Programme

### Key Facts

- The Residential Sector accounts for 16,4 per cent (2000) of final energy demand;
- Much of this energy is consumed in the form of biomass in the rural areas, but an increasing amount of electricity is used in middle and high income homes and as the national electrification programme reaches more users
- Savings can be anticipated in thermal energy demand from the incorporation of energy efficiency measures (thermal insulation) in new housing, from the implementation of appliance labeling and standards and through massive education and awareness campaigns.

**Core Objectives**

- To combat pollution on health grounds;
- To mitigate the effects of Peak Demand on power capacity;
- To introduce standards for housing and labeling/efficiency standards for household appliances;
- To introduce state-of-the-art technologies.

**Approach**

- Awareness raising to communicate the cost-benefits of energy efficiency in the home;
- Introduction of appliance labeling;
- Demonstration projects to create an incentive to invest in energy efficiency;
- The approach will initially address higher income (*i.e.* higher usage) homes and state-subsidised housing incorporating energy efficiency measures as a standard feature
- The standard for energy efficient housing (SANS 283) will be made mandatory by its incorporation into the National Building Regulations.

**Table 4.4: Residential Sector Programme**

| *Output Activity* | *Measures* | *Timeframe* | *Responsibilities* |
|---|---|---|---|
| **Standard for Housing** | • Standard 283 for energy efficient housing<br>• Incorporate SANS 283 in National Building Regulations<br>• Monitoring and dissemination of results | Phase 1<br>Phase 2 | DoH, EE Experts, CSIR, Building Industry, Thermal Insulation Industry<br>DoH, Building Industry<br>DoH, Building Industry<br>DoH, DME |
| **Appliance labeling** | • Establish standards for household appliances<br>• Label household appliances<br>• Make the Label mandatory<br>• Market appliances with labels<br>• Monitor progress | Phase 1<br>Ongoing | DME, SABS, Eskom, NER, Appliance Manufacturers, retailers, servicing industry |
| **Awareness Raising Program** | • Development of specific program<br>• Implementation | Phase 1<br>Ongoing | DME, NER, Eskom DSM |
| **Efficient Lighting Program** | • Demonstration in all sectors<br>• Implementation<br>• Monitoring | Phase 1<br>Ongoing<br>Ongoing | DME, NER, Eskom DSM, DPW |
| **Non-electric Appliance Standards** | • Study of fossil and biomass-using appliances<br>• Draft standards developed<br>• Standards (mandatory for some appliances)<br>• Implementation | Phase 1<br>Phase 2 | DME, Manufacturers, Eskom, SABS<br>DME, Manufacturers, Eskom, SABS |
| **Fuel Standards** | • Studies<br>• Development of standards<br>• Implementation | Phase 1<br>Phase 2 | DME, Manufacturers, SABS<br>DME, Manufacturers, SABS |

## 5.4 Transport Sector Programme

**Key Facts**

- Transport was responsible for 27 per cent of final energy demand in 2000. Petroleum products represented 97 per cent and electricity 3 per cent of energy demand in this sector. Road transport represents 84 per cent of energy use;
- Petrol and diesel are mainly used for road transport of passengers and freight. In terms of primary energy supply, nearly 80 per cent is from imported oil–representing the largest item on South Africa's import account;
- Transport fuels represent a significant portion of the country's imports into the economy. Government is concerned that the impact of crude oil imports on the economy should be contained.

**Core Objectives**

- Build the user base for public transport through provision of quality public transport and non-motorised transport services and infrastructure;
- Incorporate international best practice into new developments for housing, government services, sports and entertainment that support the public transport objective;
- Put in place a regime of monitoring mechanisms, penalties and rewards. Conduct research to articulate the value of energy efficiency to local transport managers and customers and use results to develop target-based performance objectives for local authorities to meet in terms of public transport provision;
- Increase public knowledge and awareness of efficiency issues, including specific efficiency indicators, as well as the engendering of a "civil responsibility" regarding environmental and sustainability issues;
- Finalise an appropriate mode freight logistics policy that incorporates the energy efficiency objective

**Approach**

- Include transport fuel efficiency promotion in ongoing fiscal reform
- Fuel efficiency labelling of vehicles;
- Conduct and support fleet audits stressing regular vehicle maintenance;
- Support Taxi-Recapitalisation process by showing efficiency potential;
- Regulations, standards and codes of practice which will stimulate the supply of energy efficient vehicle technologies
- Public information programmes that would sensitise the motoring public to the benefits of efficiency measures.
- Complimentary programmes to change the country's transport infrastructure (moving from road to rail), and the demand placed upon it by users (spatial planning) can be viewed as longer term interventions

**Table 4.5: Transport Sector Programme**

| *Fundamental Goal* | *Output Activity* | *Measures* | *Timeframe* | *Responsibilities* |
|---|---|---|---|---|
| **Optimise Passenger and goods transport** | **Passenger Transport Management policy and regulatory/incentive dispensation** | • Draft appropriate goods and passenger transport policy, accompanying regulations and necessary fiscal and budgetary reform | Phase 1 | NDoT, NT, DME, SARS, Local Government |
| | **Establish local Transport Authorities (LTAs)** | • Finance establishment of local Transport Authorities<br>• Audit local Integrated Transport Plans for energy efficiency<br>• Quantify EE baseline with TAs and establish targets | Phase 2<br>Phase 3<br>Phase 3 | NDoT, DME, Local Government |
| **Introduce Energy Management Measures in Passenger Transport** | **Fee Bate** | • Vehicle licensing<br>• Differential licensing fee<br>• Implementation, monitoring and dissemination of results | Phase 1<br>Phase 2 | NT, SARS, NDoT, DME, Local Government |
| | **Efficiency labels for motor vehicles** | • Development of energy efficiency label<br>• Introduction and marketing of label | Phase 1<br>Phase 2 | DME, NDoT, STANSA, NAAMSA, SAPIA |
| | **Emission standards for vehicles** | • Development of mandatory standard<br>• Implement standard | Phase 1<br>Phase 2 | STANSA, DME, DEAT, NAAMSA, SAPIA |
| | **Promote diesel vehicles** | • Taxi re-capitalization<br>• Tax differential on petrol and diesel | Phase 3<br>Phase 2 | NDoT, DTI<br>NT, SARS, DME |
| | **Audits on vehicle fleet operators** | • Develop audit standards for fleet operators<br>• Implementation and monitoring | Phase 2 | NDoT |

*Contd...*

**Table 4.5–Contd...**

| *Fundamental Goal* | *Output Activity* | *Measures* | *Timeframe* | *Responsibilities* |
|---|---|---|---|---|
| | **Awareness Raising Program** | • Awareness and education on driving efficiently<br>• Influence learner driver curriculum | Phase 1/ongoing | DME, NDoT, AA, Local Government |
| | **Roadworthy test including emission test** | • Include emission standards in Roadworthy Certificate<br>• Develop training curriculum for vehicle inspectors | Phase 2 | NDoT, STANSA, DME, DEAT, Local Government |
| **Facilitate Energy Efficiency in Freight Logistics** | **Advance Appropriate Mode Freight Logistics Policy** | • Finalise research on impacts of shifts between road and rail<br>• Audit EE of goods moved via range of freight modes<br>• Explore implementation of an energy efficiency levy on freight movements consistent with energy impacts | Phase 2<br>Phase 3<br>Phase 3 | NDoT, DME |
| | **Exploration of renewable energy in the freight logistics sector** | • Study regenerative braking systems on electric locomotives<br>• Run pilot project to detail costs and benefits of bio-fuel mixes to road freight sector<br>• Ensure that energy efficiency criteria are included in the capital investment plan. | Phase 3<br>Phase 3 | NDoT, DPE, Spoornet, SAPIA |
| | **Intelligent transportation System data management** | • Establish ITS forum with key stakeholders in freight sector | Phase 3 | NDoT, STANSA, SAPS, freight logistics companies, IT sector |

## 6.0 Cross-Cutting Issues

### 6.1 Integrated Energy Planning

Expenditure on energy constitutes some 15 per cent of South Africa's GDP. Therefore energy efficiency is an important facet of the Integrated Energy Planning process carried out by the DME. By virtue of its size and economic importance, the energy sector periodically requires considerable investments in new supply capacity, which impacts on the economy. Integrated resource planning decisions around the world now consider not only maintaining security of supply but give full consideration to the economic, environmental and social impacts of all alternatives, such as demand-side management and energy efficiency programmes. This Energy Efficiency Strategy will be used to inform the National Integrated Resource Plan of the National Electricity Regulator as well as the National Integrated Energy Plan of the DME.

### 6.2 Renewable Energy

There are several areas of overlap between Renewable Energy and Energy Efficiency that warrant a brief discussion.

A widespread installation of solar water heating in industrial and commercial buildings and houses has the potential to defer the need for building new power plants, as the combined heating requirements of these sectors consume the energy produced by three average power stations. The main constraint on implementing a national solar water-heating programme in the Residential Sector relates to cost, which is a function of the current small market and lack of economies of scale. This lack of demand in itself is due to low public awareness of the technology or its economic benefits. Currently the cost of a domestic solar water heater would take in excess of 5 years to pay back.

However, it is a different story when it comes to large commercial installations. Because of the size, and electricity tariff at peak times, these solar water heaters are competitive with electric geysers and hotels are installing them. The main barrier again remains lack of information about the technology.

Thermally efficient housing – houses designed to save energy, can reduce household space heating requirements. The Department of Housing in collaboration with the Department of Minerals and Energy, has developed appropriate guidelines for the construction of thermally designed housing incorporating passive solar design.

The household sector requires the following measures:

- Regulation of low-cost energy efficiency measures in housing;
- Incorporating passive solar design;
- Heat insulation in homes;
- Replacement of electric geysers by solar water heaters.

Currently the Department of Housing is re-appraising its subsidised housing strategy in preference of quality rather than quantity and will revise the Housing

Code to incorporate heat insulation. A similar approach will be followed through energy efficiency standards developed by STANSA for both buildings which are naturally ventilated (such as houses) and for office buildings that are artificially ventilated.

The implementation of the measures mentioned above will clearly reduce the need for power, mostly during periods of peak demand eg. solar water heating, heat insulation, passive solar design; and can therefore be viewed as energy efficiency interventions which reduce demand. As such DME would view such projects as eligible for funding from appropriate sources, such as the Eskom DSM Fund and indeed the NER has approved an amount of money for exactly this purpose.

In terms of bio-fuels the Transport Sector presents an important opportunity for exploring energy efficiency measures in the fuel mix in road freight. Renewable power in rail freight using regenerative braking systems in locomotives will be investigated (which also provide a means of generating power for the National Grid). Government has announced a massive investment programme for Spoornet to improve its efficiencies and thus to win back customers lost to road transport. Efforts will be made to introduce energy efficiency criteria when these investment decisions are made.

Raising awareness regarding the economic benefits of energy efficiency and renewable energy is an important step in increasing the market demand for these technologies. The development of an information strategy for both energy efficiency and renewable energy is therefore an immediate short-term priority.

## 6.3 Environment and Health

The lack of infrastructure and inadequate living conditions in many areas of South Africa has meant that millions of people are routinely exposed to fuels, which emit several noxious gases, and particulates, which can be deadly. National statistics show that Acute Respiratory Illness, associated with exposure to particulates, is the second highest cause of mortality in children under the age of five.

The medium term priorities of the *White Paper on Energy Policy* include the mitigation of the negative environmental and health effects of air pollution from coal and wood use in household environments. From this, two initiatives are currently being pursued by the DME:

**Low Smoke Fuel Strategy**

This Strategy addresses the winter coal-burning households that create a pollution problem. Several options have been assessed from a cost perspective, ranging from energy management interventions such as the 'Basa Njengo Magogo Project', to developing and manufacturing cleaner fuels for use in conventional coal stoves, to using other fuels such as liquid petroleum gas, to designing houses to require less space heating by means of insulation and ceilings. The first approach is outlined briefly below.

***Basa Njengo Magogo Project***

The local name for the so-called 'Scotch' Method of lighting a coal fire by inverting the contents, so that the volatiles are burned off first, dramatically reduces the time during which a fire produces smoke and creates a slower-burning fire in a matter of minutes, and in so doing reduces energy consumption by up to 30 per cent or more (depending on the user). This is a 10-year project targeting the 1 million homes in the winter coal-burning area.

**Improved Woodstoves**

The Programme for Biomass Energy Conservation (a Southern African Development Community regional programme) is directed at energy conservation by the use of improved energy efficient woodstoves, which require a fraction of the wood normally required. Reductions from 30-50 per cent are achievable, depending on the stove efficiency and the proficiency of the user. As biomass is approximately 9 per cent of energy demand (as compared to electricity at 22 per cent), this sector is important from a national point of view.

## 6.4 The Cleaner Fuels Programme

Cabinet has approved the phase-out of leaded petrol from 2006, the reduction of sulphur in diesel to a maximum of 0,05 per cent from 2006 and that DME determine the other relevant fuel specification parameters in consultation with the relevant stakeholders. This initiative is in line with the global move towards phasing out leaded petroleum where 86 per cent of the world's supply is unleaded.

This initiative is driven primarily by human health, environmental and air quality criteria but it is also in line with the global harmonisation of standards that are employed in Europe and Japan – those countries whose automotive manufacturers dominate the South African market. The revision of standards and their deployment fits in with the measures proposed in this document to increase energy efficiency in the transport sector.

## References

1. DME, 1998, *White Paper on the Energy Policy of the Republic of South Africa.* Department of Minerals and Energy, Pretoria, December 1998.
2. DME, 2002, *Digest of South African Energy Statistics.* Compiled by A Pouris. Department of Minerals and Energy, Pretoria, 2002.
3. DME, 2003, *Integrated Energy Plan for the Republic of South Africa.* Department of Minerals and Energy, Pretoria, 2003.
4. DME, 2003, *Joint Implementation Strategy for the Control of Exhaust Emissions from Road-going Vehicles in the Republic of South Africa.* Department of Environmental Affairs and Tourism, Department of Minerals and Energy, Version 4, November 2003.
5. DME, 2004, Draft Electricity Regulation Bill. Department of Minerals and Energy, Pretoria.

6. Draft Energy Bill, 2003, version 2, June 2003. Department of Minerals and Energy, Pretoria, 2003.
7. DME in collaboration with CaBEERE–Danida, 2002, Capacity Building in Energy Efficiency and Renewable Energy,– *Energy Efficiency Baseline Study*. September 2002.
8. DME in collaboration with CaBEERE–Danida, 2003, Capacity Building in Energy Efficiency and Renewable Energy,– *Energy Efficiency Savings Projections.* April 2003.
9. DME in collaboration with CaBEERE – Danida, 2003, Capacity Building in Energy Efficiency and Renewable Energy,-*Capacity Assessment of External Stakeholders.*
10. Energy and Development Research Centre. *Policies and measures for renewable energy and energy efficiency in South Africa.* Sustainable Energy and Climate Change Partnership, 2003.
11. International Energy Agency, 2002. *Energy Balances of OECD countries.*
12. International Energy Agency, 2003. *Key World Energy Statistics.*
13. The Electricity Act (no 41 of 1987) as amended.

## Definitions and Terminology

| | |
|---|---|
| Appliance Labeling: | Labels denoting energy consumption of appliances |
| CaBEERE: | Capacity Building in Energy Efficiency and Renewable Energy Project |
| CDM: | Clean Development Mechanism |
| CFL: | Compact Fluorescent Lamp |
| $CO_2$: | Carbon Dioxide |
| CSIR: | Council for Scientific and Industrial Research |
| CV: | Calorific Value |
| DME: | Department of Minerals and Energy |
| DNA: | Designated National Authority |
| DoH: | Department of Housing |
| DTI: | Department of Trade and Industry |
| DSM: | Demand Side Management |
| EE: | Energy Efficiency |
| Energy Intensity: | Energy use per unit of output or activity |
| ESCO: | Energy Service Company |
| GW: | Gigawatt ($10^9$ Watts) unit of electric power |
| HVAC: | Heating, Ventilation and Air Conditioning |
| IEP: | Integrated Energy Plan |
| LTA: | Local Transport Authority |

NDoT: National Department of Transport
NERI: National Energy Research Institute
NER: National Electricity Regulator
$NO_x$: Oxides of Nitrogen
NT: National Treasury
PJ: Petajoule ($10^{15}$ Joules) unit of energy
RDP: Reconstruction and Development Programme
REDs: Regional Electricity Distributors
SABS: South African Bureau of Standards
SARS: South African Revenue Service
$SO_2$: Sulphur Dioxide
STANSA: Standards South Africa
VSD: Variable Speed Drive

# Chapter 5

# *Energy Needs and Impacts: The Challenge to Science and Technology*

**Ahmed Galal Abdo**

*Academy of Scientific Research and Technology and State Ministry for Scientific Research, Cairo, Egypt*

*E-mail: galalah1@yahoo.com, galalah2@gmail.com*

## ABSTRACT

Our Planet has definitely become more crowded! The world strives on finding sustainable energy resources, protecting the environment while balancing other urgencies such as stabilizing world population, sizing escalating poverty and harmonizing global cooperation. These are interconnected problems and dragged the focus of all nations hopefully for finding remediating solutions. Egypt has committed itself to a very ambitious target of 20 per cent renewable energy share in the year 2020. This goes in line with many nations and the UNESCO target. To achieve this target, Egypt will attract investments in the next 12 years in the renewable energy sector to establish about 10,000 MW, where a considerable share will come from wind parks at the Red Sea and the rest from concentrated solar and photovoltaic power plants. Funding is not the only constraint, and more importantly we have to build human capacity, secure appropriate technologies and expand our research development and innovation capabilities. At the same time, Egypt has embarked on a very challenging energy saving and efficiency improvement programs targeting an overall 6 per cent reduction (including 35 per cent reduction in the industrial sectors) in the first phase. This is equivalent to a 15 Million Tones of $CO_2$ emission reduction annually. To achieve this, international cooperation, technology transfer and human resources are crucial elements. We should consider the impact of this on sustainable development if many countries follow the same path. This paper will address the types of renewable energy technologies considered in Egypt, current resources and vision for large-and small-scale solar thermal, photovoltaic, wind and biomass energy. A comparison will be made in respect to global industrial and innovation policies in the renewable energy sector. We will also present the strategic targets that could realize a balance between energy demand, consumption and impact on environment. The scientific research focuses on renewable energy in Egypt and the cooperative effort with the world organization and institutes will also be presented. Some examples of research findings will be discussed and attention will be paid to discuss the present escalation of the global price surge in food especially grains. The attention paid to bio-fuels and its impact on the displacement of land and farm capacity, the shortage in the world stocks of grains, and the marginalization of farm regions.

***Keywords***: *Bio-fuels, Energy, Global price, Impacts, Renewable energy, Science and technology and Solar energy.*

## 1.0 World Population and Energy Resources

According to the United Nations reports (The Day of 6 Billion, 2008), the world's population has reached 6 billion by 1999; a figure that doubled since 1960. All indications show that it is still increasing and is projected to reach over 9 billion by 2050 (Figure 5.1).

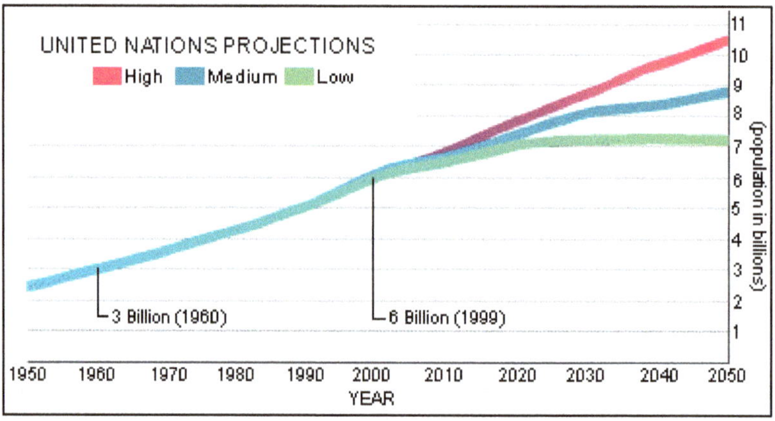


**Figure 5.1: World Population Growth (from reference 1)**

The world energy council indicated that "crude" oil is the most important primary fuel that accounts of some 37 per cent of the world's primary energy consumption (not including the biomass resources) ("2007 Survey of Energy Resources," World Energy Council). However, more than 37 per cent of the estimated ultimate recovery has been already consumed.

The resource reserve ratio (RRR) of different energy forms are: conventional oil (0.5), conventional natural gas (1.2), hard coal (5.5), brown lignite coal (5.0) and uranium (6.8). In spite of the availability of some natural resources of energy the world's demand for consumption is still rising (Figure 5.2). It is apparent that within the present century, the world has to face life without fossil fuel. On the other hand, the level of $CO_2$, a good absorber of heat because of its ability to vibrate in the infrared region of light, will probably continues to rise and contributes to the rise of Earth's

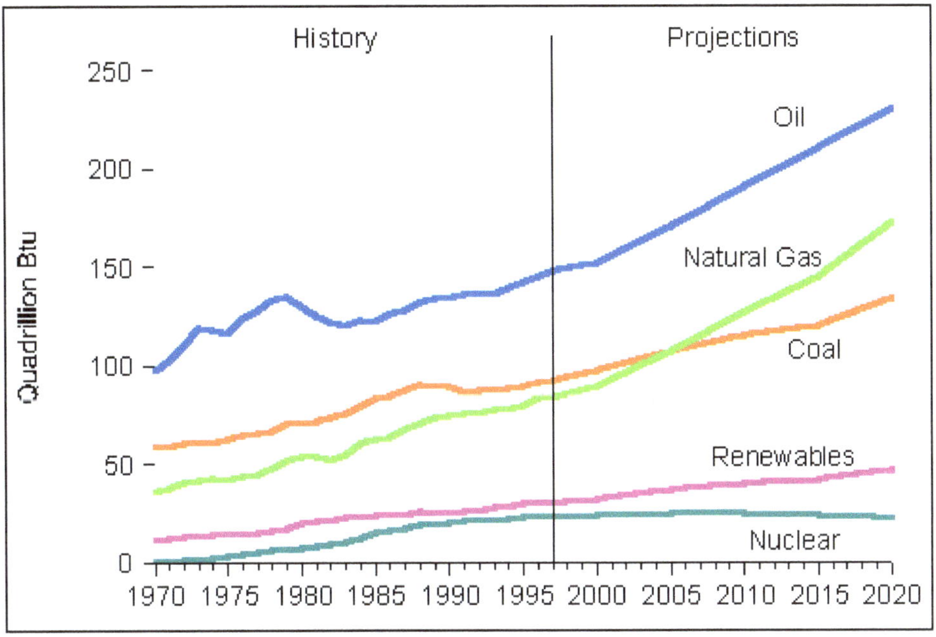


**Figure 5.2: World' Actual and Projected Consumption of Energy Resources**

temperature. Within the last couple of centuries, most natural energy resources have been consumed and present generation will likely witness the depletion of most of oil and natural gas. The total global annual energy consumption at present is over 10,000 million tons of oil equivalents and based on current projections, the global annual energy consumption rate will double within the coming two decades ("BP Statistical Review of World Energy," 2007).

The present world prices for petroleum is around 140 $/bbl (for Brent oil) and for natural gas is around 15 $/MMBtu. Meanwhile, food prices are rocketing high some of which have more than doubled through the last two years and sparked unrest in numerous countries recently. Food analysts are predicting that while protests against the rising food prices might be quelled, lowering the prices will take longer time. Several theories added that the crisis in energy prices and food are interrelated in many aspects. For instance, the spike in oil prices in recent days has pushed up fertilizer prices as well trucking food from farms to market places and shipping to export. The climate change affected by global warming has disrupted the harvesting process as witnessed by the extended droughts in Australia and Africa. The recent drive to produce biofuels, from crops as an alternative to hydrocarbons as fuel, is augmenting the shortage in food supplies. Therefore, farmers in many countries are growing crops for ethanol production influencing even the price of rice, a staple for billions of people, has spiked high to its highest price in decades. Meanwhile, the global supply of several crops reached its lowest level since decades.

## 2.0 Renewable Energy Potentials in Egypt

The following figures were collected from reference ("BP Statistical Review of World Energy," 2007) regarding the energy resources in Egypt and as compared to the world situation:

### Oil

The proved reserve mounts to 3.7 thousand million barrels (0.3 per cent of the world reserve), the daily production is 678 thousand barrels daily (0.8 per cent of the world production), and the consumption is 612 thousand barrels daily (0.7 per cent of the world consumption).

### Natural Gas

The proved reserve mounts to 1.94 trillion cubic meters (1.1 per cent of the world reserve), production is 44.8 billion cubic meters (29.3 per cent of the world production), and the consumption is 28.7 billion cubic meters (11.4 per cent of the world production).

### Coal

Egypt has limited coal resources and consumes about 1.0 million tons oil equivalent (7.8 per cent of the world consumption).

### Hydroelectricity

The consumption is 2.9 million tons oil equivalent (0.4 per cent of the world consumption).

### Primary Energy

The consumption is 58.8 million tons oil equivalent (0.5 per cent of the world consumption).

Egypt seeks to become a leader in the manufacturing and export of small/medium-technology that will render the country a leader in the MENA region. Energy is a priority for such ambition and renewable resources are set as essential for the policy agenda in consequence to the Kyoto emission reduction quotas. In its short and long-term planned strategy, Egypt has considered different forms of renewable energies: large-and small-scale solar (thermal and photovoltaic), wind and biomass

1. Pursuing economic development that is currently witnessed in Egypt necessitates paying attention to a sustainable production system based on renewable energy (RE). Egypt is targeting a RE-share of 20 per cent in year 2020 (New and Renewable Energy Authority, Egypt, Annual Report 2002/2003). The current main resource of RE in Egypt contributes 16 per cent of its electricity needs (some 2.7 GW out of total 17 GW being produced) comes from a hydro-source, the dams on the Nile. With simple projection, and expected increase in consumption that mounts 6 per cent on a yearly basis (New and Renewable Energy Authority, Egypt, Annual Report 2002/2003), Egypt will need to produce about 120 GW by the year 2050. Taking into account the limited resources of fossil fuel as previously explained, and the

confinement of hydro-electricity that will reach a high limit by the year 2020, wind and solar energies will cover the remaining needs for Egypt's energy resources. This will demand the contribution of RE of about 50 per cent. It is more likely that the resources of energy will reliably be generated from different resources to ensure more stabilized generation of electricity. While wind energy can be expected the most economically efficient at present, solar thermal generation of electricity will be of future necessity. Moreover, heat generated from the process will help desalinate sea water that will provide Egypt with another life necessity as one of the scarcest lack in the coming few decades.

Egypt enjoys the availability of both wind and solar energy resources with high quality and most importantly, geographically closer to the suppliers need. However, several issues has to be considered before deciding energy production from either or both resources that are estimated to cover all Egypt's need of energy. Those are: electricity generation cost (that includes the technology aspect for the production), landscape required for the production, controllability (such as grid distribution planning), stability (mainly in wind resources) and security issues.

Figure 5.3 shows the solar energy distribution over the Egyptian territory and Figure 5.4 shows the wind atlas of the country. It is clear that the illumination and solar intensity is sufficient to encourage the application of solar energy especially with clear skies for most of the year. The wind energy however, is limited to steady stable resources along the Red Sea at the east, Mediterranean Sea in the north and some parts of the western desert. The solar energy radiation ranges between 2000–3200 kWh/m$^2$/year over the total area of the country. Solar energy therefore demonstrates high potential for power generation amounting to about 74,000 TWh/year (Renewable Energy Sector in Egypt, Final Report by the Energy Research Centre (ERC) of the Faculty of Engineering, University of Cairo, 2006). The eastern shores of Suez Gulf have an excellent wind regime exceeding 10 m/s at about 25 m height. This region is uninhabited and can afford about 20,000 MW of wind farm production potential. The northern Mediterranean cost exhibits moderate wind speed of 5 to 7 m/s. Recently, the Egyptian Supreme Council has adopted a plan that aims to cover 20 per cent of the nation's total electricity needs using RE by the year 2020. As planned, 12 per cent of the total contribution will be shared from wind resources with a total of 7000 MW of grid-connected wind-farms. It is will thought that private sectors will be able to contribute to this ambitious plan.

The Ministry of Electricity and Energy in Egypt has submitted a new electricity act to encourage renewable energy utilization that encourages private sector involvement in the process. Egypt is in the contracting phase to implement a solar thermal power plant of 140 MW, including a solar share of 20 MW that is projected to operate by 2010. Similar plants are to be implemented in India, Morocco and Mexico and are co-sponsored by the Global Environment Facility (GEF). As indicated previously, solar concentration in northern Egypt is estimated at 2000 kWh/m$^2$ per year and reaches about 3200 kWh/m$^2$ per year in the south.

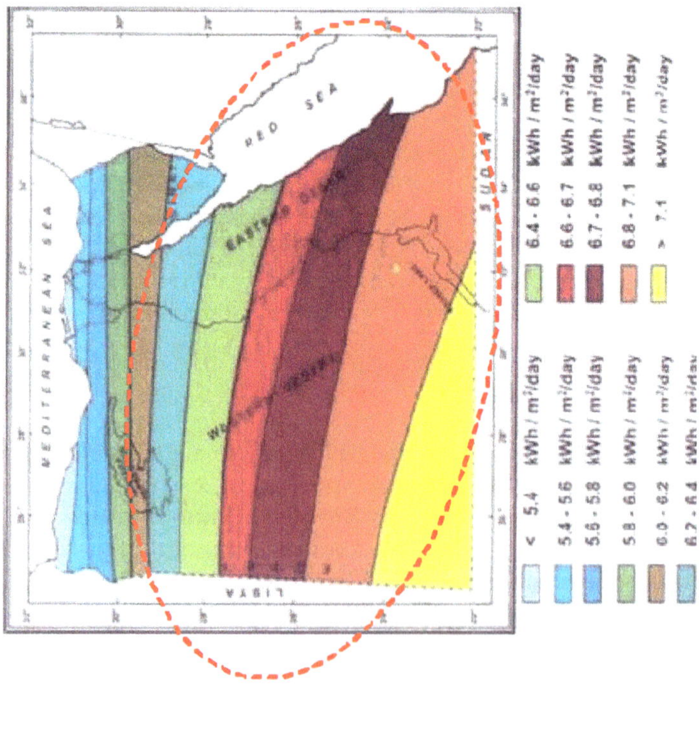


**Figure 5.3: Solar Energy Map of Egypt**

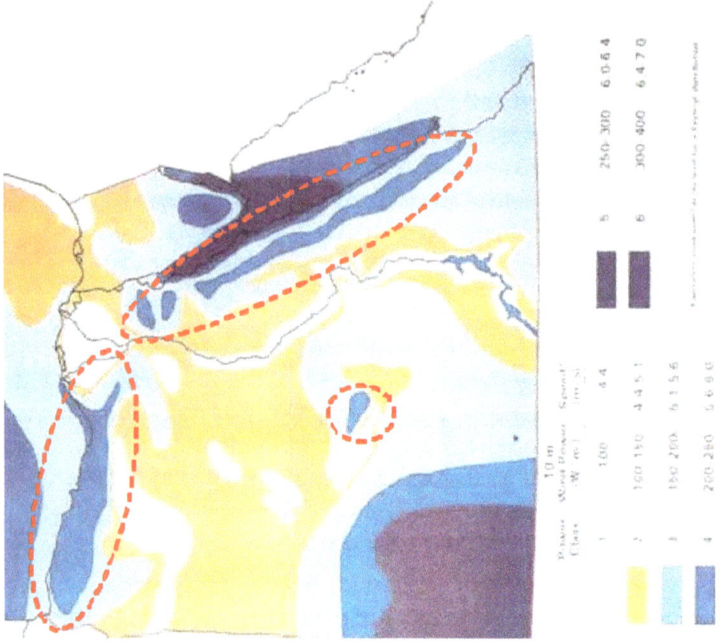

**Figure 5.4: Wind Atlas of Egypt**

New and Renewable Energy Authority (NREA) was established to act as the national lead to promote RE technologies to generate electricity on commercial scale. NREA has currently wind energy to generate electricity as its core current plan. NREA stated that Gulf of Suez is candidate to implement 20,000 MW of wind farms. A number of pilot wind farms have been installed in Zaafarana on the Red Sea, and a series of large-scale wind energy projects that will convert the previously experimental projects to produce a total capacity of 230 MW and will be connected to the national grid of which 80 MW have been recently commissioned. A target 850 MW by 2010 is set by NREA that will set Egypt to represent a hub for wind power lead in the Middle East and North Africa (MENA) area. Private enterprises are currently exploring investing in a sizeable area south of the Red Sea to establish wind farms where wind speed reaches 11 m/s.

Although an insufficient supply may be expected after 2020 for fossil fuel, several challenges are still to be overcome; namely the relatively more expensive production of RE and the high subsidy of fossil fuel products. On the national basis, Egypt is considering producing several parts either for wind or solar production that will encourage investment in RE sector. In this respect, three main scenarios for the local manufacturing capability to contribute to RE industry were suggested (Renewable Energy Sector in Egypt, Final Report by the Energy Research Centre (ERC) of the Faculty of Engineering, University of Cairo, 2006). First, locally manufacturing RE components directly using current resources of the Egyptian industry; second, RE components can be produced with current resources with help of innovation and R&D; and third, engage in joint venture with established foreign companies in such approach RE components will be produced by local industry after transfer of technology. While the first approach is likely to be immediately implemented the other two approaches will take an average of 5 to 3 years, respectively. However, each of these approaches will selectively affect the manufacturing process. For example, in the thermal solar concentrator industry, components such as glass, absorber plates, tanks, pumps, and insulation heaters can be manufactured locally through the first approach. Selective/effective coatings are rather to be manufactured through the second and/or third approach. In the case of photovoltaic (PV) components on the other hand, such as solar cells will be manufactured using the second and/or third approach while absorber plates and modules can be implemented using the first and/or second scenarios.

## 3.0 Research Prospective on RE in Egypt

Several research groups are actively involved in topics related to RE in Egypt. Those include engineers, chemists, biologists, geologists, environmentalists and physics scientists. In Egypt, some research work is directed for the study of photovoltaic of nano-structured materials (El-Barry, 2007) and solar cells conversion efficiency (El-Sayed and Abdel Rady, 2006). Other studies also included the performance of solar collectors based on latent heat storage (El-Kotb *et al.*, 2006) and solar energy storage (Shatter *et al.*, 2006). Several researchers have also considered hybrid wind/PV/fuel cell generation systems in their work (Rafat *et al.*, 2007). Numerical simulation of the limiting efficiency of the compositionally graded band gap solar cell was also studied by another group of researchers (Ahmad and El Shenawy, 2006).

Hydrogen production was also the interest of several researchers in Egypt. For example, photovoltaic arrays were used in supplying the water electrolysis system by energy requirements for hydrogen production (Galal *et a.l*, 2007). Hybrid organic/inorganic films of conducting polymers modified with phthalocyanines were used as catalytic surfaces for the production of hydrogen from alkaline solutions (Galal *et al.*, 2008). Fuel cells studies have also attracted the attention of several research groups in Egyptian universities and research centres. Electrodeposited sub-micro metals at polymeric matrices were introduced as catalytic surfaces for the direct oxidation of methanol (El-Enin *et al.*, 2008). Improvement for the performance of polymer electrolyte membrane, PEM, fuel cells as a power source was achieved by the use of new materials in the manufacture of a bipolar plate (Khalfallah and Koliub, 2007).

Wind energy research in Egypt has also attracted several researchers. An experimental investigation was performed on the effect of blade surface roughness due to dust accumulation on the performance of wind turbines (El-Sayed, 2002). Different studies addressed also the substitution potential of wind energy in Egypt. For instance, a preliminary study was conducted in Za'afarana area to assess the potential and economic impact of wind energy in the form of cost/benefit ratio to justify the substitution potential of such clean energy resource.

Scientific research and cooperation between research institutes and different stakeholders from industry and governmental authorities have been encouraged in Egypt to engage in research related to energy issues. The High Supreme Council for Science and Technology that is coordinated by the State Ministry for Scientific Research, with the involvement of several Ministries, the Science and Technology Development Fund and other organizations such as the Research and Development Innovation have announced energy-research and development related topics is one of the high national priorities in Egypt. Not only fund allocation was made available from different sources of the aforementioned organizations but also encouragements for the establishment of centre of excellence in renewable energies were suggested with the cooperation between different research centres and universities.

## Conclusions

Renewable energy should be introduced in strategic planning of science and technology. The urgent needs for energy alternatives is not only due to the depletion of current fossil fuel resources but also because of the surmount in world population. Special attention should be paid to the abuse of using bio-fuel alternatives that focuses on food crops. Egypt has engaged in an ambitious program to replace ordinary energy resources by 20 per cent renewable by the year 2020. Main focus would be on the use of solar and wind energies. Moreover, the State Ministry of Research is aiming to coordinate research efforts in universities and research institutes with the industrial organization locally and internationally for the purpose of supporting products for RE.

## Acknowledgment

The author would like to thank the State Ministry of Scientific Research and the Academy of Scientific Research and Technology of Egypt for providing the archive facilities.

## References

1. "2007 Survey of Energy Resources," World Energy Council, 2007.
2. "BP Statistical Review of World Energy," 2007.
3. "The Day of 6 Billion" UNFPA, the United Nations Population Fund, 2008.
4. Ahmad, G.E. and El Shenawy, E.T., 2006. Renewable Energy, 31 (7), 1043-1054.
5. El-Barry, A.M.A., 2007. European Physical Journal-Applied Physics, 40 (3), 241-246.
6. El-Enin, *et al.*, 2008. Journal of Power Sources, 177, 131-136.
7. El-Kotb, *et al.*, 2006. Energy Sources, Part A., 28 (12), 1113-1126.
8. El-Sayed, M.A.G. and Rady, S. Abdel, 2006. Desalination, 209 (1-3), 15-22, 2007. E.B.S. Mettawee and G.M.R. Assassa, Energy, 31 (14), 2958-2968.
9. El-Sayed, M.A.H., 2002. Energy Policy, 30 (8), 681-687.
10. Galal, *et al.*, 2007. Journal Solid State Electrochemistry, 11, 521-530.
11. Galal, *et al.*, 2008. Topics in Catalysis, 47, 73-83.
12. Khalfallah, M.G. and Koliub, A.A., 2007. Desalination, 209 (1-3), 209-220.
13. New and Renewable Energy Authority, Egypt, Annual Report 2002/2003.
14. Rafat, *et al.*, 2007. Renewable Energy, 32 (1), 21-34.
15. Renewable Energy Sector in Egypt, Final Report by the Energy Research Centre (ERC) of the Faculty of Engineering, University of Cairo, 2006.
16. Shatter, *et al.*, 2006. Energy Conversion and Management, 47 (9-10), 1264-1280.

# Chapter 6

# Present Status, Initiatives, Achievements and Future Trends Towards Energy Conservation for Sustainability

**Shantha Siri**
*National Science Foundation, Colombo, Sri Lanka*
*E-mail: shantha@nsf.ac.lk*

## ABSTRACT

Rapid depletion of fossil fuels and increasing cost of conventional forms of energy has compelled many countries to look for alternative methods of meeting their energy needs. This is more relevant and important to a developing country like Sri Lanka, which up to now heavily depends on fossil fuels. In the country's energy portfolio, almost 50 per cent is attributed to fossil fuels. The demand for power is increasing due to urbanization of rural sector and for industries. There is an annual growth more than 8 per cent in the power sector. At present, more than 35 per cent of rural sector has no access to the main electricity supply and is looking for some sources of energy which is affordable. As private sector is the engine of the economy, the government is yet to overcome the challenge of supplying an uninterrupted power supply to the industry at a competitive price. In some sectors of the industry especially the manufacturing sector, the production cost is mainly attributed to the cost of electricity. According to Ceylon Electricity Board, 17 per cent of the produced electricity is wasted due to various reasons. It will be a significant impact to the economy if the above energy wastage can be minimized. This paper explores initiatives taken by the government of Sri Lanka, its energy policy, achievements and future trends towards energy conservation for sustainability. The report also highlights the present status of energy consumption, energy portfolio and future energy demands of Sri Lanka.

***Keywords:*** *Sri Lanka, Bio fuels, Energy audit, Energy conservation, Energy efficiency, Environmental impact renewable energy and Sustainability.*

# 1.0 Introduction

Sri Lanka being a small country with moderate energy resources of its own and with no proven reserves of fossil fuels, hydro-energy and biomass (mainly fuel wood) are Sri Lanka's only indigenous energy resources. All of the country's fossil fuel needs are imported mainly as crude oil and finished petroleum products, and a very small amount of coal as of today. To satisfy the growing demand for energy, the option available for a developing country like Sri Lanka is to import its primary energy requirements as petroleum, coal and natural gas/liquefied natural gas (LNG) until such time we could locate some of these resources within our territorial boundaries.

## 1.1 Use of Imported Petroleum in the Past

In the past, the country has imported coal and used as a source of fuel in the railways, marine, transport industry and in households. The use of coal has almost ceased today except for a small amount used in the cement industry. By the turn of the centaury, the country began to import petroleum for lighting and later for motor vehicles, followed by industrial fuels like diesel oil and fuel oil to provide motive power and industrial heat.

Today hydro electricity and petroleum are the main sources of primary energy in the country. The petroleum imports and use in the country as from 1950 is given in Table 1. Between 1990 and 2000 the growth in demand accelerated and the demand for petroleum grew at a rate in excess of 11 per cent per annum on average, while between 2000 and 2005 the growth rate declined to around 8 per cent on average per annum. The decline in growth rate in this decade is partly due to high prices of petroleum. During the decade, growth in demand for transport fuels shifted to gasoline or petrol from auto diesel, while growth in demand for petroleum fuels for power generation shifted from heavy diesel to furnace oils. These are some of the fundamental changes that will affect the future supply options if no ad-hoc changes are made in the near future.

## 1.2 Importing Crude Oil

Sri Lanka imports crude oil from Middle East or the Far East. Getting crude oil from other regions is not very practical. Crude oil imported can be put to limited use only before refining. The capacity of existing refinery is 2.1 million tons per annum while the demand in the country as of 2007 is in excess of 4.1 million tons of products. Therefore, Sri Lanka has a shortfall in capacity and needs to import more than 50 per cent as refined products from the international market.

By 2015, country's imports of refined products would reach 80 per cent of our requirements if there is no addition to capacity. Between 1995 and 1997 there were two proposals made to establish export oriented refineries in Sri Lanka and both failed to materialize due to various reasons. The Ceylon Petroleum Corporation's own proposal to expand capacity of the Sapugaskanda refinery was finally studied in 1998/1999 and was kept in abeyance due to non-viability of the investment following the Far East economic crisis. The project was reactivated in 2001 when the investment became viable, but it failed due to unavoidable circumstances. A private sector proposal to invest on an expansion scheme for the refinery did not find favor in 2005.

Expanding the capacity is due to initiate with the assistance from Iran government. With the change in demand pattern for products in the last two years and the increase in demand for fuel oil, there may be opportunity to phase out the investment if the required product quality can be met for power plants. For the expected production from an expanded refinery by doubling the capacity and the different options available. Any hydroskimming option will need to give way for a conversion facility if demand for fuel oil drop with installation of coal based power plants or gas replaces fuel oil.

## 2.0 Energy Supply

Energy supply in Sri Lanka is mainly based on three primary resources, namely, biomass, petroleum and hydroelectricity. In 2004, hydro-electricity production in the country accounted for 710.71 kTOE (thousand tonnes of oil equivalent) while the biomass-based energy supply was 4,513.3 kTOE. Approximately 4,131.9 kTOE was provided by imported crude oil and finished petroleum products such as diesel and liquefied petroleum gas (LPG). Additionally, the non-conventional resources (mainly wind) provided 3.6 kTOE of primary energy, giving an aggregate primary energy supply of approximately 9,359.5 kTOE. Primary energy contributions in 2004 to national energy supply were 48.2 per cent from biomass, 44.2 per cent from crude oil and petroleum products, 7.6 per cent from hydroelectricity and other renewable sources. Figure 6.1 gives energy contribution to national energy supply as percentage by resource. The use of non-conventional energy resources in Sri Lanka is of a relatively

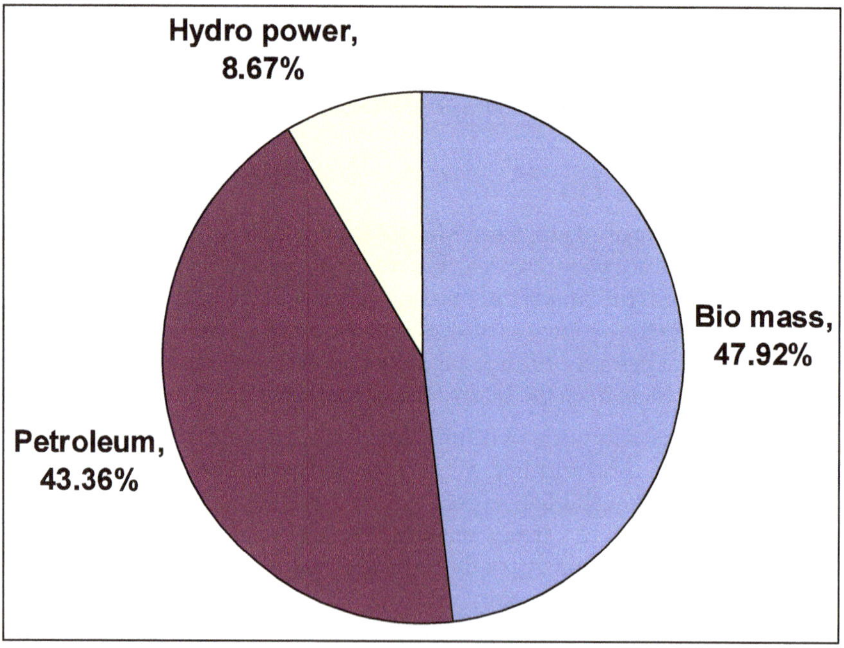


**Figure 6.1: Percentage of Energy by Resource**

smaller scale and therefore its contribution is presently of low significance in the macro energy picture.

### 2.1 Energy Demand Growth

With the increasing demand for energy to provide for the country's economic and social development, total primary energy demand is expected to increase to about 15,000 kTOE by the year 2020 at an average annual growth rate of about 3 per cent. Electricity and petroleum sub-sectors are likely to record higher annual growth rates of about 7-8 per cent. Hydro electricity production and biomass-based energy supplies, which are the only large-scale indigenous primary energy resources available in Sri Lanka, are expected to increase only marginally in the near future. This is mainly due to limitations in further hydropower development owing to lower economic viability of exploiting the remaining large hydropower sites and limited use of biomass with gradually increasing standard of living of the population. This means that the country's incremental primary energy requirements need to be supplied mainly by imported fossil fuels in the medium term. In the longer term, possible development of indigenous petroleum resources and accelerated development of non-conventional renewable energy are likely to make a significant change in Sri Lanka's energy portfolio.

## 3.0 Energy Related Organizations in Sri Lanka

### 3.1 Ceylon Electricity Board

Ceylon Electricity Board (CEB) generates supplies of electricity. It also constructs, maintains and operates the necessary works for the generation of electricity by all means and for the inter-connection of Generating Stations. It also distributes and sells electricity in bulk or other wise.

### 3.2 Lanka Electricity Company (Pvt) Ltd

Lanka Electricity Company (Private) Limited (LECO) commenced business operations on 1st June 1984 and services over 376,000 Customers in the Western and Southern coastal belt townships between Negombo and Galle.

### 3.3 Institution of Engineers Sri Lanka

The Institution of Engineers Sri Lanka (IESL) was established for the general advancement of the science and practice of engineering in all its branches and for the purpose of effectively carrying out its objectives according to rules agreed to by its members.

### 3.4 National Engineering Research and Development Centre (NERD)

The National Engineering Research and Development Centre of Sri Lanka was incorporated in 1974 under the State Industrial Corporations Act No.49 of 1957. It has emerged as one of the premier engineering research and development centres in the country. It has placed particular emphasis on activities capable of improving the quality of the life of the people in the country, especially the rural and low-income groups. It has also paid special attention to small-scale entrepreneur development projects, some of the research areas in which the centre has been active in the past are: Renewable energy, Provision of electrical power in the rural areas, Cost effective building technologies, Energy management and environmental engineering, Energy efficient lighting systems in rural homes etc.

## 3.5 Lanka Transformers Group

Lanka Transformers Group (LTL) began operations with the manufacture of transformers, but the company diversified into many fields of engineering such as switchboard manufacturing, construction of Grid Substations and Transmission Lines, Hot-dip Galvanizing, Power Generation, Steel fabrication, Mini-Hydro Power Development, IT Solutions and Energy Services.

## 3.6 University of Moratuwa

University of Moratuwa provides its services as a centre of higher learning, research and consultancy in Engineering, Architecture and other allied Professional disciplines.

## 3.7 Practical Action South Asia

Working together with the communities, Practical Action South Asia aims to build the technical skills of poor people in the region enabling them to improve the quality of their lives and that of future generations.

## 3.8 Ceylon Petroleum Corporation

The Ceylon Petroleum Corporation was set up as a state enterprise by Act. No.28 of 1961 in Parliament and further amendments carried out subsequently. The Ceylon Petroleum Corporation is engaged in importing, refining, supplying and distributing of Petroleum products.

## 3.9 Lanka IOC Private Limited (LIOC)

Lanka IOC Limited (LIOC) is a subsidiary of Indian Oil Corporation Ltd in Sri Lanka, which is the only public oil company other than the state-owned Ceylon Petroleum Corporation (CPC) that operates retail petrol stations in Sri Lanka. It has been incorporated to carry out retail marketing of petroleum products, bulk supply to industrial consumers, building and operating storage facilities at the Trincomalee Tank farm, etc., thereby not only providing energy security and supply stability for Sri Lanka but also providing world class standards of service, particularly in the retail sector in the Emerald Island.

LIOC is committed in serving the discerning customers of Sri Lanka. LIOC holds one-third share in Ceylon Petroleum Storage Terminals Ltd, the Common User Facility in Sri Lanka for storage and distribution of petroleum products. LIOC is making phased investments to provide world-class quality petroleum products and services to delight the Sri Lankan customers.

## 3.10 Shell Gas Lanka Ltd

This Company is engaged mainly in distribution of petroleum products, including LPG.

## 3.11 Sri Lanka Sustainable Energy Authority

Sri Lanka Sustainable Energy Authority offers its professional services to all energy users for analysis of energy consumption data.

## 3.12 Ministry of Power and Energy

Government ministry overseeing the power and energy sector in the country

## 3.13 National Energy Policy and Strategies of Sri Lanka

The National Energy Policy of Sri Lanka spells out the implementing strategies, specific targets and milestones through which the Government of Sri Lanka and its people would endeavor to develop and manage the energy sector in the coming years in order to facilitate achieving its millennium development goals. Specific new initiatives are included in this policy to expand the delivery of affordable energy services to a larger share of the population, to improve energy sector planning, management and regulation, and to revitalize biomass as a significant resource of commercial energy.

Institutional responsibilities to implement each policy element and associated strategies to reach the specified targets are also stated in this document. Ministry of Power and Energy has discussed the draft document with a wide group of stakeholders, obtained the views of members of the public and made the necessary amendments before publishing this National Energy Policy and Strategies of Sri Lanka.

There are nine elements in the energy policy namely providing basic energy needs, ensuring energy security, promoting energy efficiency and conservation, promoting indigenous resources, adopting an appropriate pricing policy, enhancing energy sector management capacity, consumer protection and ensuring a level playing field, enhancing the quality of energy services and protection from adverse environmental impacts of energy facilities.

The government of Sri Lanka has targeted providing 85 per cent of total households with electricity through extension of main grid by 2015 (Table 6.1) while facilitating off-grid electrification for 8 per cent of total households by 2015 (Table 6.2). It also has planned to minimize rely on oil and to increase hydropowers and NCRE in its energy portfolio (Table 6.3).

**Table 6.1: Medium-term Targets for Electrification of Households through Grid Extension**

| *Year* | *Total Households to be Provided Access to the Grid* |
|---|---|
| 2003 | 65 per cent (actual) |
| 2006 | 75 per cent |
| 2010 | 80 per cent |
| 2015 | 85 per cent |

**Table 6.2: Medium-Term Targets for Off-Grid Electrification of Households**

| *Year* | *Total Households Using Off-Grid Electricity Systems* |
|---|---|
| 2003 | 2 per cent (actual) |
| 2006 | 4 per cent |
| 2010 | 6 per cent |
| 2015 | 8 per cent |

**Table 6.3: Fuel Diversity and Security**

| *Year* | *Electrical Energy Supplied to the Grid as a Share of the Total* | | | |
|---|---|---|---|---|
| | *Conventional Hydroelectric* | *Maximum from Oil* | *Coal* | *Minimum from NCRE* |
| 1995 | 94 per cent | 6 per cent | 0 per cent | |
| 2000 | 45 per cent | 54 per cent | 0 per cent | 1 per cent |
| 2005 | 36 per cent | 61 per cent | 0 per cent | 3 per cent |
| 2010 | 42 per cent | 31 per cent | 20 per cent | 7 per cent |
| 2015 | 28 per cent | 8 per cent | 54 per cent | 10 per cent |

# 4.0 Government Initiatives and Achievements Towards Energy Conservation

## 4.1 Mega Projects

Two mega projects have been implemented by the government to meet the increased demand for energy. One is to construct a coal power plant having a capacity of 900MW and the other is to construct a thermal multi-fuel operated combined power plant with a capacity of 300MW. A national level committee on "bio-fuels" has also been appointed in order to carry out a situational analysis and make recommendations to the government on opportunities and feasibilities of manufacturing and using bio-fuel.

## 4.2 Establishment of Sri Lanka Sustainable Energy Authority

National Energy Policy of Sri Lanka has assigned responsibility of achieving several policy objectives to Sustainable Energy Authority. Among these, ensuring energy security, increasing indigenous energy and improving energy efficiency rank highest in priorities of the Authority. The vision of the organization is "An Energy Secure Sri Lanka".

Recently, a competition was organized by Sri Lanka Sustainable Energy Authority to recognize and reward most outstanding public sector institutions in energy saving. In this competition, the National Science Foundation won the first place under the medium scale upper category. It has been shown that a substantial amount of energy has saved as a result of the initiatives taken by the government. According to the

Ceylon Electricity Board, it has been able to save 88MW as a result of energy conservation steps taken by the government institutions and the general public.

## 4.3 Alternative Energy Sources for Sustainable Development

Extensive research on energy generation using biomass has been carried out by various organizations and individuals. First biomass power generation plant was established in recent past in Nuwara-Eliya district. The other has been set up by a private sector company. Both those projects had planned to utilize the sustainably grown fuel wood (SGF) such as *Glyricidia, ipil ipil, acasia* by the community of the respective areas together with other input materials found in the locality. It has been estimated that Rs. 15 billion of foreign exchange could be saved per year if 20 per cent energy requirement is met by biomass energy by year 2008. Otherwise Rs. 15 billion would be required for oil imports. If progressive development is undertaken, the savings are predicted to reach Rs. 46 billion annually by year 2010.

It has been declared by the government that *Glyricidia* as the fourth plantation crop in Sri Lanka. Research and development in biomass energy is being conducted by the Bio Energy Association of Sri Lanka and the Ministry of Science and Technology with private sector assistance. Proposals for the implementation of the policy on development of biomass as a renewable energy source for Sri Lanka have been forwarded to the Government by the Bio Energy Association of Sri Lanka (BEASL). It proposes to the government the policy that needs to be considered for action plans and achievement targets for the utilization of ample biomass resources in Sri Lanka.

## 4.4 Research and Development

The National Science Foundation (NSF) is the premier research funding organization in Sri Lanka. Research activities is being coordinated by a Research Committee on "Water and Energy" at the NSF. During past couple of years, it has facilitated carrying out following research activities in the field of energy by local R&D institutions.

1. To develop an efficient and cost effective treatment system for the serum water discharged in the process of latex centrifugation and also to use the biogas generated in the process of treatment as an energy source.
2. Geological and Geophysical Investigations for Geothermal Energy in Sri Lanka.
3. Laboratory model study and a survey to select a suitable area for ocean wave energy plant in the South and Southeast coast of Sri Lanka.
4. Estimation of daylight availability in Colombo and Energy saving potential through the use of natural daylight in building design.
5. A study of Collecting and Storing of Solar Energy in salt Pan Solar Ponds in Sri Lanka.
6. Electronic devices for energy saving design and prototype fabrication.
7. Nano-crystaline cuprous oxide thin films for solar energy applications.

8. Analysis of Energy saving potential by increased use of daylight in open office spaces in Sri Lanka.
9. Development of novel materials and devices based on polymers/natural rubber, semiconductors and graphite for energy conversion, storage and other applications.
10. Solar energy in tea drying process.
11. Apart from the NSF funded research, the followings are some of the ongoing research projects all over the country carried out by various institutions.
12. Case Study on Off-Grid Renewable Energy in Sri Lanka.
13. Economic and environmental impact of micro-hydro-and biomass-based electricity.
14. Energy Analysis of Steam Boilers in Sri Lanka.
15. Experimental Study on Ethanol and its Blends with Gasoline as a Motor Fuel Sector: Petroleum.
16. Fuel wood Fired Cook stoves in Sri Lanka and Related Issues.
17. Performance Evaluation of an Affordable Solar Dryer for Drying Crops.
18. Policy analysis.
19. Sri Lanka Energy Supply Status and Cross Border Energy Trade Issues.
20. Sustainable biomass production for energy in Sri Lanka.

## 5.0 Future Trends

### 5.1 Forecasting Future Demand for Petroleum Products

Forecasting future demand is a difficult task requiring many inputs from a number of agencies and some of the parameters taken into consideration may not behave in the manner expected. Forecasting demand for petroleum is a difficult task as the uncertainties that exist in respect to the fuel requirements for the power sector. Installing gas turbines based on heavy diesel or diesel plants too expensive and cost of supply goes up continuously. At times even with a breakdown of one of the links of the supply chain, shortages and disruptions in supply of petroleum can occur resulting in accusations and allegation of inefficiencies. Table 6.4 gives actual demand in first half of this decade while in Table 6.5 forecasted demand for the next 10 years with the information available.

It is forecasted based on both historical demand growth patterns and some information available from the power-generating sector. It is assumed that there will not be any heavy reliance on distillate diesel fuel for power generation except what has been committed so far. Due to the high price of heavy diesel, thermal generation has shifted to cheaper furnace oils. It is also assumed that new power plants would be based on furnace oil in the next 10 years as there is no firm decision on the future of the coal-based power plants (Chandrasekara, 2007). According to the forecast the aggregate demand for petroleum products would reach 5.5 million metric tons by 2015.

**Table 6.4: Petroleum Products Consumption in the First Half of the Decade (In '000 MT)**

| *Product* | *Year* | | | |
|---|---|---|---|---|
| | *2000* | *2002* | *2004* | *2006* |
| LPG | 115 | 137 | 148 | 152 |
| Petrol | 224 | 285 | 437 | 584 |
| Kero/Jet | 478 | 439 | 504 | 512 |
| Diesel Transport/Industry | 1258 | 1345 | 1443 | 1479 |
| Diesel Power Gen. | 431 | 429 | 483 | 269 |
| Naphtha Power Gen | 0 | 56 | 96 | 124 |
| Furnace Oil Industry | 269 | 253 | 207 | 239 |
| Furnace Oil Power Gen. | 467 | 503 | 530 | 733 |
| Total | 3300 | 3452 | 3883 | 4184 |

**Table 6.5: Projected Future Consumption of Petroleum (in ' 000 MT)**

| *Product* | *Year* | | | | |
|---|---|---|---|---|---|
| | *2008* | *2010* | *2012* | *2014* | *2016* |
| LPG | 158 | 164 | 171 | 178 | 185 |
| Petrol | 665 | 736 | 828 | 930 | 1045 |
| Kero/Jet. | 527 | 542 | 559 | 578 | 590 |
| Diesel Transport/Ind. | 1568 | 1663 | 1764 | 1817 | 1983 |
| Diesel Power Gen | 300 | 300 | 300 | 300 | 300 |
| Naphtha Power Gen. | 120 | 120 | 120 | 120 | 120 |
| Furnace Oil Industry | 257 | 273 | 289 | 307 | 326 |
| Furnace Oil Power Gen. | 830 | 1030 | 1030 | 1030 | 1030 |
| Total | 4218 | 4419 | 4833 | 5065 | 5317 |

## 5.2 Future Options

It seems that options available for Sri Lanka in the short term are very limited. There is no upstream oil industry in Sri Lanka and consequently there are no oil and gas fields located. The option available is to import oil and or gas (natural gas). However, in long term, there are options such as harnessing biomass for electricity generation (gasification), mini hydropower, and utilization of solar energy. It has been estimated that 10 per cent efficient solar panel extent of 18 km$^2$ will fulfill the electricity demand in Sri Lanka. Consistent support for R&D and for development of science and technology will facilitate achieving envisaged goals in terms of available options.

## 5.3 Impact of a Future Discovery of Oil and or Gas within Sri Lanka Territory

Sri Lanka is embarking on an exploration program in the Gulf Mannar where potential hydrocarbon bearing structures have been identified. It is expected that in the next few years, international companies or foreign state entities would explore in the Gulf of Mannar. Exploration is expected to be carried under a Production Sharing concept. However, one can safely argue that whether the presence of oil has been confirmed based on a systematic investigation and scientific facts. Even though the oil is presence, Sri Lanka does not have the exploration technology.

Then there is an option to sell off the crude and continue to import the crude oil to which the refinery is designed. If natural gas is discovered instead of oil then obviously gas would come ashore using underwater pipelines, to the West coast of Sri Lanka. Distribution pipelines constructed will distribute to where gas can be utilized. The gas could initially be used in the combined cycle power plants already in place and a minimum of 600 MW capacity will be available by 2012. Next conversions to gas would be the Diesel power plants using fuel oil located close to the west coast and here we would have about 250 MW of capacity available by 2012.

## 5.4 Solar Air Heating Systems for the Industry

The electricity is being used for heating and drying purposed by many industries. Solar air heating systems can be used for heating and drying purposes since the technology is already being used. Simple technologies have been successfully used world over for many decades to utilize solar energy for air heating applications such as building heating, drying, and chemical process industry. However, no significant utilization of solar air heating is evidenced in Sri Lanka as yet (Arachchi, 2007).

The tropical climate of the country with abundant sunshine and a multitude of industries utilizing low to medium grade heat energy provide vast opportunities for the utilization of this benign source of energy. Typical industrial use of hot air in Sri Lanka include; drying/withering of tea, rubber drying, drying of agricultural crops including spices, food drying, curing of ceramic ware, drying of paint work, and drying of timber.

Most of these applications require hot air within the temperature range of 35 to 100°C. A study on the tea sector has shown that the technical potential of replacing conventional energy in drying with solar energy is in the range of 30–55 per cent in most of the tea growing regions. Solar air pre-heating collectors operating at 50 per cent thermal efficiency have been shown to deliver this performance. A study by Laing Design and Development Centre, UK considered a combined solar heat collection and a modified energy efficient tea drying process and concluded that 480m$^2$ air collector system was adequate in supplying 80 per cent of the heat requirements of a factory processing 800 kg of green leaf per day over 8 hours. A similar potential for conventional energy replacement can be expected from most other industries utilizing low to medium temperature process air.

## 5.5 'Green' Factory

In the global context, environmental concerns have been increased and there will be a higher demand for products manufactured using green energy (renewable energy). This opportunity has been identified proactively by some of the export-oriented companies in Sri Lanka.

Brandix Group, one of Sri Lanka's largest apparel exporters says that it has spent 2.5 million dollars to redesign a 30-year-old factory to meet 'green' factory standards. The conversion achieved a 45 per cent saving in energy, cut water consumption by nearly 60 per cent, improved solid waste management and reduced carbon emissions. Their 130,000 square-foot casual-wear factory has a rainwater harvesting system contributing 15 per cent of the 64,000 litres of recycled water used daily. According to Brandix Group Director A J Johnpillai, the redesigned factory will reduce its carbon footprint by as much as 75 per cent. The factory is one of a series of initiatives across the group to reduce its carbon footprint by at least 35 per cent by 2012.

## 5.6 Carbon Trading

Initiatives have been taken by the Ministry of Environment to grasp the benefits of Carbon Trading, which is being evolved globally.

# 6.0 Conclusions

As a developing country, Sri Lanka will face many challenges arising due to world energy crisis. It has a potential to emerge from the crisis effectively with proactive measure, which has already been taken. However, it is safe to argue that the country cannot exclude 100 per cent from the imported petroleum fuels, as the country is yet to find such resources within its own territory.

The country has feasible alternatives such as bio mass, solar energy and hydro power which could be developed through consistent support from the government for relevant R&D and enhancement of the science and technology capabilities in the country.

Taking proactive measures with proper planning towards energy conservation and development and promotion of harnessing indigenous energy sources will help face challenges from energy crisis effectively ensuring a sustainable development. Further, steps towards gradual increase of NCRE component in the energy portfolio will lead to emerge as energy secured country in the future.


# Acknowledgement

I acknowledge the Board of Management of the National Science Foundation with sincere thanks for encouraging me to be engaged in energy conservation activities at the organization and facilitating me to acquire new knowledge in the field.

I extend my sincere gratitude to National Engineering Research and Development Centre and its engineers who engaged in the Energy Audit carried out at the NSF. I would like to thank Sri Lanka Sustainable Energy Authority for providing information with regard to list of approved organizations for carrying out energy audits.

I also thank researchers and scientists in Sri Lanka who assisted me in clarifying some of the estimated figures with regard to energy.

## References

Arachchi, K. W., 2007. Solar air heating systems for the industry: a new approach for their design and Development. SLEMA Journal, 13 (1), PP. 17-28.

Chandrasekara, D., 2007. Options for Sri Lanka (Oil and Gas). SLEMA Journal, 13 (1), pp. 12-16.

http://thakshana.nsf.ac.lk/wwwisis/rgra/form.htm (Accessed 4th June 2008).

http://www.energy.gov.lk/energypolicy/energypolicy.asp (accessed 4th June 2008).

National Energy Policy and Strategies of Sri Lanka.

Research Grants, Online Database, National Science Foundation.

## Abbreviations

NSF: National Science Foundation
LNG: Liquefied Natural Gas
LPG: Liquefied Petroleum Gas
kTOE: Thousand Tonnes of Oil Equivalent
NCRE: Non Conventional Renewable Energy
SGF: Sustainably Grown Fuel wood
MT: Metric Tons
SLEMA: Sri Lanka Energy Managers Association

# Chapter 7

# *Energy Policy for Mauritius: Analysis, Prospects and Challenges*

**P.M.K. Soonarane**
*Ministry of Public Utilities, Port Louis, Mauritius*
*E-mail: psoonarane@mail.gov.mu*

## ABSTRACT

As a small island developing state (SIDS), Mauritius is at the crossroads of its future socio-economic development, particularly in the context of the looming energy crisis. Against such a backdrop, Mauritius is actively putting in place a number of energy policies and strategies to be able to confront the daunting challenges lying ahead. Such strategies are focused on both the demand and supply side of energy. The strategies are firstly geared towards regulating the efficiency of new electrical appliances used by households and the imposition of appropriate code of practice for industrial and commercial utilization of energy. In addition incentive frameworks are being put in place to encourage the use of efficient technologies such as energy efficient lamps for lighting on a very wide scale. The supply side strategy is underpinned on the use of renewables. Mauritius has a long and rich experience in the use of biomass, particularly bagasse which is a byproduct of the sugarcane industry. Bagasse based electricity is poised to increase further in the coming years. In addition, Mauritius is proposing to use ethanol produced from molasses for part-substitution of gasoline while research in coordination with other international organizations is ongoing on the gasification of bagasse that could potentially double the amount electricity that can be generated from bagasse. Mini-hydro potential is being given a boost, while the development of a major wind-park will complement the already acquired experience with small wind generators to improve the energy mix of the country. Mauritius is also giving a new drive for the wider use of solar water heaters for substitution of electricity and gas for domestic water heating. Furthermore, a new framework to allow households to generate their own electricity from either solar or wind and sell any excess to the national grid is being put in place.

***Keywords:*** *Policies, Strategies, Renewables, Bagasse, Solar, Wind, Ethanol and Conservation.*

## 1.0 Introduction

With the price of oil hovering around $135 a barrel, the vulnerability of Mauritius as a small island developing state (SIDS) in terms of its reliance on oil cannot be overstated. Compared to 2006, there has been a 17 per cent increase in 2007 of the energy imports from Rs 19 billion to Rs 22.3 billion, with obvious adverse effects on the country's balance of payment. In 2007, energy accounted for 18 per cent of total imports and this is likely to increase in 2008 with the sustained increase in the price of oil. Given that oil reserves are expected to last some 40 years or so as predicted by many experts, Mauritius is at the crossroads in terms of developing an energy policy and strategy for the medium to long term that would reduce as far as possible and practical its reliance on oil. Such a strategy should be multi-pronged, reckoning with the constraints that the country has in terms of its geographical location, availability of other less volatile and better distributed reserves of fossil fuel such as coal, energy resources available locally and the availability and competitiveness of technologies in the marketplace for their conversion, level of energy demand and the ongoing macro-economic re-orientation in the light of the challenges and threats to the existing pillars of the economy with the slow down in global growth.

This paper provides a background of the institutional set-up of the energy sector in Mauritius, analyses the current electricity mix in Mauritius, makes a diagnosis of the policy drivers that enabled the achievement of such a mix, and discusses new energy policy drivers and initiatives, in the light of the experience acquired, that are necessary to allow the country to successfully meet the daunting and serious challenges it is facing in the context of the current near-to-crisis situation in oil supply in the world market against the backdrop of macro-economic difficulties. As regards the transportation sector, the paper examines the way forward in the context of ongoing transport policies and initiatives in tandem with the linkage for energy demand in the sector, particularly in the light of the dedicated bus lane project opted by Government and the use of sugarcane based bio-fuels in motor vehicles.

## 2.0 Current Legal and Institutional Framework in the Energy Sector

The Ministry of Public Utilities (MPU) is responsible for the design and implementation of energy policies. It is primarily responsible for the power sector, as well as water and wastewater sectors, and oversees the power utility, the Central Electricity Board (CEB). The Ministry of Commerce oversees the petroleum industry, including the State Trading Corporation, which is responsible for the import of all petroleum products. The import and distribution of coal are not regulated. The principal legislations governing the sector include the: (i) Electricity Act of 1939 (amended in 1991); (ii) Electricity Regulations of 1939; (iii) Central Electricity Board

Act (1964); (iv) Petroleum Act of 1970 (amended in1991); and (v) Consumer Protection (Price and Supplies) Act of 1998.

The CEB, which generates power, complemented by generation from independent power producers (IPPs), is the sole supplier of electricity to consumers in the country. Currently, the CEB also acts as the electricity regulator, but this situation will change by mid-2009 when an independent regulator is expected to be in place. The Ministry of Commerce is responsible for the enforcement of the Petroleum Act and the Consumer Protection (Price and Supplies) Act. The Petroleum Act primarily includes provisions for the granting of prospecting and mining leases for petroleum, although there are as of to date no known indigenous sources of petroleum. The Consumer Protection (Prices and Supplies) Act empowers the Ministry of Commerce to control the import and prices of petroleum products. Gasoline, gas oil, kerosene and fuel oil are all imported through the State Trading Corporation, and the Ministry of Commerce controls the retail price of these products by adding excise duty, wholesale and retail margins to the CIF cost. LPG is imported by the Energy Storage Company Limited, a private company. The price of jet fuel is not regulated. The petroleum products, imported by the State Trading Corporation, are distributed by four multinational companies, including Indian Oil. Coal is imported by the private sector and is used by the IPPs in the sugar industry to generate electricity during the off-crop season and by the textile industries in their boilers for steam production.

The MPU also manages an Energy Fund which has received significant financial resources in budget 2008-09 to provide support to the development of Renewable Energy in the country. The fund would have a key role for promoting the use of renewable energy through the provision of appropriate financial incentives and grants.

## 3.0 Analysis: Demand and Supply

### 3.1 Imported and Local Supply

Mauritius has no known oil, natural gas or coal reserves, and is therefore heavily dependent on imported energy carriers. The country imports approximately one million tons of petroleum products, comprising 380 cst and 180 cst heavy fuel oil, gasoline, diesel and LPG, and 650,000 tons of coal annually. The balance is met from local and renewable sources, namely bagasse, a by-product of sugarcane processing, and hydro.

Mauritius, being a tropical country, has a very good solar regime, which gives it a potential average annual solar radiation value of some 6 kWh/m$^2$. One of the means used to tap this source of energy is through the use of solar water heaters. Although solar water heaters are manufactured locally in addition to imports, easy access to the electricity grid and reliable power supply combined with relatively high prices of the solar water heaters have restrained demand, that too despite the low interest loan facilities available to the public at large from the Development Bank of Mauritius (BBM) since the 1990's for the purchase of such units.

The wind regime in Mauritius is very good in some areas with an annual average speed of 8.1 m/s at 30m above ground level. Pilot projects in the mid 1980's were not successful as the wind turbines were damaged by cyclones. However, with the progress in the design of turbines, Government is putting emphasis on the use of this renewable source of energy on a large scale.

Biomass energy in Mauritius consists mainly of bagasse, fuel wood and charcoal. Bagasse is the most plentiful primary energy resource used by the sugar industry to meet all their energy requirements in terms of heat and electricity generation. In addition, surplus power from bagasse is fed into the national grid. There is still room to increase this contribution through the ongoing centralization programme of the sugar industry. As regards fuel wood and charcoal, their use for cooking is not significant, while LPG is widely used by households for cooking.

Hydropower potential has been almost fully tapped in Mauritius, except for mini-hydro at a couple of sites, and there is competing demand for the existing water resources. There are eight hydropower plants with a combined installed capacity of 59 MW. However, due to seasonal rain conditions and limited storage capacity, only two of the hydro plants can generate all year round during peak hours, whilst the other six generate as and when water is available, mostly during the period January to March.

### 3.2 Electricity Demand

Total electricity generation in the island in 2007 was 2160.2 GWh. Hydro contribution was only 83.3 GWh, below the average hydro generation in a normal year of about 100 GWh, and the remaining was thermal out of which 850.2 GWh was from heavy fuel-oil in diesel plants, 2.8 GWh from kerosene in gas turbines for peak shaving 879.8 GWh from coal and 346.8 GWh from bagasse. The coal and bagasse-based electricity, totalling 1226.6 GWh, was generated by the sugar industry.

Forecast shows that in 2025 about 4500 GWh of energy would have to be generated to meet demand, with a peak power demand of more than 700 MW. As it will be discussed later, bagasse would be able to contribute some 600-900 GWh to that requirement depending on the success of the implementation of measures in the wake of the accelerated centralization of the sugar industry. Coal, on the other hand, would substitute oil to some extent thereby significantly reducing the country's dependence on oil.

## 4.0 Prospects and Challenges

### 4.1 Government's Energy Policy

The energy policy for the short and medium-to-long-term of Government is four-fold. Firstly, availability, security and diversity of supply is of topmost priority; secondly, affordability is the key to Government's policy to ensure socio-economic development of every region of the country; thirdly, energy conservation is another central plank of Government's energy policy; and fourthly, reform of the power market and industry, which includes the opening of certain areas of the electricity market to the private sector, is a vital thrust of Government's policy.

There are a number of challenges lying ahead in the design and implementation of key strategies to achieve the stated policy aims and objectives of Government. One of such challenges is the need to achieve the right energy mix strategy to be able to ensure both security and diversity of supply. In the 1990's more than 70 per cent of the country's electricity requirements were met from oil. This made the country very vulnerable in view of the volatility of the prices of oil products, more so during times of crisis such as the last two gulf wars and the current near-to-crisis with the price expected to even reach $200 in the near future. The dependence on oil for electricity supply has been reduced to some 39 per cent today through enhanced use of renewable energy sources such as bagasse and hydro and the use of coal as a complementary fuel to bagasse during the sugarcane off-crop season. The share of bagasse in total electricity generation in 2007 was about 16 per cent and that of coal about 40 per cent. Hydro contribution was about 4 per cent.

With a view to further diversifying the country's electricity base away from oil, it would be ideal to have a better distribution of the share of renewables and coal in the energy mix, that is a contribution in the medium to long term of 40-45 per cent from renewables, and about 40 per cent from coal and the remaining from oil, principally for part of semi-base load supply and for peaking. While hydropower projects, except for mini-hydro, cannot be further developed as there are no other economic sites, bagasse has the scope for further development for electricity generation.

## 4.2 Electricity from Bagasse

Energy generation has been vital to the sugar industry, the more so as about 50 per cent of investments in a typical factory is linked to the boiler and the turbo-alternator. Such investments can be optimized if power generation for supply into the national grid is included. In such an arrangement, the sugar factory can obtain steam and electricity required for its operation free, and in return the power plant uses the bagasse without any charge. This has been one of the cornerstones of Government strategy to optimize bagasse for power generation and at the same time modernize the sugar industry, in addition to diversifying its revenue for its sustainability in the context of the ongoing phased reduction in the guaranteed price of sugar from the European Union, 5 per cent in 2006, 17 per cent in 2008 to cumulatively 36 per cent in 2009.

Significant power can only be obtained if a factory handles large volumes of cane. This implies that power generation from the sugar industry in Mauritius, in particular from bagasse, can be improved if there are fewer mills, but bigger in terms of cane handling capacity. Given the phased reduction of the guaranteed price of sugar with the European Union (EU), the sugar industry is compelled to accelerate its centralization programme as elaborated in the Multi Annual Adaptation Strategy (MAAS) 2006-2015 agreed with the EU for support in the context of the restructuring of the industry. In that context, potentially only four sugar factories will stay in operation in the medium to long term, one each in the North, East, West and South of the country. Active discussions are currently being held between the stakeholders on the strategy for power generation from bagasse and coal from the sugar industry. With appropriate investment, the extraction of an average of more than 120kWh of energy from one ton of cane can be easily achieved. Given that the country produces about 5.6 million tons of cane annually, the total amount of energy that can be generated from bagasse in the coming years would be more than 600 GWh that is 50 per cent more than the current level of generation from this source. The spreader stoker boiler with the condensing-extraction steam turbine (CEST) is the preferred technology by the industry for achieving the said output of energy from bagasse.

In the longer term, it is quite possible that Mauritius may have to resort to higher bagasse yield type of sugarcanes to further improve the contribution of bagasse-energy in its energy mix. Such a strategy can be beneficial to both the energy and agricultural sectors, in particular in the context of the phased reduction in the guaranteed price of sugar by the EU. Mauritius has the real possibility, with the right balance between energy and sugar production, to make both the renewable energy and the agricultural sectors more viable and sustainable side by side. Bagasse will, therefore, continue to play a vital role in the country's energy mix, not only in the short-to-medium term, but also in the longer-term with attendant beneficial effects to the agricultural sector, least to say the potential that bagasse offers to ensure the long-term viability and sustainability of sugar production by still more added value of energy from sugarcanes with higher yield of bagasse. In that context, the Mauritius Sugar Industry Research Institute (MSIRI), an institution specialized in sugarcane research, is playing an important role in developing the right varieties of sugarcane. A 50 per cent increase in the amount of bagasse available from an increase in the fibre content of cane for example would allow the generation of approximately 900 GWh of electricity from bagasse annually that is about 40 per cent of current total consumption. In addition, The MSIRI is also working in close collaboration with international institutions for the setting up of a prototype gasifier in Brazil. However, funds to the tune of $ 25 million required for the project has not been secured yet.

## 4.3 Electricity from Hydro

As indicated earlier, the hydropower potential in the country is almost fully tapped. However, there are still a couple of sites for the installation of mini hydro units. In this context, a 350 kW mini hydro power station will be constructed at an existing dam for supply of potable and irrigation water, where an outlet pipe had already been provided during the construction of the dam structure. Electricity will be generated when water will be released to a downstream reservoir via a feeder canal. A similar mini-power station will be constructed on the same feeder canal, where there is sufficient head to install a hydro turbine. The total cost of the two projects is estimated at Rs 70M.

## 4.4 Electricity from Wind

Mauritius for the major part of the year is exposed to windy conditions of the South East Trade Winds, and this makes it a favorable candidate for wind energy exploitation. Government therefore intends to encourage the development of wind power in Mauritius and the outer island of Rodrigues. As stated earlier, Government policy is to substitute as far as possible imported oil for power generation, and wind power development is considered as a priority. In this context, Government relies on the assistance of the Government of India. India is one the few countries in the developing world which has made great strides in the use of wind power. Its assistance is helpful to give the much-needed boost to the wind technology in Mauritius. A Memorandum of Understanding has been signed with the Government of India for renewable energy development in Mauritius, under which Mauritius, *inter-alia*, can benefit from technical assistance in the various fields of renewable energy.

A study in the mid 1980s, demonstrated that there are potential sites in Mauritius and in the island of Rodrigues for the setting up of wind farms. Mauritius and Rodrigues are, however, exposed to very high winds during tropical cyclones. These cyclones occur during the summer season and may be harmful to wind turbines. The highest recorded wind gust in Mauritius was about 280 km/hr that occurred in 1975. Wind turbines have, therefore, to be of such a design that they are not damaged in cyclones. It has to be noted that the CEB has already installed a wind park of three 60 kW units at Trefles in Rodrigues Island. These units can be easily tilted during cyclones to avoid damage.

The Ministry of Public Utilities in collaboration with the Public Private Partnership (PPP) Unit of Ministry of Finance and Economic Development (MOFED) will soon launch a RFP to develop a wind farm at a site called Bigara in the central area of the country with very good potential on a Build Own Operate scheme with the installation of bigger wind turbines for a capacity of 25-40 MW. A transaction adviser will be appointed to advise and assist the PPP Unit and MPU in the procurement process till the commissioning of the wind farm. Given the initial high costs involved in the development of renewable energy projects, Government will enlist the financial participation of development partners and rely on potential contribution from the sale of CERs applicable to such project to arrive at an affordable purchase price of electricity by the CEB from the wind farm.

As regards Rodrigues, the CEB will shortly proceed with the installation of two 275 kW wind turbines at Grenade estimated to cost Rs 40 M. Over and above the two units, with the financial assistance of Government it is proposed to install five additional units estimated to cost about Rs 100M. The objective of the CEB is to increase the number of wind turbines at Grenade to ten in the medium term.

## 4.5 Solar Energy

With regard to solar energy for water heating, a Solar Water Heater Scheme was introduced by the Development Bank of Mauritius (DBM) in the early 1990's. The scheme aimed primarily at providing soft loan facilities to households for the purchase of a solar water heater. So far, it is estimated that approximately 25,000 households are equipped with solar water heaters. A recent study carried out by the Ministry of Public Utilities has shown that the main barrier for further market penetration of solar water heaters is its high price. However, in view of the rising energy prices, Government has revamped the scheme with a direct grant of Rs 10,000 to households from July 2008 for the purchase of a solar water heater with a view to making it more attractive and achieving greater use of such units in the country. In addition, this measure will help to reduce the peak demand on the CEB system.

As regards solar electricity, a pilot project comprising a 76.8 kWp grid-tie photovoltaic system was implemented by Government in 2002 with a view to assessing the technology in the country. The technical and operational feasibility of the technology has been successfully established. However, the generation cost of the electricity is still unfavorable compared to other conventional and renewable sources. Nonetheless, to increase further the use of solar energy in the country from a long-term strategic perspective, it will henceforth be compulsory for promoters of development projects to make maximum use of solar energy for water heating in buildings and photovoltaic systems for outdoor lighting. Compliance with this new policy will be ensured through the introduction of appropriate conditions to that effect in the EIA license for new building projects. The target group includes new hospitals, hotels and IRS projects. Moreover, private households will also be encouraged through appropriate incentives to install photovoltaic panels to meet their own electricity needs and to supply any excess to the CEB grid.

## 4.6 Municipal Wastes and Landfill Gas

Discussions are underway between the CEB and a private promoter for the setting up of a 20 MW incinerator that would burn about 300,000 tons of municipal waste to provide some 144 GWh of electricity annually. In addition, the CEB is proposing to purchase about 2 GWh of electricity generated from a 3.5 MW plant that would use landfill gas.

## 4.7 Renewable in General

One of the primary constraints in the large-scale commercial use of renewable energy resources is the difficulty to match supply with demand pattern. Furthermore, energy from renewable technologies such as solar, sea wave, and tidal remain still relatively costly compared to conventional alternatives, although such a situation may reverse if the price of oil were to still increase and reach levels as high as $200 a barrel. Conscious of the need for a reliable supply of electricity to a growing economy, cost is obviously an important decision factor. Fortunately, we may just be reaching the point where renewable sources will progressively become economically attractive in comparison to generation from conventional sources. It is in this context that a grid code is being prepared to establish the protocols and standard to encourage households to generate electricity from either solar pv panels or wind turbines for their own consumption and sale of any excess to the national grid.

## 4.8 Electricity from Coal

The policy of Government with regard to coal utilization is that this source would be a major carrier in the country's energy mix for two reasons. Firstly, it allows bagasse to be a reliable source for year-round electricity generation. Otherwise, a power plant using solely bagasse would remain idle during the off-crop season. Secondly, coal is the only other fossil fuel available with proven reserves that can last much longer than oil, in addition to their even distribution around the globe, making it less prone to political and other associated risks, and therefore less volatile in terms of prices. Given the huge investment in the developed world to make coal a still cleanlier and environment friendlier fuel, Mauritius would be able to use such available technologies in the marketplace in its new strategy.

## 4.9 Energy Conservation/Demand Side Management

As regards the demand side strategy of Government, it is considered the right time at this juncture to further public awareness of the impacts of energy consumption on natural resources as well as the demands that consumption patterns place on electricity infrastructure, the more so as the economy in Mauritius continues to mature. To this end, the Ministry of Public Utilities has launched a comprehensive and sustained energy savings campaign targeting users in various category groups. The campaign would also involve the inculcation of a culture of energy savings among school children, which would bear its fruits in the medium-to-long term.

Also, Mauritius is faced with a highly variable daily demand pattern: system load averages about 200 MW during overnight periods, to reach nearly 370 MW during the evening peak hours between 18.00 and 20.30 hrs. The CEB must have sufficient production capacity to meet these peak hour demands, yet much of this capacity remains idle for long periods each day and on weekends. In the national interest, regular appeals are made to industrial and commercial customers and the public at large to limit their electricity consumption during these peak hours. If customers shift their electricity consumption to hours outside this peak period, the maximum demand will be lowered and system generation, transmission, and distribution assets will be more efficiently used. For example, the CEB has been encouraging industrial customers to modify their processes so that production cycles use a multi-shift or batching mode in which electricity-intensive activities are performed outside of peak hours, and to curtail non-essential activities such as automatic water pumping, electric water heating, and washing machine use during peak hours. Furthermore, the CEB is looking into the possibility to introduce time-of-use tariffs, which would encourage off peak consumption.

The use of compact fluorescent lamps (CFL), also known as energy saving lamps, will be promoted as one of the conservation measures to reduce the use of fossil fuels for electricity generation. In this context, the CEB will procure one million CFLs for resale to its customers at half the price. Each customer will be allowed to purchase a maximum of 3 CFLs. The project reconciles several objectives namely the savings of energy, reduction of lighting load, reduction of GHG, the reduction of the evening peak demand, lowering the electricity bills of customers and finally securing carbon credits from the market.

A GEF funded project on *Removal of Barriers to Energy Efficiency and Energy Conservation in Buildings* will be implemented over the next four years. During the first year of the project, priority will be given to the drafting of an Energy Efficiency Bill including secondary legislation to make mandatory the labelling of energy intensive domestic electrical appliances. This will give the opportunity to potential customers to choose between equipment, which generally have a low price at purchase but are not energy efficient, and energy efficient appliances which can be more expensive at purchase but have low running costs as they are energy efficient. The aim is to give customers the opportunity to make the judicious choice of their appliances.

Summer time will be adopted by Mauritius on a pilot basis whereby the clock would be moved forward by one, over the period 01 November 2008 to 31 March 2009. Besides reducing energy consumption, the summer time is expected to reduce peak demand by some 15 MW. The summer time will become a permanent feature if it proves successful.

The Ministry of Environment in collaboration with UNEP is actively developing a five-year project called 'Development of National Programme on Sustainable Consumption for Mauritius'. Resources use Efficiency with focus on energy is one of the five components of the programme. In this context, a number of energy conservation projects and initiatives will be identified for implementation over the next five years.

## 4.10 Transportation Sector

The transportation sector, which accounts for about 50 per cent of the consumption of imported oil products, poses more challenges in terms of fuel substitution. Such a situation arises from the dominance of traveling by motor vehicles for personalized traveling and buses for mass transportation.

Government has a declared policy of substituting 20 per cent of the gasoline utilized in cars by ethanol produced from molasses, a by-product from sugar processing. Such an initiative has benefits on two fronts in that it would reduce the reliance of the sector on oil with attendant benefits to the economy, and secondly it would help to boost up the revenue of the sugar industry that would offset part of the loss of revenue from the sale of sugar to the EU.

Government would proceed in phases with the E10 project at first, which comprises 10 per cent gasoline substitution. Positive results were obtained on the use of E10 in twenty five cars over a period of three months. The bio-fuel was available at only one filling station. Under Phase II, it is proposed to make E10 available at 6 filling stations with appropriate incentives for users to shift to the bio-fuel mixture. Furthermore, the sustained high price of oil is likely to make hybrid cars more popular in Mauritius in the coming decade.

In addition, a dedicated bus lane would be put in place within the next three years along the main urban route between the capital of Port-Louis and the town of Curepipe. It is expected that the energy intensity in this dedicated lane would improve and motorists traveling by motor vehicles for personalized traveling would have the necessary incentives to opt for the bus, which would see significant reduction in journey times.

# 5.0 Conclusions

Government is implementing a wide range of measures for diversifying the country's energy mix away from oil. With regard to electricity, the strategy is to promote the use of local and renewable sources as much as possible. In that context, a number of initiatives are already being put in place for enhancing the energy potential with regard to bagasse and wind technology with a view to achieving the target of 40 per cent of the country's electricity supply within the next decade, inclusive of hydro power, in the short to medium term. In the longer term, the country would have to resort to the cultivation of new varieties of sugarcane with higher fibre content to generate more bagasse for conversion into electrical energy. In addition, with the introduction of a grid code over the next six months or so. Government will encourage households to generate electricity from either pv panels or wind turbines for their own consumption and the sale of any excess to the national grid. Government is also keeping track of progress in new renewable technologies in terms of their development and cost. For instance, although tapping of ocean thermal energy (OTEC), tidal and sea-wave energy is still at a stage of Research and Development, the possibility of the application of these technologies in Mauritius with the assistance of friendly countries is being kept in perspective.

As renewable energy can never on its own provide a complete solution to the country's vulnerability linked to oil dependence. Government multi-pronged strategy is also geared to the utilization of coal through the use of efficient, clean and environment-friendly technologies. As regards the transportation sector, 20 per cent use of ethanol for substitution of gasoline is a strategy being put in place for a reduction of the country's dependence on oil. Furthermore, in the coming decade the use of hybrid cars are likely to become more popular among car users. The implementation of a dedicated bus lane along the major urban corridor is likely to shape the future of traveling in the country with people in urban areas leaving their cars for buses with attendant benefits in terms of improved energy intensity in the sector.

## References

1. Digest of Statistics, Central Statistical Office, www.gov.mu/portal/site/cso.
2. Budget Speech 2008-09, www.gov.mu.
3. CEB Annual Report 2006,ceb.intnet.mu

## Abbreviations

CEB: Central Electricity Board
CIF: Cost, insurance and freight
CER: Certified Emission Reduction
CEST: Condensing-extraction steam turbine
CFL: Compact Fluorescent Lamp
CSO: Central Statistical Office
Cst: Centistokes
DBM: Development Bank of Mauritius
E10: 10 per cent Ethanol mixture with gasoline
EU: European Union
GEF: Global Environment Facility
GHG: Greenhouse Gases
GWh: Gigawatthour
kW: kilowatt

LPG: Liquefied Petroleum Gas
MAAS: Multi Annual Adaptation Strategy
MOFED: Ministry of Finance and Economic Development
MPU: Ministry of Public Utilities
MRC: Mauritius Research Council
MSIRI: Mauritius Sugar Research Institute
MW: Megawatt
OTEC: Ocean Thermal Energy Conversion
PPP: Public Private Partnership
RFP: Request for Proposal
Rs: Mauritian Rupees
UNEP: United Nations Environment Programme
UoM: University of Mauritius

# Part II

# *Cleaner Production*

# Chapter 8

# *System for Preventive Energy Saving and Cleaner Production Innovation Development and Implementation*

**Hab. Jurgis Staniskis**

*Institute of Environmental Engineering, Kaunas University of Technology, Lithuania*

*E-mail: www.apini.lt, justa@ktu.lt*

## ABSTRACT

Role of industry in the process of sustainable industrial development is obvious. It relates to changes in production processes, products and services aimed at reduction of impact to the environment in the entire life cycle perspective that result in improvement of environmental, economic and social performance of enterprises. To ensure sustainable industrial development, systematic application of the following measures is needed (Staniskis *et al.*, 2005):

1. Cleaner production (CP),
2. Environmental and integrated management systems,
3. Product related measures of sustainable industrial development (*e.g.* eco-design, life cycle approach),
4. Sustainability reporting.

CP should be an essential part of any comprehensive environmental management system at an enterprise or national level. In many cases the adoption of CP improvements can reduce or even eliminate the need for end-of-pipe investments and therefore can have both environmental and economic benefits. Experience shows, that often companies having identified cost-effective and technically feasible CP options, may still not be able to make necessary CP investment to realize the financial benefits and environmental advantages. Financing of CP projects varies among countries and types of the projects. Domestic and international efforts to strengthen environmental financing still face a number of serious obstacles, many of which are related to profound economic, political and social problems. The Institute of Environmental Engineering (APINI) (EU Centre of Excellence in Sustainable Industrial Development) initiated implementation of CP concept in Lithuania and is the main institution in the country working in this area. In 1992-2007, with support provided by different donors and in co-operation with different foreign and local partners APINI supported more than 150 Lithuanian companies to implement CP projects. In 1997, the Institute of Environmental Engineering has developed a system for preventive innovation development / implementation (APINI SPIN). The system consists of the pool of experts (on preventive innovation generation, financial engineering, implementation and monitoring), the financing source – soft credit line at Nordic Environment Finance Corporation (NEFCO) and industrial companies. The system is based on company's material and energy flows, and properly evaluated environmental costs based on Environmental Management Accounting (EMA). The methodologies employed in the system are flexible, can be applied to different company levels and enable decision–maker to obtain information in terms of the intended economic and environmental purposes. The system has been used for development of 141 innovations in 69 Lithuanian industrial companies. APINI experience was successfully disseminated in Africa, South East Asia, Central America and CEE. APINI SPIN System was awarded by ENERGY GLOBE 2007 National Award. The paper provides an overview of activities and experience of APINI in development of preventive innovations in industry and lessons learned in terms of strengthening corporate commitment in sustainable development.

***Keywords:*** *Sustainable industrial development, Preventive innovation, Material and energy balance, Environmental management accounting, Financial engineering.*

## 1.0 The APINI Method for CP Innovation Development

The below described logical steps of CP assessment are embedded in an organised procedure. This procedure is instrumental in organising the CP efforts, informing the necessary stakeholders within the company and bringing together the persons that can develop, evaluate and implement CP opportunities. CP assessment is most often divided into five phases (Staniskis and Stasiskiene, 2002):

1. *Planning and organisation*: CP planning is a systematic method for identifying options to reduce or avoid the generation of waste. The process:
    - Ensures the selection and implementation of the most cost-effective CP options;
    - Consistency of CP objectives and activities with those identified in the organisation's broader planning process;
    - Facilitates broader investment analysis and decision-making (such as capital budgeting and purchasing);
    - A documented CP plan may be a condition for receiving financing or insurance at more attractive rates.
2. *Pre-assessment*: The prime objective of the phase is the selection of one or a few assessment focuses. It requires a preliminary identification and

evaluation of the CP potential at the plant level. While doing so, first inventories of obvious options as well as a preliminary estimate of the waste generation costs are made.

3. *Assessment*: The phase focuses on the production process that involves waste streams, and is based on examination and re-evaluation of the production processes. The re-evaluation consists of "source identification" followed by "cause diagnosis" and "option generation".
   - For the *source identification*, an inventory is made of the material flows entering and leaving the company, with the associated costs. This results in a process flow diagram, allowing the identification of all sources of waste and emission generation.
   - Material flow analysis (MFA) is a systematic assessment of the flows and stocks of materials within a system defined in space and time. It connects the sources, pathways and intermediate and final sinks of a material.
   - In general, an MFA begins with the definition of the problem and of adequate goals. Then relevant substances and appropriate system boundaries, processes, and goods are selected. Next, mass flows and stocks are calculated and uncertainties considered. The results are presented in an appropriate way to visualize conclusions and to facilitate implementation of goal-oriented decisions.
   - The next step *(cause diagnosis)* is based on the material and energy balance and investigates the factors that influence the volume and composition of the waste and emissions generated. Evaluation of the relative importance of each of the possible waste generation causes is performed.
   - The purpose of the *option generation* is to create a vision of how to eliminate or control each of the causes of waste and emission generation.
4. *Feasibility studies*: These studies have to prove whether each of the options is technically and economically feasible and whether each option contributes to the environmental improvement. The level of detail in the feasibility studies needs to be tailored to the nature of the option, since options may vary from simple operational improvements, the use of alternative materials to installation of advanced equipment.
   - Technical evaluation consists of two interrelated parts. First, it should be evaluated whether the option can be put into practice, *i.e.* a check of the availability and reliability of equipment, the effect on product quality and productivity, the expected maintenance and utility requirements and the necessary operating and supervising skills. Second, the changes in the technical specifications can be converted into a projected material balance, reflecting the input and output material flows and energy requirements after implementation of the CP option.
   - The economic evaluation consists at least of data collection (regarding investments and operational costs, as well as benefits), choice between

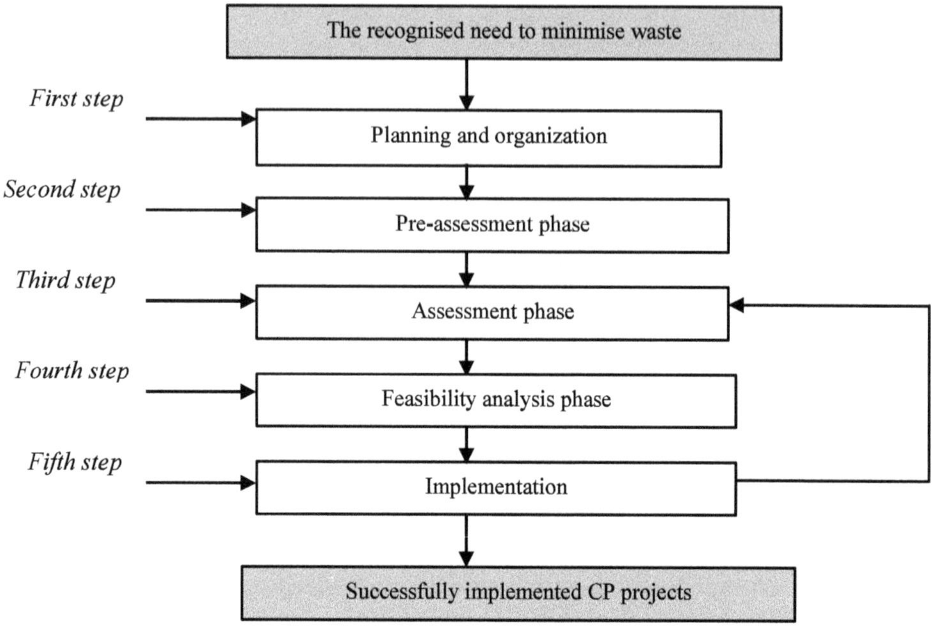


**Figure 8.1: Cleaner Production Procedures**

evaluation criteria (payback period, net present value (NPV) or internal rate of return (IRR)) and feasibility calculations. The economic data collection is based upon the results of the technical evaluation. In order to properly incorporate the long-term economic advantages of CP, it is highly recommendable to apply Environmental Management Accounting (EMA) principles to the economic evaluation (especially for high cost options).

- It is suggested that all significant environmental costs (and impacts) should be traced to the responsible product (as opposed to the use of overhead accounts). That is, material tracking could be used for materials costs and other related product costs (such as energy, water use, and so on), as well as for waste streams. Where tracking does identify costs, these should be clearly identified with the tasks or processes generating the costs.
- The objective of environmental evaluation is to determine the positive and negative impacts of the option on the environment. It is of little value if an option reduces one problem and creates another, potentially a worse one. In order to be comprehensive, an environmental evaluation must take into account the whole life-cycle of a product or service.

  The following steps can be used:

  1. Evaluate changes in: (i) quantity and (ii) toxicity of wastes/emissions at all stages in their life-cycle (at least in terms of raw materials, production, use and disposal).

2. Evaluate changes in energy consumption throughout the life-cycle.
3. Determine whether the environmental effects will be shifted to other media.
4. Determine whether there will be a change in the re-usability of this or other waste streams.
5. Determine whether the degradability of the wastes/emissions will be changed.
6. Determine the extent to which renewable raw materials can be used.
7. Determine the extent to which energy consumption can be reduced.
8. Determine the extent to which renewable energy sources can be used.

- *Implementation and continuation*: The feasible prevention measures are implemented and provisions taken to assure the ongoing application of CP. The development of such ongoing programme requires of monitoring and evaluation of the results achieved by the implementation of the first batches of prevention measures. The expected result of this phase is threefold: (i) implementation of the feasible prevention measures; (ii) monitoring and evaluation of the progress achieved by the implementation of the feasible options and (iii) initiation of ongoing CP activities.

## 2.0 APINI–SPIN CP Financing Scheme

The technical appraisal of the project in the APINI–SPIN methodology is focussed on:

- Ensuring that the project is technically feasible, technical solution is cost-effective, and that no experimental technologies are being applied;
- Confirming that the environmental and economic benefits are achievable;
- Establishing that the time scales are reasonable; and confirming that the procurement approach is acceptable and that the price estimates are realistic.
- The processing facilities are complex systems of unit operations and streams. Consequently, their environmental impact cannot be optimally mitigated by simple end-of-pipe measures. Instead, it is crucial to gain global insights into how mass flows throughout the process, and to use these insights as a consistent basis for developing cost-effective cleaner production solutions (Kliopova and Staniskis, 2006).

Over the past decade, significant advances have been made in treating different production processes as integrated systems. This holistic approach can be used to enhance and reconcile various process objectives, such as cost effectiveness, yield enhancement, energy efficiency and cleaner production. In this context, a particularly

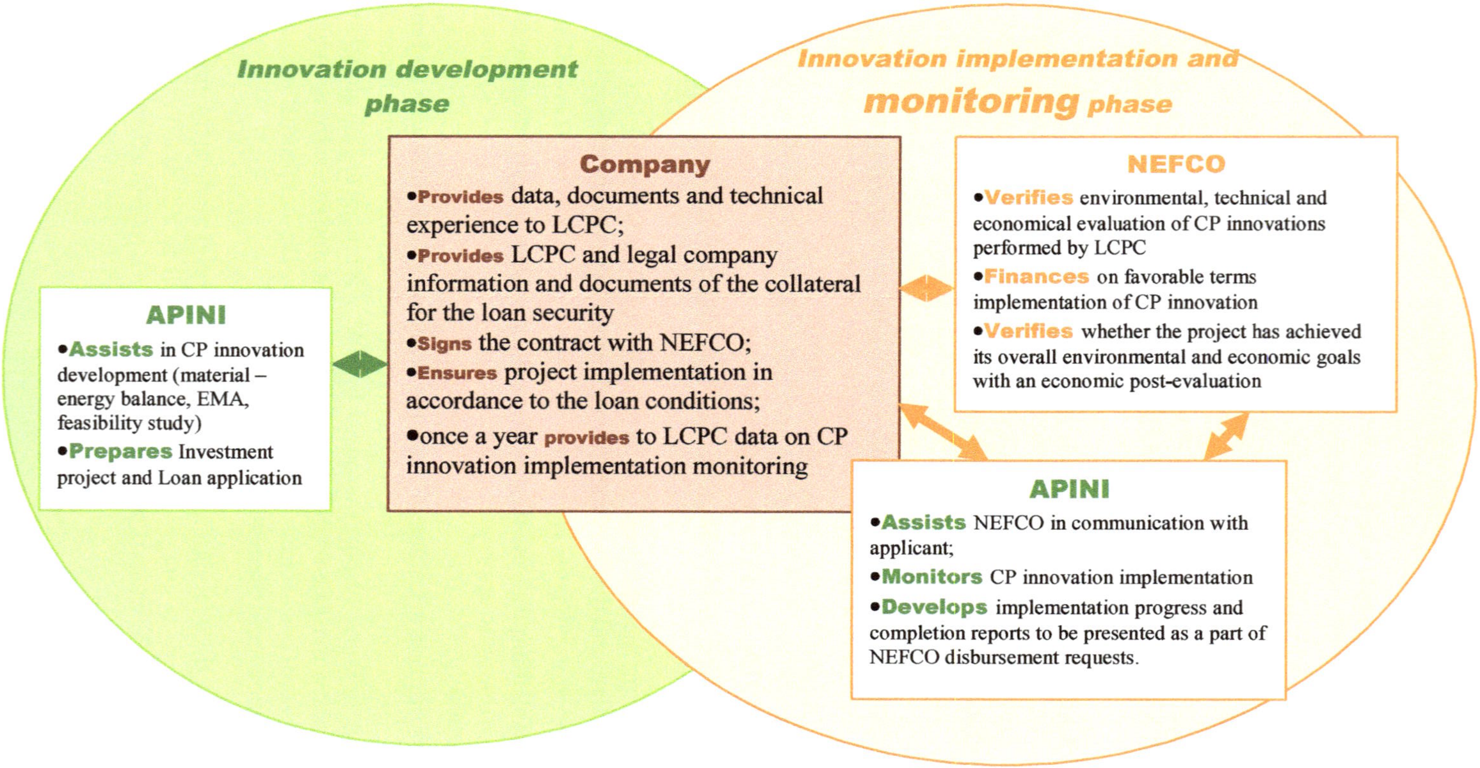


**Figure 8.2: APINI–SPIN "Soft" Credit Line for CP Innovations Financing**

powerful concept is mass integration, which deals with the optimum routing of streams as well as allocation, generation and separation of species.

In the APINI–SPIN method, material flow analysis is taken as a basis for problem identification, *i.e.* evaluation of CP potential at the plant level, a preliminary estimate of waste generation costs, in-depth analysis of selected assessment focuses (quantification of the volume and composition of various waste streams and emissions as well as a detailed understanding of the causes of these waste streams and emissions).

The systematic investigation of material flows and stocks of anthropogenic systems provides a new approach to production, economics and the environment. Material flow analysis links anthropogenic activities with resource consumption and environmental loading.

In case the important flows and stocks of materials used by humans are uniformly analysed using the methodology of material flow analysis and the information obtained is linked to powerful databases on material sources, pathways, intermediate stocks and final products, then more efficient use of resources will be possible. This motivates management of the companies and reallocates investments from end-of-pipe solutions to preventive measures and to get significant environmental and economic benefits.

The actual investments in cleaner production take place in the last phase of the assessment procedure. However, difficulties encountered while preparing and implementing CP investments may have been caused in earlier stages of CP assessment (due to, for instance, an insufficient depth of the diagnosis of the process, waste generation sources and causes, the lack of quantification of the final benefits from low and no-cost options implemented in earlier stages).

Many companies simply include all environmental protection costs into their general overhead costs, together with the top management salaries, advertising costs and all other costs that were not traced back to individual production processes. At a time when environmental compliance costs were marginal and profits high, this might have been reasonable. But within increased environmental awareness, strong competition and the need to improve production efficiency, especially with regard to material efficiency, the cost of tracking and tracing material flows throughout the company are by far outweighed by the improvement potentials identified and realised (Arbaciauskas and Staniskis, 2004).

## 3.0 Implementation Results

Cleaner production measures have mainly been used to reduce energy and water consumption. In terms of innovation type, most of investments have been used for process optimisation and technology change. Detailed information about CP activities in selected industry sectors is presented in Table 8.1.

Results from implemented CP innovations are summarised in Table 8.2.

**Table 8.1: Implementation of CP innovations in Lithuanian industry**

| *Industry Sector* | *Number of Enterprises* | *Number of CP Options Analysed* | *Number of Implemented CP Measures* | *CP Investments, EUR* | *Savings from implemented CP Measures, EUR/Year* |
|---|---|---|---|---|---|
| Textile industry | 14 | 42 | 39 | 2 734 000 | 2 474 000 |
| Food industry | 13 | 27 | 25 | 2 027 000 | 1 365 000 |
| Chemical industry | 6 | 15 | 14 | 435 000 | 493 000 |
| Machinery production | 5 | 5 | 5 | 1 033 000 | 389 000 |
| Production of radio, TV and telecommunication equipment | 2 | 6 | 6 | 1 478 000 | 613 000 |
| Furniture production | 6 | 10 | 10 | 1 030 000 | 421 000 |
| Wood industry | 3 | 6 | 6 | 1 431 000 | 1 067 000 |

**Table 8.2: Results from Implemented CP Innovations**

| | |
|---|---|
| Number of companies | 69 |
| Number of implemented preventive innovations | 141 |
| Environmental results (yearly): | |
| El. energy consumption reduced | 30 365 MWh |
| Heat energy consumption reduced | 60 510 MWh |
| Waste amount reduced | 86 700 t |
| Chemicals consumption reduced | 850 t |
| Air emission reduced | 79 500 t |
| Drinking water consumption reduced | 297 500 $m^3$ |
| Diesel consumption reduced | 387 000 l |
| Natural gas consumption reduced | 5 883 000 $m^3$ |
| Petroleum consumption reduced | 656 800 t |
| Wastewater amount reduced | 622 500 $m^3$ |
| Industrial water consumption reduced | 468 900 $m^3$ |
| Economic benefit: | |
| Total investment in preventive innovations | 16 529 000 EUR |
| Yearly savings from preventive innovations | 9 605 000 EUR |

Some key learning points concerning cleaner production are as follows:

1. CP approach reduces pollutant generation at every stage of the production process and thus minimises or eliminates wastes that need to be treated at the end of the process;
2. CP is not a one-time event, it is a continuous process;
3. CP can be achieved through good operating practices, process modification, technology changes, raw material substitution, and redesign and/or reformulation of product;

4. Effluent treatment, incineration, and even waste recycling outside the production process are not regarded as cleaner production, although they remain necessary activities to achieve a low environmental impact;
5. The economic advantages of CP are that it is more cost-effective than pollution control. The systematic avoidance of waste and pollutants increases process efficiency and improves product quality. Through pollution prevention at the source, the cost of final treatment and disposal is minimised;
6. The environmental advantage of CP is that it solves the waste problem at its source. Conventional end-of-pipe treatment often only moves the pollutants from one environmental medium to another;
7. The barriers to CP are mostly related to human rather than technical factors. The end-of-pipe approach is well known and accepted by industry and engineers. Existing government policies and regulations often favour end-of-pipe solutions. There is a lack of communication between those in charge of production processes and those who manage the wastes that are generated. Managers and workers, who know that the factory is inefficient and wasteful, are not rewarded for suggesting improvements;
8. Considering that CP addresses the problem at several levels at the same time, introduction of an industry/plant level programme requires the commitment of top management and a systematic approach to cleaner production in all aspects of the production process.

In spite of the results presented above, industry very rarely considers cleaner production initiatives as an option for improving productivity in parallel to a reduction of the impact on the environment. End-of-pipe solutions are still commonly applied if environmental demands are to be met.

Experience from CP projects implemented in Lithuania has shown that it is difficult to ensure the follow-up of CP activities on a company level after the project end. Therefore, a decision has been made to integrate CP and environmental management systems (EMS) and to start the implementation of preventive environmental management systems:

1. Firstly, cleaner production is not a continual and normal practice for industry;
2. Secondly, EMS does not guarantee that CP will be applied or even that an environmental performance above regulatory requirements will be obtained, *i.e.* EMS focuses on the quality and strengths of procedures and not the actual outcome of these procedures regarding pollution abatement and improved resource productivity.

## 4.0 Conclusions and Considerations

The implementation of sustainable development is mainly based on two fundamental and sometimes conflicting approaches. One approach is indirect, which generally assumes that system level reforms and capacity building are sufficient to motivate industry development toward sustainability. It focuses on reforms of tax policies, the use of market-based instruments, education and other tools that create an external environment, which motivates decision-makers to choose more sustainable alternatives. Another approach is direct; it focuses on specific intervention processes, technical assistance and direct financing to industry and organizations. In reality, it is assumed that indirect efforts must be supplemented by direct ones, *i.e.* that industry needs a helping hand to progress.

Until now, the direct approach remains one of the most common features of sustainable development programmes and in particular Cleaner Production programmes. Many CP programmes, especially in Lithuania, have shown that they produce positive net benefit. At the same time it is very difficult to prove if the investment in the programme has created environmental benefits exceeding those that could be achieved through other alternatives with similar funding. It is very complicated to determine what really works at the indirect level, because indirect interventions take a long time to show any effect. At the direct level, financing is an important component in promoting CP programmes and in most cases takes the largest share of capital for a particular programme.

Thus, financing sustainable development should focus on changing the conditions that reduce sustainability. Development organizations in this case have two options to help. They can help industry by using existing financing mechanisms, for instance, private and commercial banks, leasing, equity financing, etc. and/or have to create new financing mechanisms. The action is extremely important for all organizations promoting Cleaner Production and other preventive strategies.

The Organization for Economic Cooperation and Development (OECD) has studied many aspects of CP promotion and financing and has targeted on the achievement of a Basic Capacity Level in the country. The logic of CP financing, as recommended by OECD, follows several key steps. First, financial institutions must be taught to understand and appreciate CP investments so that they are acceptable to financial institutions. But this is not enough, because financial institutions are hard to change, and many smaller industrial companies do not meet the basic requirements for getting a commercial credit. The second major step in CP financing is creation of special financing mechanisms to encourage CP.

It must be recognized that many existing CP promotion programmes worldwide are missing the step of Basic Capacity Level and education of financial institutions. Before a company can apply for financing, it must know how to professionally prepare an investment proposal and make a good business case to a financial institution. This implies understanding the principles of environmental cost identification and assessment and environmental management accounting. This is the developmental step that must precede all efforts to promote CP financing. Hopefully, the present monograph will contribute to overcoming this barrier.

Besides that, environmental costs assessment and environmental management accounting are a company's internal tools for capital budgeting of environment-related investments. The capital budgeting process refers to the internal allocation of capital to actions, *i.e.* allocation within the boundaries of the company. For this purpose companies appear to use a combination of indicators to evaluate capital investment. However, the actual effect of environmental performance indicators on the decision-making process is poorly designed. Nevertheless, the accounting aspects of environmental management have received more attention, inducing the development of new tools of management. It is important to stress that to overcome the shortcomings of existing management accounting systems and techniques, new accounting systems must be developed or those traditional redesigned. Additionally, the external factors that influence the costs and benefits of environmental investments should be studied. Examples of these factors are (Staniskis and Stasiskiene, 2002):

- Taxes (low/no VAT) on CP technologies and cleaner products;
- Subsidies on environmental investments;
- Tax credits for environmental investments;
- Reduced capital cost by reducing the risk premiums.

There are many reports on company-based case studies, which demonstrate that environmental costs frequently have a significant share in total operational costs.

Our experience has confirmed that if any financing of CP is to be provided, a revolving facility is a more appropriate mechanism. Such facility may offer low interest rates, grace periods or other incentives to borrowers. But it is not necessarily easier to get financing from a revolving facility than from a commercial bank. A revolving facility might even ask for additional information or participation of the borrower in specific, for instance Cleaner Production, programmes or other support systems.

At the same time, it should be noted that a revolving facility is very difficult to design well, and it can be easily undermined by changes in the macro-economic or political environment. They do not solve the problem of unprepared applicants who are not yet able to formulate a good proposal.

In such cases, non-financial support services are expected to be offered by other service providers. Although there may be differences of opinion on the subject, there is no doubt that a suitable synergy between financial services and business development services (BDS) can make credit facility schemes more effective and produce more successful lending programmes. The majority of the companies in the developing world need some help to convert ideas into a technically and financially acceptable business plan. BDS can facilitate a possible partnership with investors but only if there is sufficient data, records and information on the business and the investment proposal.

Some banks try to solve this problem by organizing their own advisory and consultancy services. Inevitably, this is costly but apart the cost, there is a potential conflict of interest between those who help to draw up the business plan and the loan approval process which has to remain objective and independent. The real help that BDS can give is often before an approach is made to the financial facility.

Generally, the borrower has to go elsewhere to find a BDS provider: a private consultant, a business association or some private or public business centre. Working with BDS providers can reduce the transaction costs of a bank in handling a loan request. Credit facility may also benefit from working with BDS providers in recommending enterprises to get external outside advice when encountering the problems that result in poor repayment. However, BDS providers must not become debt collectors (Kliopova and Staniskis, 2006).

Finally, as a number of financiers have commented and different studies have shown, there is much more capital for qualified projects than there are qualified projects. In the end, the financing of investments by capital through loans of a financing facility heavily depends on (Staniskis *et al.*, 2008):

- The creditworthiness of companies. They have to meet conditions on liquidity, solvency, and profitability;
- The application itself – the financial costs and benefits of Cleaner Production investments should be identified properly;
- The existing relationship between a facility and a company.

It can be expected that as society becomes more aware of the environmental problems, shareholders might increasingly consider the environment in their behaviour. The number of investment facilities that incorporate ecological criteria into their decisions, as well as the amount of capital they control is increasing. This phenomenon can be explained by a rising number of environmentally conscious investors as well as by expected financial return on investments. The quality of environmental management and the degree of implementing Cleaner Production are recognized as a good indicator of the overall quality of business management. Accordingly, investment facilities are beginning to research these drivers as one of the elements in their financial analysis.

## References

Arbaciauskas, V., Staniskis, J.K., 2004. The Sustainable Industrial Development: Reality and Visions. Technological Choices for Sustainability. Ed. Subhas S.K., Glavic P., Jain R. ISBN 3-540-21131-4 Springer, p. 91-99.

Kliopova, I., Staniskis, J.K., 2006. Evaluation of Cleaner Production Performance in Lithuania Industry. Journal of Cleaner Production. ISSN: 0959 – 6526. Elsevier Science, Vol. 14, p.1561-1575.

Staniskis, J.K., Stasiskiene, Z., 2002. Promotion of Cleaner Production Investments: International Experience. Journal of Cleaner Production ISSN: 0959-6526 Elsevier Science, Vol. 11/6 p. 619-628.

Staniskis, J.K., Stasiskiene, Z., Jasch, Chr., 2005. Assessment of Environmental Costs for Sustainable Industrial Development. ISBN 9955-09-959-3, Kaunas, Technologija. p.260.

Staniskis, J.K., Stasiskiene, Z., Jasch, Chr., 2008. Cleaner Technologies: Environmental Management Accounting, Investment Appraisal and Financing. ISBN 978-1-60456-614-7 Nova Science Publishers inc., New York, 345p.

# Chapter 9

# *Cleaner Production and Energy Conservation for Sustainable Development in Zambia*

***John Lukonde Chongo***
*Department of Science and Technology–Technology Section,*
*Ministry of Science Technology and Vocational Training,*
*P.O. Box 50464, Lusaka, Zambia*
*E-mail: jlchongo@mstvt.gov.zm*

## ABSTRACT

Zambia is largely a mining country with abundant deposits of copper, cobalt, emeralds, coal, amethyst, gold, lead and zinc. Historically, the industry has been a pillar of Zambia's economy and continues to do so today. It is also the country's largest foreign exchange earner. However, by its nature, mining is a highly disruptive activity. It requires large tracts of land, consumes significant quantities of water and leaves a lot of waste materials. To maintain sustainable development in the country in the sight of economic development, Cleaner Production was introduced in Zambia in 1997. The Cleaner Production Project has achieved a number of goals. The CP Project was funded by Norwegian Agency for Development (NORAD) under the Industrial Pollution Prevention Programme (IPPP). In 1994, the United Nations Industrial Development Organizations (UNIDO) and the United Nations Environment Programme (UNEP) joined forces to help introduce Cleaner Production in developing countries and economies in transition through the National Cleaner Production Centres (NCPCs) programme. The UNIDO/UNEP programme is aimed at fostering dialogue between industry and government and enhancing investments for transfer and development of Environmentally Sound Technologies (EST). Since then, the concept of Cleaner Production (CP) has rapidly been gaining ground worldwide as a strategy to increase profits while reducing pollution and waste and using energy and raw materials more efficiently. The goal of CP is to avoid or at least minimize generating pollution at source–which frequently cuts costs, reduces risks, and identifies new opportunities. Over the last ten years, twenty-four National Cleaner Production Centres (NCPCs) have been established under the programme in developing countries and economies in transition of which nine are in Africa. The programme focuses on understanding the forces that drive consumption patterns around the world and how to translate those findings into tangible activities for business, governments and other stakeholders. Much of the current thinking on environmental protection focuses on what to do with wastes and emissions after they have been created. Throughout history industry has employed different strategies for environmental management. Before 1970, dilution and disposition methods were widely used while filtering was the method of choice for the 1970s. Reuse and recycling strategies were options of use in the 1980s and since the 1990s the world has since more preventive strategies such as CP and Environmental Management Systems. The cleaner production project in Zambia has achieved a number of goals. Some of these have been the training of 50 personnel from industry in CP technologies and the dissemination of information to stakeholders who include decision makers from selected industry and government institutions.

## 1.0 Introduction

Zambia is a Land Locked country covering an area of 752, 614 sq km (2.5 per cent of Africa). It is centrally placed and shares a border with Eight Countries. It is a country rich in natural resources, some of the finest in Africa. Minerals, farmland, forests, wildlife and water resources all provide the foundations for an economical development of the country. Zambia's has tropical type of climate; modified by altitude; rainy season (October to April). Its terrain is mostly high plateau with some hills and mountains. The Country's Natural Resources includes Copper, cobalt, zinc, lead, coal, emeralds, gold, silver, uranium and hydropower. Its Ethnic groups are of African 98.7 per cent, European 1.1 per cent, other 0.2 per cent. Administratively, Zambia is divided into 9 Provinces and 72 districts. Total population is 11.502 million with an annual growth rate of about 2.9 per cent. Most of the population is clustered along the line of Rail, which is highly urbanized due to economic activities. See the Zambian map (Figure 9.1).

The country is well endowed with a wide range of indigenous energy resources and is self sufficient in the supply of energy except in the case of petroleum. Though the Country is not a producer of Oil and Gas, recent studies have shown indications of Hydrocarbon potential. Zambia's main energy sources include biomass, hydropower, coal and renewable energy.

Biomass in the form of charcoal and wood fuel is the most dominant source of energy as only 20 per cent of the population has access to electricity. Electricity, Petroleum and coal are mainly used in the mines and industry. Like many other countries in the Sub Saharan Region, Wood fuel accounts for over 80 per cent of Zambia's total energy supply. Estimates for the other energy sources are 10 per cent for electricity, 8 per cent for petroleum, 2 per cent for coal while the use of renewable energy such as solar, wind and geothermal minimal or still untapped as is in the case of geothermal.

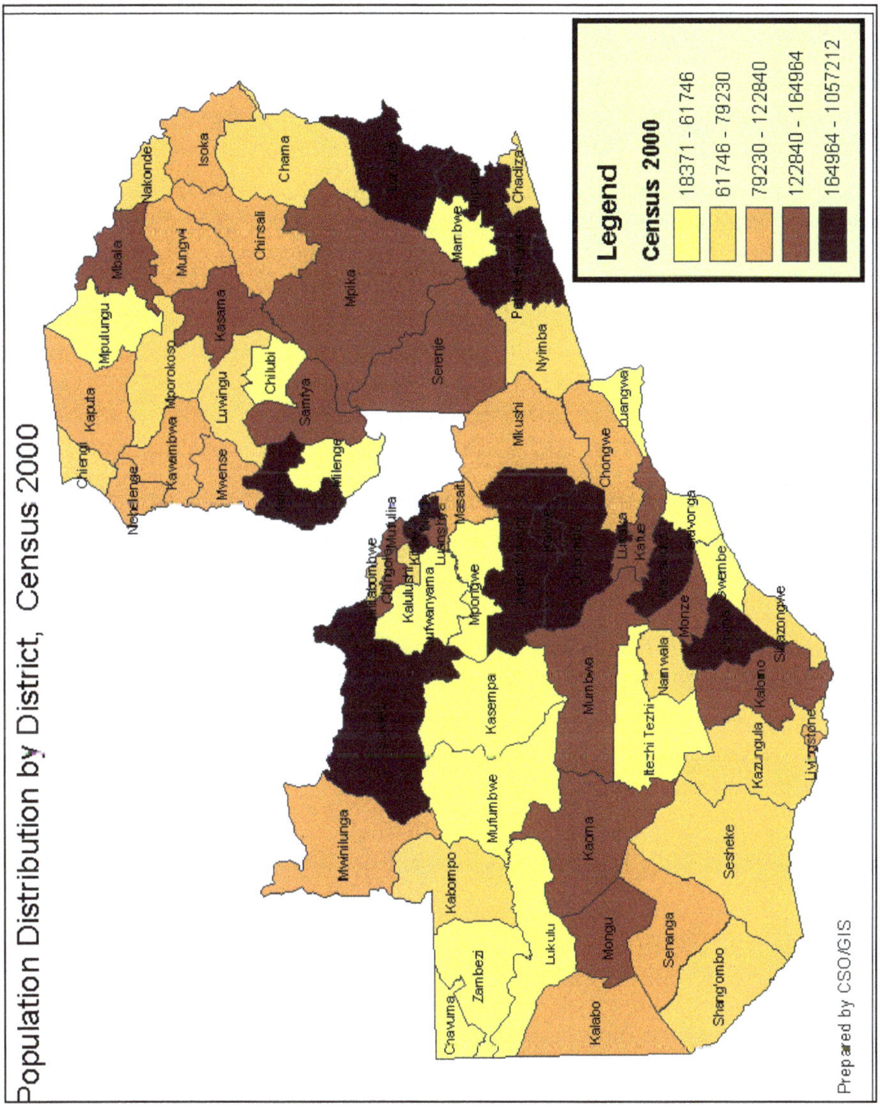


**Figure 9.1: Map of Zambian**

In Zambia, Lusaka and Copper belt provinces are the most urbanized areas. However, these are even the most highly polluted regions is the Zambia. The two provinces have a high concentration of industrial chemicals because of the manufacturing industry. These industries include pesticides and chemical manufacturing, wood preservation, metal plating and finishing, paper processing, battery, paint, food processing, leather ceramics, dye-textile, and petro-chemicals, to cite a few. The hazardous wastes produced by these industries are responsible for polluting many of the country's bodies of water. The mining and industrial activities has caused a significant environmental impacts including large waste generation, emissions to air, water and land, as well as excessive energy consumption.

Therefore, one of the most important issues in industries and mining is waste management to minimize the long-term environmental effects. Industrial wastes have a profound effect on the surrounding ecosystem. If they are not chemically stable, they can be a permanent source of pollution to the natural water system. This results in alterations in quality, quantity and availability of surface water or ground water. Such changes can affect the spawning grounds of fish and the habitats of bottom-dwelling creatures. Acid drainage may be the most widespread threat to aquatic species. Such effects may lead to extinctions, or may restrict access to species that local communities depend on, such as fish, vegetables, and medicinal plants.

## 1.1 Energy Conservation in Zambia

Energy is irreplaceable in the sustainable development of a country as it drives the engines of economic activities serving as a powerful tool in uplifting people's livelihoods.

Availability of affordable and reliable energy is critical in attracting sustained flows of investment thereby contributing to quality living standards of the people.

Sadly, the energy shortages being experienced in the SADC region, has become a source of grave concern to governments as the power deficits have potential to reverse the significant economic gains scored so far.

Power utility companies in a bid to minimally satisfy demand, have had to resort to load shedding much to the chagrin of consumers both domestic and commercial.

The status of hydro power generation capacity in the region is insufficient considering the economic progress being experienced and there is need to explore alternative low cost sources of energy to address the current shortages. Power shortages, power deficit, power outages, power black-outs, load shedding, have become common in the SADC region. Private sector participation in energy conservation is thus very cardinal.

Zambia has abundant hydroelectric resources and meets most of its energy needs from its own hydroelectric stations, which are operated by the state-owned Zambia Electricity Supply Company (Zesco). Zambia exports electricity to its neighbors, especially Tanzania and Kenya.

To lessen on the burden of hydroelectric resources, Zambia is now encouraging the use of renewable and alternative energy sources. At the moment renewable energy sources are increasingly being used though it still remains insignificant in terms of contributing to the total national energy supply. It is in this regard that promotion of renewable energy is one of the strategies aimed at sustainable energy development in the Fifth National Development Plan (FNDP) and in the National Energy Policy.

Renewable energy sources being promoted in the country include solar (thermal and photovoltaic), mini/micro hydroelectric, biomass (agricultural wastes, forestry waste, industrial/municipal organic waste, energy crops and products and animal waste) Geothermal and wind.

In Zambia, Petroleum, coal and electricity form the major commercial energy while the firewood and charcoal remains the dominant sources for domestic energy.

Solar energy is viewed as a prime tool for Southern Africa's rural (off-grid) electrification programs, which have been slowed by the high costs of grid extension services.

Zambia's government has encouraged investment by eliminating all import duties on solar panels and waiving the otherwise obligatory annual license fees for solar energy projects.

### 1.2 Cleaner Production in Zambia

The introduction of Cleaner Production (CP) in Zambia in 1997 has highly contributed to effective and efficiently running of the Zambian industry. The Environmental Council of Zambia (ECZ) in conjunction with Norwegian Agency for Development (NORAD) under the Industrial Pollution Prevention Programme (IPPP) coordinated a CP Project.

The theory and practical applications of CP were disseminated to Zambian industry through a series of workshops and in-company visits. As a supplement to the basic CP training, additional courses in environmental management systems (EMS) and business development for financing CP projects have been delivered. As part of capacity building, more than 15 CP experts have been trained through the train the trainer training. In addition a seminar on how to introduce CP in the planning process of developmental projects by integrating CP with the Zambian Environmental Impact Assessment (EIA) regulations was carried out.

The CP Project successfully conducted seven training programmes. Measurable environmental and economic benefits have been recorded in Zambia.

## 2.0 Results for Cleaner Production Project in Zambia

The CP Project achieved a number of goals. The project conducted seven training programmes from 1998 to 2004. The project trained over 50 personnel from industry in CP technologies and the dissemination of information to stakeholders who included decision makers from selected industry and government institutions. In addition, measurable environmental and economical benefits were recorded.

The following matrix summarizes the results for the CP project:

**Table 9.1: Total Average Costs Vs. Total Average Yearly Savings in CP Programs**

| Training Programme | Companies | Participants | | | Average Cost (USD) | Average Savings (USD) |
|---|---|---|---|---|---|---|
| | | Institutions | ECZ staff | Industry | | |
| CP1 (1998) | 8 | 5 | 3 | 29 | 50,000 | 530,000 |
| CP2 (1999) | 16 | 2 | 4 | 30 | 450,000 | 1,180,000 |
| CP3 (2000) | 11 | 3 | 3 | 24 | 27,000 | 63,000 |
| CP4 (2001) | 9 | 3 | 2 | 21 | 47,000 | 65,000 |
| CP5 (2002) | 8 | 2 | 2 | 23 | 45,000 | 251,000 |
| CP6 (2003) | 5 | 2 | 2 | 16 | 5,000 | 12,000 |
| CP7 (2004) | 14 | 5 | 5 | 21 | 8,000 | 84,000 |
| Total | 71 | 15 | 21 | 143 | 632,000 | 2,185,000 |

As can be seen from Table 9.1, Zambian companies made a lot of savings after applying Cleaner Production methods in their operations and appreciated the important of cleaner production mechanisms.

Some of the specific example for the CP program in Zambia, that were carried out in the year 1999 are shown in Table 9.2.

## 3.0 Benefits of Cleaner Production

As can be seen on some of the examples for the case studies above, Zambia has benefited from the cleaner production concept in the following ways:

- Decreased costs of raw materials and energy
- Decreased waste management costs
- Improved public image
- Enhanced productivity
- Improved product quality
- Increased profit margin

According to the United Nations Environmental Programme (UNEP), it is estimated that 70 per cent of all current wastes and emissions from industrial processes can be prevented at source by the use of technically sound and economically profitable procedures.

Further to this, it is estimated that more than 50 per cent of the waste volumes can be avoided by simple management measures and minor process changes!

## 4.0 Cleaner Production Challenges

Southern Africa which includes Zambia faces various environmental problems, including pollution of water supplies, deforestation, desertification, pollution associated with oil and gas development, soil erosion, decreasing quality and quantity of water, poor sanitation services and poor urban conditions and dramatic decline in biodiversity throughout the region.

**Table 9.2: The Second Zambian–Norwegian Cleaner Production Program (1999)**

| *Company* | *Projects/Options* | *Savings/Year, US$* | *Payback Period* | *Environmental Improvements* |
|---|---|---|---|---|
| Bitumen Products | 1. Bitumen recovery | 1,376 | Immediate | 1. Reduction of land pollution |
| | 2. Waste material analysis | 5,395 | Immediate | 2. Reduction of land pollution |
| | 3. Modification of water pumping system | 403 | Immediate | 3. Improved working environment |
| | 4. Drum cleaning | 8,510 | Immediate | 4. Improved working environment |
| BP (Z) Ltd | 1. Oil sample recycling | 1,910.00 | Immediate | 1. Reduced oil in the effluent |
| | 2. Used oil disposal | 18,125 | Immediate | 2. Reduced land and water Pollution |
| | 3. Compressor utilization | 3,215.00 | 15.5 months | 3. Efficient energy use |
| Central Breweries | 1. (i) water usage minimization<br>(ii) beer leakage<br>(iii) Cooling tower | 1,920 | Immediate<br>Immediate<br>Immediate | 1. (i) Effluent reduction by 50%<br>(ii) Effluent reduction by 50%<br>(iii) Effluent reduction by 75% |
| | 2. Energy recovery | 5,644 411 | 2.4 months | 2. Reduction of coal consumption by 12.5% |
| | 3. Use of already milled mealie-meal | 16,000 | Immediate | 3. Safe working environment |
| Chilanga Cement | 1. Change of input material-coal from peas and nuts to fines | 270,000 | Immediate | 1. Reduced coal dust pollution |
| | 2. Using used oil blended with diesel | 57,773 | Immediate | 2. Reduced ground water pollution for kiln preheating |
| | 3. Kiln 3 overhaul | 390,000 | 1 year | 3. Reduced $CO_2$, SOX and NOX emissions |
| | 4. Product change-Blockmate | 229,196 | 4 months | 4. Reduced $CO_2$, SOX and NOX emissions |
| Kafironda Explosives | 1. Replacement of citric acid with 70% acetic acid | 2,554 | 1 day | 1. Less solid waste |
| | 2. Installation of electric heaters | 1,120 | 1 day | 2. Less solid waste |
| | 3. Reuse of ANS spills | 234 | 1 month | 3. Less effluent discharges |
| | 4. Use of $NH_4NO_3$ solution instead of | 9,183,437 | 1 month | 4. Less solid and air pollution $NH_4NO_3$ solids |

*Contd...*

**Table 9.2–Contd...**

| *Company* | *Projects/Options* | *Savings/Year, US$* | *Payback Period* | *Environmental Improvements* |
|---|---|---|---|---|
| ZCCM-Nchanga Division | 1. Flocculant handling and mixing | 22,995 | 1.3 days | 1. Reduced land and water pollution |
| | 2. Recirculation of SX-T/H Solution | 1,252,680 | 1.22 days | 2. Less polyacrilamide levels in water |
| | 3. Crude pumping and treatment | 2,208,250 | 2.94 days | 3. Reduced water contamination |
| ZCCM-Konkola Division | Reduce TSS in process waste water | 19,500 | 7 days | Reduced land pollution |
| ZCCM-Nkana Division | 1. Pre-heater off gas at acid plant | 55,660 | 3 months | 1. Reduced $CO_2$ emissions |
| | 2. Retrofitting of gas cooling and cleaning | 4,329,817 | 5 months | 2. Reduced $SO_2$ emissions system at cobalt plant |

In order to help in answering to the above challenges Zambia has proposed to establish a sustainable CP Centre (CPC), which will follow up on the success and continue the economic and environmental improvements in Zambian industry.

## 5.0 Conclusions

There is urgent need to safeguard the environmental from further deterioration, reduce pollution and build better ways of managing resources for the future so that the much needed benefits of development can be wide spread and long lasting.

Zambia has put in place a number of programs aimed at conserving energy and promoting cleaner production. The country is promoting more efficient technologies such as improved stoves, use of alternative renewable energy sources *e.g.* solar, gel fuel, more efficient firewood management (drying wood, splitting wood, etc) and improved kitchen management (*e.g.* reducing time for cooking process, better ventilation at cooking place).

Allow me to end my presentation by thanking NAM S&T Centre for invite Zambia to participate and learn from other on cleaner production and energy conservation for sustainable development in countries.

Zambia subscribes to NAM S&T Centre as a member Country. It became a member of the NAM S&T Centre in April 1993. Since joining the NAM S&T Centre, over twenty (20) scientists/experts from different institutions in Zambia have attended various scientific events such as international workshops, Conferences, Training programmes etc, organized by the Centre in other Countries. Zambia managed to host two very important activities for NAM's a scientific event in the form of an African Regional Conference in January 2006 and the 2nd Bureau Meeting of the 11th Governing Council for Nam S&T Centre from 5th to 7th May 2008 at Chrisma Hotel in the tourist city of Zambia, Livingstone.

## References


1. http://www. copperbeltenergy.com/Zambia.html.
2. http://www. fes.org.zm/publications/energyreport.pdf.
3. http://www.erb.org.zm.
4. http://www.hedon.info/goto.php /The Zambia charcoal/Industry.
5. http://www.necz.org.zm.
6. Journal of Cleaner Production, Volume II, Number 2, March 2003, pp. 141-146(6).
7. Ministry of Commerce, Trade and Industry, (2005). Commercial Trade and Industrial Policy, Lusaka.
8. Ministry of Science Technology and Vocational Training, (2006). Fifth National Development Plan, Zambia.

# Chapter 10

# *Status on Cleaner Production and Energy Conservation in Kenya*

***Mobagi Dinah***
*National Environment Management Authority, Kenya*
*E-mail: dianasoila@hotmail.com*

## ABSTRACT

Promoting programmes and technologies for the cleaner production in Kenya is a great challenge due to poverty and inadequate capacity institutionally, individually and at organizational level. The sources of energy in Kenya include: Petroleum, electricity, gasoline, Solar, Wind, Biomass, Tidal energy on the Indian Ocean coast. Other potential sources are nuclear power and natural gas. Emissions such as $SO_2$ are approximately 5.545 ug/m$^3$ per capita, $NO_2$ (2.464 ug/m$^3$ per capita) and $CO_2$ (0.3 tonnes per capita). These levels are far below the maximum allowable levels of 50 ug/m$^3$ for $SO_2$ and 100 ug/m$^3$ for $NO_2$ respectively; an advantage, as Kenya can trade in emissions at the international level. Putting in place energy saving systems has adequately been planned for though implementation is a bottleneck. However in the industrial sector, energy savings of up to 25 per cent are possible in steam systems, by improving the efficiency of steam boilers, better steam distribution, and use recovery of waste heat and condensate. Motor systems, the largest users of electricity in industry, are often oversized, resulting in lower efficiency of operation. Potential savings are of the order of 20-50 per cent through motor system efficiency improvement. The Electricity Regulatory Board (ERB), established under the Electric Power Act, No. 11 of 1997, is empowered to enforce environmental and safety standards in power generation and ensure that electrical apparatus and works meet the standards set by the Kenya Bureau of Standards (KEBS) or, where no such standards exists, with the relevant international standards approved by Kenya Bureau of Standards. There is considerable potential for technology improvements in Kenya, for example in smaller and lighter cars, low or zero emission vehicles, or hybrid cars. Energy efficiency in the transport sector is increasing gradually. Poverty will remain to be the leading factor making most Kenyans not to switch to cleaner fuels, improved firewood cookstoves such as the kuni mbili (maendeleo) model and other cleaner production technologies. Measures for energy efficiency improvement as part of the government's national policies, strategies and priorities for achieving social and economic development, and protection of the environment have been put in place. Efforts at policy and technical interventions in atmospheric and indoor pollution are aware that improved charcoal and firewood cookstoves have demonstrated their potential to save 10-50 per cent of fuel. Policies geared to provide incentives for agro-forestry and village woodlots are being launched. Tax reduction on goods commonly consumed by the poor is one important policy measure that the Kenya government has been pursuing to eliminate the value of added tax on kerosene, cooking gas. There is commitment to improvement to access of modern energy at affordable prices for all sectors of the economy. Kenya energy policy aims at ensuring adequate, quality, cost affective and affordable supply of energy to meet development needs, while protecting the environment. The industrial development policy is based on the strategy for industrial transformation by the year 2020, sessional paper No. 2, 1997. Its vision is to bring about sustained economic growth and rapid employment generation through industrialization. The Government has been pursuing policies and practices intended to achieve energy conservation and efficient use, for instance, the Kenya association of manufactures (KAM) together with the ministry of Trade and Industry has engaged the Government to address the efficient use of energy. It has initiated the GEF-KAM Industrial Energy Efficiency Project since late 1990's.

*Keywords: Cleaner production, Energy conservation, Energy efficiency, Energy management, Energy sources, Poverty Reduction Strategy and Renewable energy.*

## 1.0 Introduction

In Kenya, reliable, affordable and sustainable energy is critical. The high cost of energy is one of the biggest bottlenecks to economic activity in Kenya (KIPPRA, 2005). Available data shows that the cost of electricity in the country is four times that of South Africa, the main competitor in the region, and more than three times that of China (KIPPRA, 2005). The problem of high cost is compounded by unreliability of supply. On average, Kenyan companies lose 9.5 per cent of production because of power outages and fluctuations. This excludes the losses from damaged equipment because of power interruptions, which could be up to Ksh 1 million for a company in 2001. Inefficiency in energy use is also one of the factors impeding the competitiveness of the country's products in international markets and a hindrance to adoption of cleaner production.

Access to adequate energy services is essential in both rural and urban areas. Lack of such access reduces the potential for achieving major structural changes in rural economies, required for income-generating activities and poverty alleviation. Access to electricity in rural areas is only 4 per cent compared to the national average of 15 per cent (Kamfor, 2002), one of the lowest in developing countries. Yet the Government of Kenya has been spending, since 1973 huge amounts of money under the Rural Electrification Programme. There is thus huge unmet demand in the country, particularly so in rural areas. Kenya heavily depends on electricity as a source.

Demand for electricity exceeds generation capacity during peak periods. The supply problem in Kenya is particularly acute during droughts when dam reservoirs are low or when some of the hydroelectric generating plants are out of service (Kamfor, 2002). There is also considerable loss of power during transmission and distribution, partly due poor technology.

Insufficient energy in the country has led to increased use of other energy sources that are not clean. The use of wood fuel, charcoal, dried dung and other farm residues has resulted into worsening indoor air pollution. In rural areas the lowest income household's urban areas depend on firewood. This together with other utilization of vegetation has posed great environmental challenges in the country. Kenya's main productive sectors (agriculture, fisheries, mining, and timber) depend on raw materials provided by the natural environment (Economic Survey, 2005). Water pollution has increased with emissions on an upward trend over the years. There are also cases of concern in the concentrations of heavy metals such as lead, mercury, zinc and cadmium, in the countries to water bodies.

Increase in air pollution from Carbon dioxide emissions have lead to global warming, (0.2 tonnes per capita) in 1990 and (0.3 tonnes per capita) in 2000. Accumulation of solid wastes is a common feature in the country. However there is potential for the use of solid wastes in urban centres to produce sustainable energy. The Kenya National Cleaner Production Centre (KNCPC) established by UNDP and the government of Kenya through the Kenya industrial research and development institute (KIRDI) has been championing the mainstreaming of cleaner production in enterprises as a management, economic, environmental and quality improvement tool. The National Environment Management Authority (NEMA) is among the significant government bodies that develop and monitor environmental regulations that articulate cleaner processes in the country.

## 2.0 Energy Sources in Kenya

There are three main sources of energy in Kenya. These are wood fuel, petroleum and electricity, accounting for 70 per cent, 21 per cent, and 9 per cent of total energy use respectively. Renewable energy is also becoming important although it remains insignificant in the country's overall energy mix.

### 2.1 Electricity

The major sources of electricity are hydro, geothermal and thermal power. The installed power capacity, in June 2005, was 1155.0 MW. The breakdown was: hydropower at 677.3 MW, oil thermal power at 344.2 MW, geothermal power at 128 MW, and wind power at 0.4 MW. The effective capacity was 1066.9 MW (KPLC, 2005). The key players in the power sector are KPLC, KENGEN, ERB, MOE, and independent power producers (IPPs). KPLC is 48.4 per cent government-owned and is the only licensed public electricity transmitter and distributor. KENGEN accounts for more than 82 per cent of the country's total installed generation capacity. The Government of Kenya faces several challenges in improving electricity generation and distribution to meet increasing industrial and residential demand. Some of the major challenges include:

1. Connecting a minimum of 150,000 new consumers per year to the electricity grid, including expansion of the Rural Electrification Programme.
2. Reducing the cost of electricity in Kenya, which is higher than costs of regional economic competitors in South Africa and Egypt.
3. Investing in upgrading of the national electricity grid to provide constant high quality power especially to industrial consumers.
4. High consumer tariffs due to operational inefficiencies and high taxes (*e.g.* VAT at 16 per cent).

## 2.2 Petroleum

Petroleum fuels are the most important source of commercial energy in Kenya, and are mainly used in the transport, commercial and industrial sectors. The country relies entirely on imported petroleum products. Imports of petroleum accounted for 16 per cent of the total import bill in 2002 and consumed 31 per cent of the country's foreign exchange earnings from merchandise exports. Consumption of petroleum products was 2.4 million metric tonnes, with per capita consumption at 76.2 kg. Projected growth in this demand, with Kenya's anticipated economic recovery of 2 per cent per annum (Ministry of Energy, 2004).Kenya's oil imports have not seen major shifts apart from sharp increases during the power crisis of 1998-2000 when the country imported extra tonnes of oil to meet increased demand. In 1998, the country imported 2157.7 tonnes of crude petroleum and 1387.8 tonnes of petroleum fuels. This declined to 1493.4 and 1023.5 tonnes respectively in 2002, mainly due to a strengthening of hydroelectric generation and easing of the power crisis.

The petroleum sector has been liberalized and the only direct government involvement in the petroleum industry is in the oil refinery it co-owns through the Kenya Petroleum Refineries Ltd (KPRL) with three private companies (Shell, BP and Chevron) on a 50-50 equity basis, and the oil storage facilities at Kipevu, capable of holding 1.5 million barrels. Kerosene as a cooking and lighting fuel is important for the poor in rural and urban areas and has in some cases served as a substitute for wood fuel. Hence, any efforts to increase kerosene consumption will undoubtedly relieve pressure on wood use. The government has often used tax reduction or non-increase for kerosene for this purpose and also as a poverty mitigation measure. Key emerging concerns in this regard are the impact of kerosene on the levels of indoor air pollution, the consequent health impacts on the poor and the adulteration of other fuels with kerosene. The IPP is currently supplying 44 mw of power. The market structure for the petroleum sector is not adequately competitive and the cost of energy is therefore high. Regulatory functions in the petroleum sector are shared among various players including the MOE, provincial administrations, local authorities and the Kenya Bureau of Standards (KEBS). The Petroleum Institute of East Africa (PIEA), a voluntary membership institution patronized by major oil companies, plays a key role in capacity building and awareness creation.

## 2.3 Biomass Fuels

Wood fuel has remained the most important source of energy in Kenya, meeting over 70 per cent of the country's total energy consumption needs. Eighty per cent of the population depends on it (Mugo and Kituyi, 2002). It provides 90 per cent of rural households' energy requirements and 85 per cent in urban areas. This state of affairs has major implications on sustainable development. Unsustainable harvesting, given the lack of efforts in reforestation and on-farm planting of wood lots, has often led to soil degradation, deforestation and associated diseases. According to the Poverty Reduction Strategy Paper for Kenya for the period 2001-2004, demands for wood fuel outstripped supply, and the country faced a wood fuel deficit of 4.1 million tonnes in 2005.About 47 per cent of the Kenyan households use charcoal. Some 82 per cent of urban households use the fuel compared to only 34 per cent of households in rural areas. Total charcoal production is about 2.4 million tonnes (or 67 million bags of 36 kg each). Charcoal continues to be harvested from trust lands and gazetted forests, an annual business worth Ksh 17 billion (Ministry of Energy, 2004). Improved charcoal production technology has had minimal impact on recovery and production. Some of the charcoal in Kenya are dubbed environmental friendly and exported but accurate figures are unavailable. Biomass is seen as the poor people's source of energy (Kituyi and Mugo, 2002).

### 2.4 Other Renewable Energy Resources

These include solar energy, windmills, power alcohol and biogas. Programmes for their increased use have been formulated and are intended to supplement the non-renewable ones. Since they are renewable, these sources of energy have the potential to contribute to social, economic and environmental dimensions of sustainable development. Other potential sources of energy in Kenya are nuclear power and natural gas. The contribution of renewable energy sources (other than biomass) to the overall energy supply is minuscule. However, with concerted efforts, renewable energy may be significant in the years ahead.

## 3.0 Potential for Improving Energy Efficiency

While access to energy in Kenya is still very limited, significant potential and opportunities exist for improving energy in all sectors of the economy, the industrial sector. In this sector, food, beverages and tobacco, paper and paper products, printing and publishing, chemicals as well as petroleum, rubber and plastic products are among the major consumers of energy. The major end users of energy in any of the above – mentioned categories are steam and motive power. Energy savings of up to 25per cent are possible in steam systems, largely by improving the efficiency of steam boilers, better steam distribution, and use and recovery of waste heat and condensate. Motor systems, the largest users of electricity in industry, are often oversized, resulting in lower efficiency of operation. Potential savings are of the order of 20-50 per cent through motor system efficiency improvement.

### 3.1 Current Situation in the Industrial Sector in Kenya

In the 2004, the industrial sector in Kenya consumed an estimated total of 514 million tonnes of oil equivalent. This constitutes about 18.2 per cent of the total commercial energy consumed in the country in that year. It has been estimated that in the industrial sector, savings in excess of Kshs. 2.5 billion are possible. The rate of economic growth in the country has, meanwhile, picked in the period between 2003 and 2006 and is expected to continue at that level or to grow at an even faster rate in the years ahead, with attendant increase in the demand for petroleum and petroleum products. All these should set the stage for higher savings in energy consumption in the industrial sector. Increasing local as well as international competition is causing Kenyan enterprises to look at their energy use critically. Potential energy savings exist even in highly efficient, well-managed industrial plants.

The industrial development policy is centered on the Industrial Transformation to the year 2020 (Sessional Paper No.2, 1997)[5] whose vision is to bring about sustained economic growth and rapid employment generation in Kenya through industrialization, and the part the government must play in its implementation. Among the critical policy strategies to be pursued include investment to rehabilitate and expand physical infrastructures, specifically those aimed at improving energy efficiency. New management approaches will also be established for improving enterprise competitiveness and productivity; and to strengthen the capacity of both the public and private sector actors to manage development resources within the constraints of the country. The latter will have particular focus on assisting the private sector to develop efficient and competitive industries. The sessional paper decries the persistence of inadequate sources of energy coupled with energy –efficient technologies. This has escalated the cost of manufacturing industrial products in Kenya, thus rendering them less competitive internationally. The issues should therefore be addressed as a matter of urgency.

An important priority area is to expand and improve Kenya's infrastructures, *i.e.* roads, railways, air and water transport, telecommunication services, pipeline transport and, course, energy efficiency. The critical role of the infrastructures is to spur growth and support a conductive environment.

## 3.2 Energy Management in the Boiler and General Steam System

Energy management in the boiler and general steam is an area that raises much interest in Kenya since it yields numerous opportunities for very significant energy savings.

A boiler is the energy-transforming device, which converts the inherent energy available within various fuels such as coal, light and heavy petroleum fuels, and biomass (agricultural and forestry residues) into other forms of energy. The efficiency, of a boiler comprises of the combustion efficiency is obtained by using petroleum as a fuel, since virtually all of that fuel is transformed into energy, with negligible residue. Coal and biomass on the other hand contain higher percentages of ash, and some carbon is generally lost with the ash residue. Heat transfer efficiency, on the other hand is determined by the residual energy disappearing with the exhaust, or flue gas. The higher the flue gas temperature, the higher is this loss. When biomass is used as fuel, a higher loss is incurred because part of the heat is absorbed by the moisture content in the biomass, which then evaporates and escapes with the flue gas. Steam generated in a boiler plant must be conveyed through a piping system to equipment where the heat energy is required.

The following are some of the most opportunities in Kenya for better energy management in the boiler and general steam system.

1. Steam conditioning by removal of non-condensable gases and solids;
2. Condensate formation and its returns to boiler;
3. Managing the parameters of steam quality to ensure that they remain appropriate for the process;
4. Piping systems for steam distribution and condensate return;
5. Boiler efficiency determined by such aspects as fuel oil conditioning, atomization etc;
6. Co-generation.

### 3.3 Other Technologies for Improving Energy Efficiency

Apart from those explained above, there are other technologies and techniques used to improve the efficiency in industries, hotels, hospitals and other institutions that use heat energy pumps in air conditioning;

1. Air drying and purification for compressors;
2. Combined heat and power (CH+P);
3. Co-generation (Co-Gen);
4. Energy efficient chillier systems;
5. Heat recovery in foundries and kilns, etc;
6. Used of scrap metal instead of ore;
7. Continuous casting of primary metals; and
8. Conversion from wet to dry process

## 4.0 The Kenya Government Policies on Energy Efficiency and Conservation

### 4.1 Formulation of the Energy Policy

Kenya experienced unprecedented power shortages in 1999 and 2000, which had a major impact on the country's economy. The shortages also underscored the need to write an overall national energy policy. Consequently, MOE formed a Sector Working Group in 2001 with the mandate of producing a Sessional Paper on energy. The group comprised of representatives from departments and parestatals from the Ministries: the Department of Renewable Energy, KPLC, KENGEN, the National Oil Corporation of Kenya (NOCK), the Kenya Pipeline Company (KPC) and ERB. The policy has since been partly implemented.

### 4.2 Energy Policy Aims and Weaknesses

Kenya has set in action a number of policies in the past to address energy issues in support of its development challenges. The national Energy Policy under assessment has a number of broad objectives including ensuring the adequate, reliable, cost effective and affordable supply of energy to meet development needs, while protecting and conserving the environment. The specific objectives are:

1. Provide sustainable quality energy services for development
2. Utilize energy as a tool to accelerate economic empowerment for urban and rural development
3. Improve access to affordable energy services
4. Provide an enabling environment for the provision of energy services
5. Enhance security of supply
6. Promote development of indigenous energy resources
7. Promote energy efficiency and conservation as well as prudent environmental, health and safety practices.

## 4.3 Integrated Assessment of the Energy Policy with Focus on the Transport and Household Energy Sectors

The current energy Policy has addressed poverty, gender, biomass fuels (charcoal production), energy processing marketing, and renewable energy. The Policy has deliberations on provisions to decongest and improve quality roads, target road improvement as a fuel reduction option, mass transport in cities; construction of ring roads, flyovers and by-passes; repair of dilapidated roads; promotion of use of bicycles; gradual elimination of very old vehicles from the roads; quality monitoring of smoke and vapour emitted from vehicles; and annual vehicle inspection programmes.Some of these are special provisions from the Traffic Act, Kenya Roads Board Act, and the First National Communication to the United Nations Framework Convention on Climate Change (UNFCCC).

The governments current policy of assisting industry in energy conservation measures through energy audits is very likely to be continued and strengthened Furthermore its envisaged that the government t will assist industry in awareness among the private sector, on the benefits arising from investments that lead to improved energy efficiency (EE). Also, to develop a cadre of professionals who have been soundly trained in the design and implementation of EE projects.

One likely development is establishment of demonstration projects to clearly show the benefits from investment in conservation and fuel substitution in energy production. In addition, financial institutions and donor will be encouraged to provide funds for EE investment projects.

Solar hot water systems can, in most instances, satisfy the demand for hot water even in very large buildings. Proper building design can also reduce or eliminate the need for air conditioning systems, particularly in the mild highland climate seen in Nairobi, Nyeri and Embu. A likely government action in this area will be to encourage the use of solar water heating natural ventilation in all new government buildings. The policy will most likely insist that new government buildings incorporate these features where possible, and will encourage existing building to retrofit.

## 4.4 Renewable Energy in Contribution to Context of Energy Conservation and Efficiency

Renewable energy can play an important role as alternative source of energy to more polluting energy source such as petroleum and biomass. Increased usage of renewable energy sources is particularly important because given the fact that a significant portion of Kenya's natural resources are, by their nature, renewable in nature. These renewable sources include:

1. Hydro-power generation potential of river and streams
2. Geothermal fields within Rift valley
3. Solar energy
4. Wind energy
5. Biomass energy
6. Tidal energy on the Indian Ocean coast.

The policy options available to government for reducing the country's over-dependence on wood fuel (leading to deforestation and environmental degradation) and other biomass sources include:

1. Reduced operating costs and hence increased profits;
2. Improved access to sustainable biomass and fuelwood supplies and technologies, and their commercialization;
3. Progressive shifting to socially acceptable alternative energy sources such as solar energy, wind, mini-and micro-hydro power installations, and steam generation using biomass;
4. Widespread promotion of energy efficiency and conservation as a means of reducing demand for fossil fuels; and
5. An appropriate pricing policy for commercial forms of energy, to shift demand towards modern forms of energy albeit in sustainable varieties and appropriate technologies such as manner; and supporting electricity services based on grid extension in electrical supplies and other appropriate technologies.

As previously indicated, solar energy for lighting and water heating, wind generation of electricity and mini-and micro-hydro power generation schemes for electricity, constitute Kenya's primary renewable energy sources, followed by steam generation from biomass or co-generation.

For the of EE investments projects, the minister for finance is responsible for most matters relating to all kinds of investment in Kenya, including within the energy sector. While the minister for Energy handles other energy-related matters, relevant ministries and other specialized Government agencies pertaining to EE investments are quite handy.

## 4.5 The Power Act No. 11 of 1997

The national regulatory authority responsible for all matters pertaining to the generation, transmission and distribution of electric power in Kenya is the Electricity Regulatory Board (ERB), established under the Electric Power Act, No. 11 of 1997.

The electricity regulatory board is empowered to enforce environmental and safety standards in power generation and ensure that electrical apparatus and works meet the standards set by the Kenya Bureau of Standards (KEBS) or, where no such standards exists, with the relevant international standards approved by Kenya Bureau of Standards.

### 4.6 The Standards Act (CAP. 496)

Kenya Bureau of Standards (KEBS) is a statutory organization whose duty is to formulate standards and codes of practice among other responsibilities.

### 4.7 The Environmental Management and Co-ordination Act

Environmental matters are regulated under the Management and Co-ordination Act, 1999. Investors in industrial energy efficiency projects have to comply with provisions relating to environmental impact assessment reports, unless the project is considered as normal factory improvement and audits, as per the Act.

### 4.8 The Petroleum Act

Even though petroleum prices are no longer price-controlled, the petroleum Act regulates the importation, transportation, storage and retailing of petroleum products in Kenya. This Act has an impact.

## 5.0 Cleaner Production Initiatives in Kenya

The KNCPC is mandated to spearhead production and sustainability in Kenya by assisting the enterprises, local authorities and the general public to reduce water and energy consumption, emissions of pollutants and waste volumes. KNCPC was established in July 2000 with funds from UNDP through UNIDO. Enforcement of EMCA 1999 has greatly enhanced demand for cleaner production in the country. A total of 500 CP experts have been trained in CP theory, philosophy and practice. KNCPC employs the following tools with lead agencies to enhance and promote cleaner production: cleaner production assessments, environmental auditing, energy auditing, EIAs, environmental management systems, environmental cost accounting, lifecycle assessments and eco-design, corporate environmental reporting, green procurement, hazard assessment, awareness raising, in-plant assessments, training, policy analysis and information dissemination.

In 2004, 1.93 Tg (1Tg = 1012 g) of $CO_2$ from gasoline and 2.94 Tg $CO_2$ from diesel were emitted. Other emissions include $CH_4$, $N_2O$. There are other emissions such as $SO_2$ (5.545 ug/m$^3$ per capita) $NO_2$ (2.464 ug/m$^3$ per capita) and $CO_2$. (0.3 tonnes per capita). Current levels are far below the maximum allowable levels of 50 ug/m$^3$ for $SO_2$ and 100 ug/m$^3$ for $NO_2$ respectively, an advantage as Kenya can trade in emissions at the international level. Moreover, any increase in emission levels is not expected to be dramatic with NEMA, increasingly stringent on the enforcement of environmental standards. A number of factors explain the composition and magnitude of the above pollutants. The first is traffic congestion, which translates into loss of time and fuel.

Much of the road network in Kenya needs rehabilitation because it is in poor condition. The roads also carry over ten times the amount of traffic they were designed for UNEP (2005).

## 5.1 Pollution from Household Energy Consumption

Production and consumption of solid fuels (firewood and charcoal), and liquid fuels (kerosene and LPG), are a key source of a range of gaseous and particulate pollutants. These include $CO_2$, CO, $N_2O$, NOx, $CH_4$, NMHCs and PMs. These gases and aerosols have a range of impacts on human health and the environment. $CO_2$, $CH_4$ and $N_2O$ are greenhouse gases (GHGs) with potential effect on global warming, while others such as NOx, $CH_4$ and CO are precursors to important secondary products such as tropospheric ozone, which has serious impact on vegetation and human health particularly in high altitude regions. PMs and CO is also well known respiratory health hazards. Unsustainable charcoal and fuel wood production can also contribute to deforestation and desertification. Although quality of data is still lacking, there is indication of rapidly changing demand for these fuels in rural and urban areas. Demand for charcoal has dropped from 2.91 million tonnes in 1997 to 2.48 million tonnes in 2000 and to 1.6 million tonnes in 2004–05. The use of forestry and agricultural residues shows an increase from 1.37 million tonnes in 1997 to 2.9 million tonnes in 2000. Demand for firewood decreased 47 Gross heating values of air-dried maize cobs is 18.9 MJ/kg compared to 20.0 MJ/kg of air-dried wood slightly from 15.42 million tonnes in 1997 to 15.1 million tonnes in 2000. The use of kerosene and LPG for cooking continues to grow steadily in both rural and urban markets. For instance, Kenya had fewer than 50,000 household LPG cylinders in use in 1995, confined to a few key urban areas. However, by 2002, over 700,000 cylinders were in use throughout the country. The growing use of kerosene was in part attributable to convenience and competitive pricing especially as charcoal price was rising.

Several factors are responsible for the above levels of emissions from households. The use of inefficient fuel conversion technologies, including poor stove engineering characteristics, is one of the factors. Other parameters include moisture content, type of fuel, and type of food being cooked, among others. Efforts at policy and technical interventions in atmospheric and indoor pollution must be aware that improved charcoal and firewood cookstoves have demonstrated their potential to save 10-50 per cent of fuel, but are yet to reach the majority of the market. Similarly, efficient charcoal kilns with potential recoveries of 30-45 per cent are not well known. Over 95 per cent of the kilns used in the country are inefficient, with 8-20 per cent conversion efficiency. The cost of alternative clean energy options is too high for most poor households to afford. Although the cost of renewable energy technologies such as PV systems have fallen significantly over recent years, they remain uncompetitive compared to alternatives such as kerosene. LPG is out of the reach of most of the population due to tariffs and duties (18 per cent VAT and import duty on LPG, and 18 per cent VAT on cylinders and appliances). Poor energy policies and legislation also explain emission composition and magnitude.

Unsustainable wood production practices are another factor determining the level of emissions from households. Charcoal production for nearly 80 per cent of urban households destroyed over 50,000 hectares of the country's dryland savannah woodlands and rangelands in arid and semi-arid regions in 1989. Trees felled from clearing land for agriculture and settlement is another source of charcoal. The forested area also continues to diminish each year. Charcoal production is more intensive than wood gathering, continuing until wood resources at a site are exhausted. The demand for biomass fuels in 2000 stood at 35 million tonnes, of which only 15 million tonnes were sustainably supplied. Despite the potential of agroforestry to supply firewood from most parts, this option is yet to be fully exploited owing to land tenure problems. A range of actions however, for implementation is on going to promote clean production and minimize energy waste in the country as hereby outlined.

1. The Government began preparations along the stretch of proposed by-pass paths in February 2004. About Ksh 5 billion is needed for the construction of the three by-pass roads around Nairobi. Total collections from the fuel levy fund rose from Ksh 1.5 billion in financial year 1994/1995 to Ksh 9 billion in financial year 2004/2005. About Ksh 9 billion will be required annually to maintain 65 per cent of the 64,000 km network. In addition, about Ksh 45 billion is required in a one-off operation to rehabilitate the remaining 35 per cent of the classified network that is in poor state and therefore not maintainable.
2. Innovation on the part of the KPRL in November 2005 led to the improvement of existing infrastructure, including changes produce two grades of unleaded gasoline *i.e.* 91 octane premium and 87 octane regular grade (J. Murutu, personal communication). This was with a view to beat the 31 December 2005 deadline set by the Dakar Declaration.
3. The Kenya Pipeline Company, the only white oil product pipeline within East and Central Africa, has performed its mandate fairly well by transporting over 90 per cent of oil products consumed in Kenya. To maximize its potential, pipeline extension from Eldoret in Kenya to Kampala in Uganda is at an advanced stage, with the actual construction having begun in 2005. This mode of fuel transportation has replaced a significant number of freight trucks that would have had to traverse the city of Nairobi daily, compounding the congestion and air pollution challenge.
4. Overall efficient implementation of all the projects proposed would lead to a total reduction of $CO_2$ of 8.3 Tg by the year 2030, or about 37 per cent below the anticipated Business As Usual emissions. Apart from interventions that reduce the number of inefficient vehicles on the roads, the other option to bring about significant reductions in GHG emissions involves improving the quality and scale of road networks in the country. It was estimated that an hour spent in traffic jams in Nairobi wasted, per car, 1.26 litres of fuel. Transport sector intervention is well recognized as a measure for climate change mitigation by the First National Communication to the United Nations Framework Convention on Climate Change (UNFCCC). A big challenge lies in implementing existing policies and laws, including the Traffic Act. A significant reduction in $CO_2$ is expected as a

result of road improvements. Other key pollutants are also expected to reduce significantly. Various government departments are already implementing some of the measures.

5. MOPND concluded negotiations with a German railway investor to construct a Ksh 300 billion-railway line linking Nakuru in Kenya and South Sudan.
6. A South African company, Sheltam Rail, which recently was issued a concession to run the Kenya-Uganda railway, plans to invest a total of US$80 million. However, nothing is earmarked for the city commuter segment of the railway. The Kenya Railways has undergone three decades of decay to the point where it now needs about Ksh 40 billion to upgrade its infrastructure and rehabilitate the railway.
7. Implementation of the Road Maintenance Levy Fund (RMLF) Act of 1994, which consists of an automotive fuel levy and transit toll collections, is meant to go towards public road maintenance.
8. The government has zero-rated duty on bicycles thereby encouraging their use. As a result, bicycles are being used as taxis (boda boda) in many western Kenya towns. This contributes to job creation and emission reduction through foregone vehicle use.
9. Production of ethanol from sugarcane molasses is an existing enterprise in Kenya with two major firms, Kisumu and Muhoroni, already producing large volumes for industrial purposes. The policy challenge is, therefore, to make the domestic production of power alcohol competitive with motor gasoline to facilitate its re-introduction as a motor fuel blend.
10. On-going efforts by NEMA are on the development and possible enforcement of air quality and their development and monitoring on waste regulations and environmental audits.
11. Initiatives to switch to improved firewood cookstoves such as the kuni mbili (maendeleo) model stoves designed by the Maendeleo Ya Wanawake Organization, a Kenyan women's NGO, as a fuel saving device. The rising fuelwood crisis has resulted into demand for residues for cooking on an upward trend projected to 18.2 million tonnes in 2030. Maendeleo model of It requires about two pieces of wood to make a typical meal, hence the name kuni mbili.
12. Capacity building workshops on CP, biomass fuel improvement, technology improvement.

## 6.0 Summary and Conclusions

The achievement of cleaner production and energy conservation for sustainability requires the government to streamline coordination of activities by different government departments, harmonize energy efficiency, pollution prevention policies and laws scattered among sectoral policies and laws. Implement climate change adaptation measures in the UNFCCC. The government should source and mobilize resources for capacity building and technology transfer, monitoring and enforcement for cleaner production projects, guidelines and regulations to have long lasting impacts. There should be energy efficiency skew in the implementation of all the cleaner production strategies since they often have strong elements of cleaner production.

## References

1. Electricity Regulatory Board, ERB, 2005. Retail electricity tariffs review policy: 2005. Nairobi: Electricity Regulatory Board.
2. Government of Kenya, 1996. Sessional Paper No. 2 of 1996 on industrial transformation to 2020. Nairobi: Government Printer.
3. Government of Kenya, 1999. Sessional Paper No. 1 of 1999 on national policy on water resources management and development. Nairobi: Government Printer.
4. Government of Kenya, 1999. Sessional Paper No. 6 of 1999 on environment and development. Nairobi: Government Printer.
5. Government of Kenya, 2002. First National Communication to the UNFCCC. Ministry of Environment and Natural Resources.
6. Government of Kenya, 2003-2007. Economic Recovery Strategy for Wealth and Employment. Nairobi: Government Printer.
7. Government of Kenya, 2004. Sessional Paper No. 4 on Energy. Ministry of Energy. Nairobi: Government Printer.
8. Government of Kenya, 2005. Sessional Paper No. 2 of 2005 on development of micro and small enterprises for wealth and employment creation for poverty reduction. Nairobi: Government Printer.
9. Government of Kenya, 2005. Sessional Paper No. 9 of 2005 on Forest Policy. Nairobi: Government Printer.
10. Kamfor, 2002. Study on Kenya's energy demand, supply and policy strategy for households, small scale industries, and service establishments. Final Report. Ministry of Energy.
11. Kituyi E., 2002. Lost opportunities: Wood fuel data and policy development concerns in Kenya. Energy Policy Opinion Paper, WSSD Edition, Nairobi: African Centre for Technology Studies.
12. Ministry of Energy, 2004. The Energy Bill, 2004. Nairobi: Ministry of Energy.
13. Renewable Energy Technology Assistance Programme RETAP, 2000. GEF-SGP Eco-Schools Project Report. Renewable Energy Technology Assistance Programme, Nairobi.
14. Renewable Energy Technology Assistance Programme RETAP, 2001. Energy-saving stoves in the Mt.Kenya region, Kenya.
15. UNEP, 2005 Air Pollution and Africa. Report of the Secretary General for the CSD-14. (E. Kituyi, ed.).
16. Republic of Kenya; "sessional paper No. 2 on industrial transformation to the year 2020"; 1997.
17. Republic of Kenya; "sessional paper No. 4 on energy"; November 2004.

# Part III
# *Renewable Energy*

# Chapter 11

# *Bio-Energy for Sustainability of Energy in Indonesia*

**S. Tursiloadi*, T.B. Bardant and K.C. Sembiring**

*Research Centre for Chemistry, Indonesian Institute of Sciences, Kawasan Puspiptek, Serpong, Tangerang, Indonesia*

**E-mail: tursilo@gmail.com*

## ABSTRACT

Brief description in this article is putting the Indonesian's problems in energy development at a clear perspective. These problems can be categorized in two major challenges. The first one is increasing the energy consumption rate and the second one is reducing the Energy Intensity. These are not easy problems to solve since Indonesian topography is an archipelago. This topography made energy distribution system through cable and wire, like the system applied in Japan, Europe and America, will requires very expensive investment. This unique topography forced the Indonesian Government to perform his own research for providing and distributing fuel and energy consumption which ended to a proven applied technology for its application in national scale. Indonesia is developing several bio-energy projects in order to maintain the country's sustainability of energy. Indonesian unique topography gives many opportunities for developing several kind of bio-energy at once. On the other hand, this archipelago topography is a big challenge for distributing fuels from one place to others through out Indonesia. This situation stimulates many unique ways in fulfilling local energy demands that developed by local culture. Some of these cultures were touched by applied technology. Above these local actions, Indonesian government also carries on deep concerns in developing sustainable bio-energy. Most of it aim to scale up energy productions and requires more hi-technology. This spirit of development based on several researches about Indonesian potential energy source range from in-situ source like geothermal and ocean wave to the ex-situ source like bio-diesel. Some industrialist takes Indonesian government policy as good business opportunities and had already made some investment in new developed bio-energy industry in Indonesia. The Research Centre for Chemistry of Indonesian Institute of Sciences is one of research centres that carry out research programs in developing bio-energy. Several research collaborations had already done with other countries such as Japan, Korea and Germany and also with local institutions and industries. This article gives an overview about Indonesian developments in bio-energy as a guiding reference for anyone who is interested to learn more about Indonesian potential energy resources and make collaboration projects with Indonesia in developing it. In conclusion, this article tells about Research Centre for Chemistry of Indonesian Institute of Sciences proposed action for solving the Indonesian's problems in energy development.

***Keywords:*** *Bio-energy, Sustainable energy.*

## 1.0 Introduction

Indonesia archipelago lies from 95° to 141° East longitude and 6° north to 11° South latitude. Its territory was bordered by Indian Ocean in the west side and south side, by the Pacific Ocean in the east, then by the South China Sea in the north side. Within this territory, Indonesia has total area about 9.8 million square km, consisting of land area just about 1.9 million square km (1/5 of the total area). The rest of the area is territorial waters, which includes the exclusive economic zone of about 7.9 million square km (4/5 of the total area). It consists of 17316 islands, of which only about 6,000 are inhabited. The five main islands are Sumatra, Kalimantan, Java, Sulawesi and West Papua considering the land size and population number. Indonesian population had already reaches 240 million people and will still growing. This huge number requires a very hard duty from the government to fulfilling the three basic needs for live, which is food, water, and energy.

In the mean time, energy demands are fulfilled from mostly fossil fuel. Quoting from Business Week July 4, 2008 "Indonesia's energy minister, Purnomo Yusgiantoro, must feel funny at times going to OPEC meetings. Indonesia is the only OPEC member that isn't a net oil exporter, and yet it continues to pay millions of dollars in annual dues to keep its membership. Over the past 16 years, Indonesia's oil production has declined 1.5 per cent a year, while consumption has grown at 5.4 per cent annually. Today, Indonesia imports about 60,000 barrels of oil a day, compared with exports of 27,000 bbl. daily last year."

Goldenberg and Suarez had done empirical approach in studying the relationship between energy consumption and HDI. HDI is a new tool developed by United Nation Development Program for measuring achievement of a country development which covers three dimensional parameters, lifelong and healthy (measured by expected age for new born baby), education (measured by number of adults that can read and education system) and prosperity (measured by Gross Domestic Product per capita). The results show that there is positive correlation between those two variables. They also show that energy consumption per capita per year that was proven to be capable in supporting prosperity, with a main condition that the consumption is well distributed, is 1.0-1.33 kW per capita or 5.15 – 6.85 Barrel of equivalent (BOE) per capita per year. With that consumption rate, the HDI could be increased to 0,8. Consumption rate above that value will not significantly affect the increase of HDI.

Indonesian energy consumption rate for these last few years is about 3.42 BOE per capita per year. This consumption rate is way below the lower level that pointed above. Besides gave more proven for the empirical studies, this fact had led us to a conclusion that Indonesia need to work hard for increasing their energy consumption rate. On the other hand, there is also another interesting fact that Indonesian Energy intensity is 21 MJ/ US$GDP. This value is way above Japan which only 4 MJ/ US$GDP. This fact is also led us to a conclusion that for increasing 1 US$ GDP per capita, Indonesian needs to expend energy more than five times that the energy required by Japanese.

Brief description above is putting the Indonesian's problems in energy development at a clear perspective. These problems can be categorized in two major challenges. The first one is increasing the energy consumption rate and the second one is reducing the Energy Intensity. These are not easy problems to solve since Indonesian topography is an archipelago. This topography made energy distribution system through cable and wire, like the system applied in Japan, Europe and America, will requires very expensive investment. This unique topography forced the Indonesian Government to perform his own research for providing and distributing fuel and energy consumption which ended to a proven applied technology for its application in national scale.

## 2.0 Renewable Energy and National Energy Policy on Renewable Technology

The Government of Indonesian (GOI) acted fast for solving the problem by announcing a national plan to reduce the country's dependency on oil. The GOI aims to increase the use of new and renewable resources to produce energy from 0.2 per cent currently to 4 per cent by 2025. The target is certainly achievable given Indonesia's massive potential renewable energy (RE) resources. Potential supplies of geothermal, solar, wind, micro hydro and biomass energy are estimated of 160 gigawatts of electric capacity. However, the development of RE has been slow because the state-of-art technology involved of this energy is more expensive than energy produced by conventional hydrocarbon-fired plants. Commercial opportunities for U.S. companies exist in equipment and technology related to RE energy development.

The National Energy Policy is issued with the main objective to create the sustainable energy development. The Government of Indonesia has formulated the general energy policy since 1980. It was motivated by the reverse oil shock that started in the early 1980's. The first energy policy was established under presidential decree in 1982 and the policy be periodically revised. The last revise was in 2003, which called "National Energy Policy" with vision to guarantee the sustainable energy supply to support national interest.

The main missions of the policy are guaranteeing energy supply to meet national interests; increasing the added values of energy sources; managing sustainable sources of energy in an ethical and sustainable manner; focusing of attention to conservation of environmental functions; providing affordable energy for low income people and less developed area; and developing domestic capacities. The main policy measures of National Energy Policy in the supply side are energy intensification and energy conservation (production optimization). In the demand side the policy consist of energy diversification and energy efficiency.

Serious energy crisis, which will be faced by Indonesia, has pushed the Government to start to give a main attention on it. The most strategic plan is utilizing the renewable energy resources recognizing the limited nature of conventional energy and the global concern on environmental conservation. Indonesian dependency on oil can be reduced by diversifying its energy supplies, several policies, laws, and regulation. Indonesian Government has support the utilization of renewable energy by establishing policies, laws, and regulations as summarized in Table 11.1.

**Table 11.1: Indonesian Policies, Laws and Regulations on Renewable Energy**

| *Policy/Law/Regulation* | *Purpose* |
|---|---|
| Government Regulation No. 5/2006 | Organize Indonesian energy regulation |
| Presidential Instruction No. 1/2006 | Supply and utilize of bio-fuel as alternative fuel source |
| Presidential Instruction No. 2/2006 | Supply and utilize of liquefied charcoal as an alternative fuel source |
| Ministerial of Economic Decree No. KEP-11/M.EKON/02/2006 | Develop a team for supplying and utilizing alternative energy |
| Presidential Decree No. 10/2006 | Develop national team to developing biofuel to accelerate the decreasing of the poor members and un-employments of societies |
| Ministerial of Energy and Mineral Resources Decree No. 051/2006 | Pre-requirements and Orientation of trade license of bio-fuel as alternative energy |
| Law no. 27/2003 | Geothermal development |
| Law No. 6/1994 issued June 2002 | Launching of Green Energy Program that intends to realize renewable supply utilization to help sustainability of the long-term national development |
| Ministerial Decree, Kep.Men No.1122/ K/30/MEM/2002 | Obligating the key players in the electricity sector to utilize small hydropower as part of their portfolio |
| Government Regulation No. 3/2005 | Increasing the role of local government, cooperatives, private companies, community organizations and individual on electricity supply |
| UU No. 22/1999 or the Local Autonomy Law policy | Institutionalization of the shifting of governance authority to local government mandating greater autonomy and responsibility to local government to manage its local electricity supply system |

## 3.0 The National Energy Policy Target

Figure of Indonesian energy resources is shown in Figure 11.1. It can be seen there that the most of Indonesian energy demand right now is supplied by oil source. Prediction of Indonesian mix energy resources in 2025 can be seen in Figure 11.2. While it have been discussed before that Indonesian oil source decline gradually and

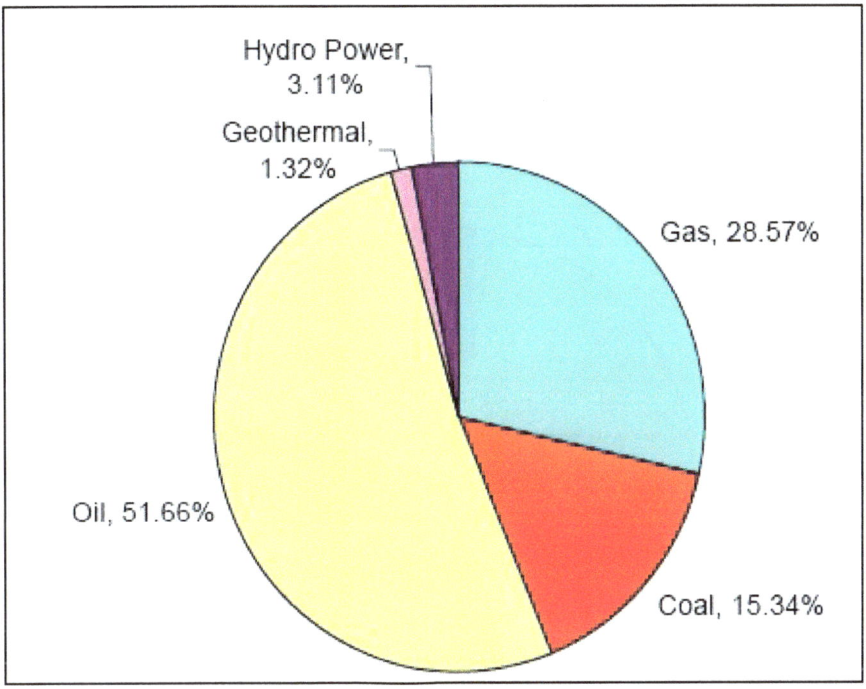


**Figure 11.1: The Current National (Primary) Energy**

predicted will be used up for the next 10 years, so that Government of Indonesia rearrange prediction of mix energy resources in 2005 consider utilization of renewable energy resources. Renewable energy resources which will be explored in 2025 are bio-fuel, geothermal, biomass, nuclear, hydropower, solar energy, wind power, and coal liquefaction. All of those renewable energy resources will replace 17 per cent of origin national primary energy mixture as can be seen in Figure 11.3.

## 4.0 Existing Indonesian Renewable Energy

The potential of Indonesian renewable energy is shown in Table 11.2. Surveys have identified 217 geothermal prospects, 70 of them may provide energy of about 27 GW. So far, the installed geothermal plant is only 800 MW. The Government has enacted the Geothermal Law No. 27 Year 2003 to promote the development of geothermal. The potential of hydropower (large, mini and micro) is theoretically about 75 GW, spread-out in more than 1,300 locations. The utilization of large-scale hydropower is 4.3 GW (just about 5.6 per cent of the country's potential), while utilization of mini-micro-hydro is about 84 MW (out of 500 MW potentials). The Ministry of Energy and Mineral Resources has established a Mini-Micro Hydro Clearing House to provide information and networking on mini-micro power development.

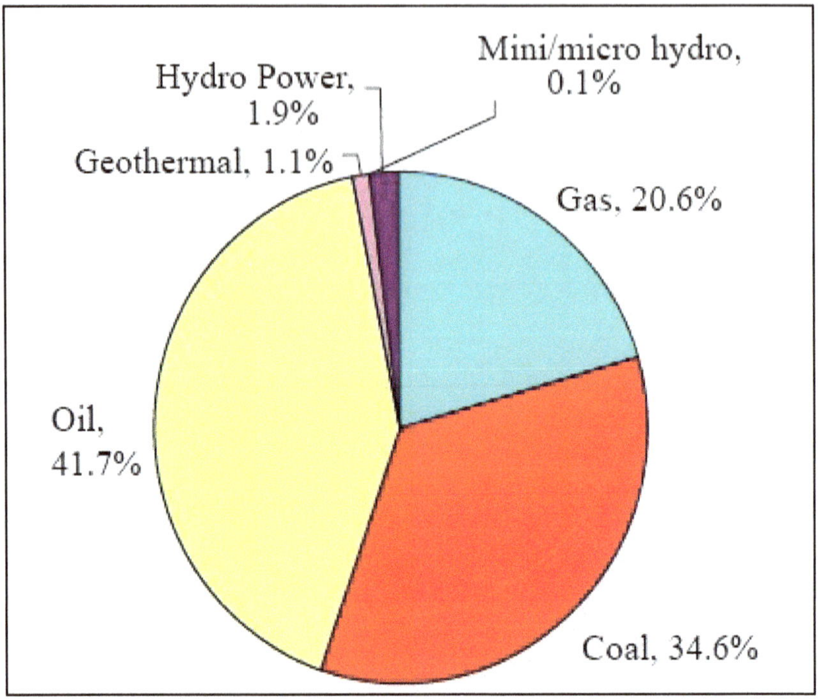


**Figure 11.2: The National (Primary) Energy Mix of 2025 without Optimization**

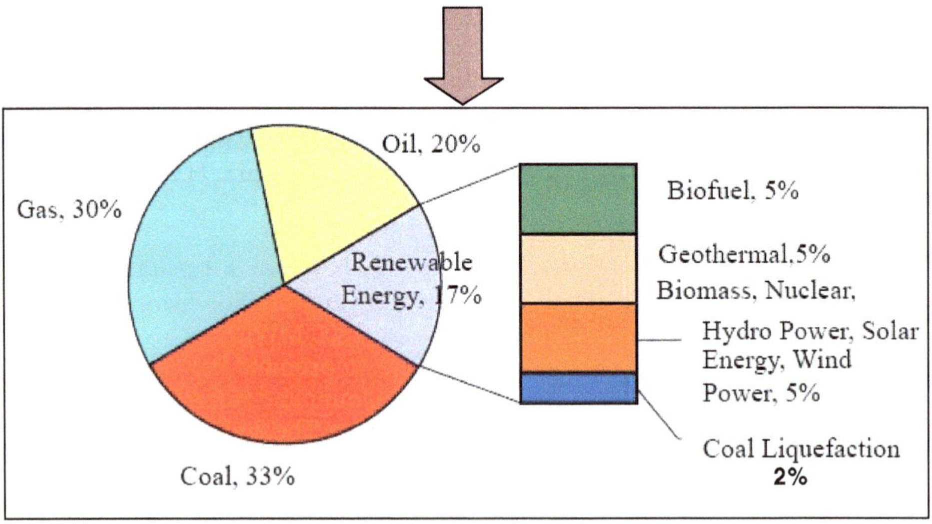


**Figure 11.3: The Target of National Energy Mix of 2025**

***Source*: Ministry of Energy and Mineral Resources, 2005**

**Table 11.2: Indonesian Potential of Renewable Energy**

| *Type* | *Potential* | *Installed Capacity* |
|---|---|---|
| Hydropower | 75.67 GW | 4,200 MW |
| Geothermal | 27 GW | 807 MW |
| Mini/micro-hydro | 500 MW | 84 MW |
| Solar Energy | 4.8 kWh/m$^2$-day | 8 MW |
| Wind Energy | 3-6 m/sec | 0.6 MW |
| Biomass | 49.81 GW | 445 MW |

*Source*: Ministry of Energy and Mineral Resources, PT PLN.

## 4.1 Hydropower and Micro-Hydro

The oldest machines for capturing the energy of moving water were waterwheels. In the days before electricity, it was common to use water wheels to provide the power for mills that ground grain or cut lumber. To start the mill, the miller simply opened a gate to let the water flow over the top of the wheel. The water wheel was connected to a massive millstone or metal saw blade through a system of gears. Water for the wheel usually came from a small dam and reservoir, called the millpond.

Hydroelectric facilities often depend on a dam to raise the level of the water in the reservoir. Water from this reservoir is allowed to fall through huge pipes to a building that houses water-driven turbines. Pressure from the falling water spins the turbines at high speed. The turbines are connected to huge generators that make electricity as they turn. This electricity is carried to cities and towns that may be located hundreds or even thousands of kilometers a way.

Utilization of hydropower on a large-scale for power plants until the year 2002 had already reached 4,208 MW. It is only around 5.6 per cent of the available resources. However, hydropower potential in Java has been optimally developed, namely around 2,389 MW or 53 per cent of the total reserve. Where as the capacity of mini-hydro and micro-hydro is around 500 MW; 84 MW of which has been utilized to support rural electrification. To provide information related on mini-micro hydro development as well as networking among all stakeholders, Ministry of Energy and Mineral Resources has established the Mini-Micro Hydro Clearing House.

Hydropower resources for electricity power generation are still under-utilized. The total estimated hydropower energy potential is about 75,000 MW including micro-hydro energy at 500 MW. Total installed capacity is 2,700 MW of which 2,500 MW is owned by PT PLN. Besides PT PLN, PT Inco, the state-owned nickel company, is building the Balambano hydropower plant in Soroako, Sulawesi. More than 200 units of micro hydro with a capacity between 50 kW and 500 kW are installed around the country. Most of the existing micro hydropower plants are installed off-grid as they are used in rural and remote areas. Currently, PT PLN has 12 ongoing projects in Papua, Nusa Tenggara, Sulawesi and Kalimantan. The projects are financed by ADB and are expected to be in operation by 2006-2007. PT PLN plans to develop 8 hydro power plants during 2008-2010 in Sulawesi, Jambi, West Java and East Nusa Tenggara.

## 4.2 Geothermal

It's understood that the earth's interior and core is at high temperature, estimated to be between 3000 and 5000°C. This heat is generated by the slow breakdown of radioactive elements, and by the immense gravitational pressures acting on the rocks and minerals of the earth's interior. Temperatures in exceed of 500 °C can be found in the Earth's crust just a few thousand meters below the surface, but geothermal heat right at the surface of the land is barely detectable. Geological surveys throughout the Indonesian region have identified 217 geothermal locations. Exploration indicated that 170 of geothermal resources are high temperature providing the country with a potential of geothermal energy about 27,189 MW. The current installed capacity of Geothermal is about 807 MW.

Indonesia, located in the "ring of fire" volcano belt, is estimated to hold about 40 per cent of the world's geothermal reserves, equivalent to a total of 27,140 megawatts (MW) of power. While the country has several operational geothermal power plants, their combined capacity currently is only 807 MW or about 3 per cent of the country's total geothermal potential. The development of the use of geothermal resources is hampered by several factors such as high investment costs, limited incentives and lack of technology. The Geothermal law issued in 2003 (Law No. 27/2003) is expected to provide a conducive atmosphere to geothermal development in Indonesia.

Currently there are 33 Geothermal Working Areas that has been issued, which 15 of them owned by the state-owned oil and gas company, PT Pertamina. In addition, the GOI offers 28 open geothermal working areas, which bring a total potential of 13,497 MW. This energy potential is expected to fulfill the geothermal development target to generate electricity of 6,000 MW in 2020 and 9,000 MW in 2025. PT PLN, the state-owned electricity company, will build an additional 16 geothermal power plants with a total capacity of 1,150 MW. Up until the first quarter of 2005, the installed capacity of PLN's Bali-Java system was 19,514 MW with around 4 per cent generated from geothermal power plants. Outside Java-Bali, from 5,573 MW installed capacity, only 20 MW is generated from geothermal plants. The existing geothermal power plants in operation are: Darajat (150 MW), Dieng (60 MW), Kamojang (140 MW), Gunung Salak (330 MW), Sibayak (20 MW), Lahendong (20 MW) and Wayang Windu (110 MW) among others. Road Map of Geothermal energy development is shown Figure 11.4.

## 4.3 Solar Energy

To utilize such potential of solar energy, two technologies have been applied, namely thermal solar energy and photovoltaic solar energy. Photovoltaic (PV) systems have emerged to be the cost effective and a suitable approach for rural electrification in the rural areas of the developing countries. Many PV systems have been installed in different parts of Indonesia during the last decade. Road Map of solar energy development is shown in Figure 11.5.

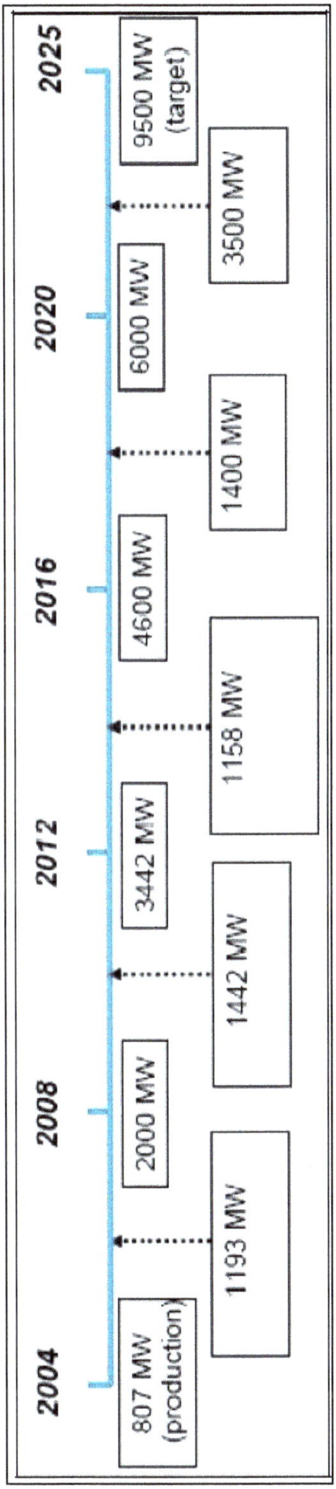


**Figure 11.4: Road Map of Geothermal Energy Development**

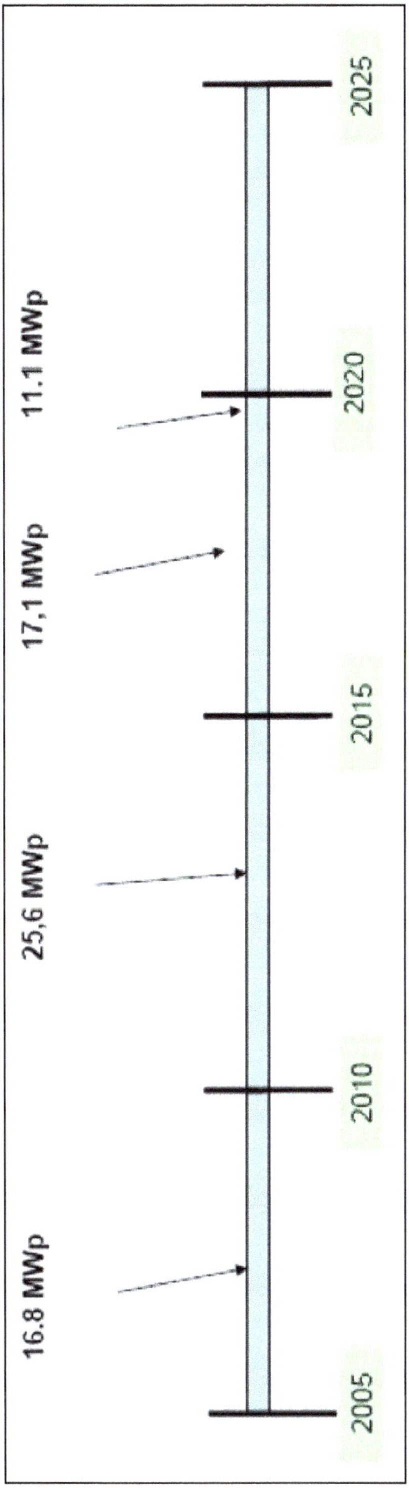


**Figure 11.5: Roadmap of Solar Energy Development**

The potential of solar energy in Indonesia is relatively good with a daily average intensity between 4.8 to 5.1 $kWh/m^2/day$. Solar thermal energy is used for solar stove, water pumping, drying equipment for agricultural product, and water heaters. The current installed capacity is only 8 MW because of its high investment and operational cost. The government plants to build 15,000 units of solar power plant in the eastern part of Indonesia, each with a capacity of 100 watts. The cost for each unit ranges from $500–$600 and will be financed from the state and PLN's budgets.

### 4.4 Wind Energy

Commercial wind energy is one of the most economical sources of new electricity available today. Wind turbines can be set up quickly and cheaply compared with building new coal-fired generating stations or hydroelectric facilities. Modern wind generating equipment is efficient, highly reliable, and becoming cheaper to purchase. The environmental impact of large wind turbines is negligible compared with an open pit coal mine or a reservoir and during their operation produce no air pollution. Because of these factors, wind energy is recognized as the world's fastest-growing new energy source. The potential of wind energy in Indonesia is relatively low with the average of wind speed around 3-6 m/second. However, in certain areas, especially in the Eastern Regions of Indonesia, wind velocity is more than 5 m/second. Around 0.6 MW has been installed and operated for lighting, water pumping and battery charging.

The potential of wind energy is relatively small with the wind speed between 3 and 5 m/second. The current limited utilization of wind energy (around 0.5 MW) concentrates on stand-alone electricity production in rural and remote areas and it is used for rural electricity, water pumping and battery charging. The local industry is able to build wind energy conversation systems up to 5 kW of capacity. PT PLN currently is developing 3 wind power plant projects in Bali (3 x 250 kW), West Nusa Tenggara (3 x 250 kW) and East Nusa Tenggara (6 x 250 kW). It is expected that these power plants will be in operation in the 2006/2007 time frame.

## 5.0 Biomass Resource from Agriculture and Forestry for Local Energy in Indonesia

Java has the largest population number, followed by Sumatra at second. Both of these islands have only a little portion of forest left. Most of industries and agricultural business take place in these islands and consumed the forest. However, from the biomass point of view, these facts lead to several new homogenous biomasses potential from agricultural business. Some of them were shown in Table 11.3.

Kalimantan and West Papua still have large area of forestry. Many international boards that have big concern in global environment put very big concern to the forest in these islands. They are doing everything they can to prevent the forest from their extinction. However, capitalism drift still pulls Indonesian government policy to give privilege for industry cutting down the forest. This fact still gives potential benefits for biomass utilization. Some of the potential are shown in Table 11.4.

**Table 11.3: Biomass from Forest Related Industries**

| *Biomass from Forest Related Industries Main Region* | *Production [million t/year]* | *TEP [million GJ/year]* | *Remarks* |
|---|---|---|---|
| Logging residues | 4.5 | 19 | Small logs factory residues were |
| Sumatera, Kalimantan | | | used free energy resource for |
| Sawmill residues | 1.3 | 13 | household cooking in local communities. |
| Sumatera, Kalimantan | | | |
| Plywood and veneer production residues | 1.5 | 16 | Together with sawn timber residues for particle board raw material. |
| Kalimantan, Sumatera, Java, Irian Jaya, Maluku | | | |

*TEP: Technical energy potential.

**Table 11.4: Biomass from Agricultural Industries**

| *Biomass from Agricultural Industries Main Region* | *Production [million t/year]* | *TEP [million GJ/year]* | *Remarks* |
|---|---|---|---|
| Rubber wood Sumatera, Kalimantan, Java | 41 (replanting) | 120 | Small logs ∅ <10 cm big and medium logs are used as fire wood in brick and roof tile industry: price 20,000–30,000 IDR/m³ |
| Sugar residues Java, Sumatera, South Kalimantan | Bagasse: 10 Cane tops: 4 Cane leaves: 9.6 | 78 | Bagasse was generally used in sugar factories (90 per cent). The use of cane tops and leaves needs further investigations. |
| Rice residues Java, Sumatera, Sulawesi, Kalimantan, Bali/Nusa Tenggara | Husk: 12 Bran 2.5 Stalk: 2 Straw: 49 | 150 | Stalk and straw were generated at the field and burn out. In some areas they were used as cattle feed or raw material for paper industry. Most of husks were just burn out. |
| Coconut residues Sumatera, Java, Sulawesi | Shell: 0.4 Husk: 0.7 | 7 | Husk were generated locally and usually left on the plantation field. Shells were largely used for the charcoal production. |
| Palm oil residues Sumatera new areas: Kalimantan, Sulawesi, Maluku, Nusa Tenggara, Papua | Empty fruit bunches: 3.4 Fibers: 3.6 Palm shells: 1.2 | 67 | Shells, fibers and empty fruit bunch were incinerated as common fuel sources for palm oil refinery factories. |

Facts in both tables had already shown that local communities around the industries had already used the biomass for energy resource. Most of them are for household cooking. In Indonesia, the energy consumption for household cooking was estimated at 0.88 $m^3$ of fuel wood/head/year or equivalent to about 17.7 MJ/ person/day (Kamaruddin, 1998). So, further development that might be very useful is increasing stove efficiency. Traditional three stone stove efficiency that were used for household cooking is still low, just about 5 per cent efficiency. Transforming them into refined wood stove such as the two pot un-insulated metal wood stove will increase the efficiency up to 27.9 per cent (Baldwin, 1987). Another simple stove can be made by using common metal oil barrel with sawdust as fuel. This stove will give 16 per cent thermal efficiency and can be used as cooking stove and as main or auxiliary heating producer in small processing unit (Kamaruddin, 2000) or in various agro-based industries. Such technology is also being used to dry sawn coconut wood in timber industry in Manado, North Sulawesi.

Another interesting biomass potential comes from essential oil extraction plant. Indonesia has 925 units essential oil extraction plant that spread in 12 provinces with area coverage about 470.941 ha. The essential oil yield is only in a range 2 – 5 per cent or dry or wet mass. For producing patchouli alcohol from local plant name nilam, for instance, just give 2 per cent yield of dry weight and only 0,26 per cent yield based on its wet weight. The solid organic waste from these extraction plants are a very interesting tropical biomass potential. However, this technique is very dependent to the sunlight and surrounding climates. The other method is by converting the waste into biogas. This method has other advantages because this method also produces organic fertilizer. A research had already done for *Mentha arvesis* or in local name serai wangi. Serai wangi is a plant for essential oil raw material. After composting the solid waste from its oil extraction plant, the organic fertilizer was used for growing the serai wangi plant. These plants gave higher yield of menthe oil than the one with commercial fertilizer (Djazuli *et al.*, 2002).

**Table 11.5: Potential for Biomass Technologies in Rural Industry**

| *Mills Category Mill Capacity* | *Capacity of CHP Technology* | *Biomass Potential for Power Generation* |
|---|---|---|
| Saw mills<br>1000–3000 $m^3$/y | 40–100 kWe | 0.6 $m^3$ wood waste/$m^3$ sawn timber<br>~ 130 kWh/$m^3$ sawn timber |
| Ply wood mills<br>40 000–120 000 $m^3$/y | 1.5–3 MWe | 0.8 $m^3$ wood waste/$m^3$ plywood<br>~ 200kWh/$m^3$ plywood |
| Sugar mills<br>1000–4000 TCD | 3–10 MWe | 0.3 t bagasse/t sugar cane<br>~ 100 kWh/t sugar cane |
| Rice mills<br>< 0.7 t/h / >0.7 t/h | 30–70 kWe/<br>100–300 kWe | 280 kg husk/t paddy<br>~ 120 kWh/t paddy |
| Palm oil mills<br>20–60 t FFB/h | 2–6 MWe | 0.2 t EFB/t FFB, 0.2 t fibre/t FFB<br>70 kg shells/t FFB, ~160 kWh/t FFB |

TCD: Tones of Cane per day; FFB: Full Fruit Bunches; EFB: Empty Fruit Bunches; CHP: Combine heat and power.

*Source*: ZREU (2000)

Biomass energy can be also converted to produce electricity, which is necessary to support our daily activities at home, office or industries. Several mill plant categories and their potential capabilities of biomass that could be converted into heat or power for electricity were shown in Table 11.3. A sawmill with 1000 – 3000 $m^3/y$ capacity can produce 40-100kWe using CHP (combined heat and power) technology while a sugar mill with 1000-4000 TCD (total cane/day) can produce 3-10 MWe.

At this time, an electric power generation using rice hull gasification was proposed by BPPT and local industry for supplying Riau province needs. This power plant is capable to produce 18 kW electricity from a 25 KAV unit with charge Rp 406.45/kWh (or about US$ 0.051/kWh) at the end user. This charge is a little higher than the targeted purchasing price of PLN (National Electric Company). This program was funded and supervised under the newly enacted Small Distributed Power Generation program (improvement of the old PSKSKSM) by the Ministry of Energy and Mineral Resources and still in progress for lowering the production cost. Further improvement was made in import duty and tax scheme of some important components of the power plant.

Many kinds of project like the one proposed by BPPT as described above was also very potential to generate domestic income and increase the prosperity of Indonesian people. However, these projects have faced many difficulties in accessing commercial funding from banks for their working capital. The government has already creating source of funds to give soft loan for these projects. The fund was collected from the state owned company, redemption of oil subsidy, etc. However, the funding is still become the main obstacle.

Serious R/D efforts on bio-ethanol had been going on in Indonesia since long time ago. Ethanol production from cassava starch was started in 1987 in Lampung with production capacity of 8,000 litres/day and 12,000 litres/day. Other attempts using other source of biomass such as sago starch, is also found in abundance in Indonesia. Some research still developing bio-ethanol production from rich-glucose content raw material. Two of Indonesian plant with local name Lontar and Gewang, which are still in palm family in its taxonomy, growth easily in eastern Indonesia and able to produce rich-glucose liquid from its flower cuts.

## 6.0 Indonesian Government Research in Bio-energy

Many commercial biomass power generations are already established in sugar and palm oil plantations. However, the market potential is still huge and waits to be developed. Figure 11.6 shows market potential for power generation and process heat using biomass energy (ZREU, 2000)

At this moment, Indonesian Government had arranged new tax policy for ethanol producers with fuel-oriented product. The government hope this policy will increase the number of bio-ethanol production plant so their goal for substituting 20 per cent gasoline consumption in 2015 could be achieved.

Indonesia is one of the biggest world producers of palm oil right below Malaysia. In order to use of this potential resource as another fuel source alternative, an Indonesian Bio-diesel Forum (FBI) has been formed which is fully supported by the

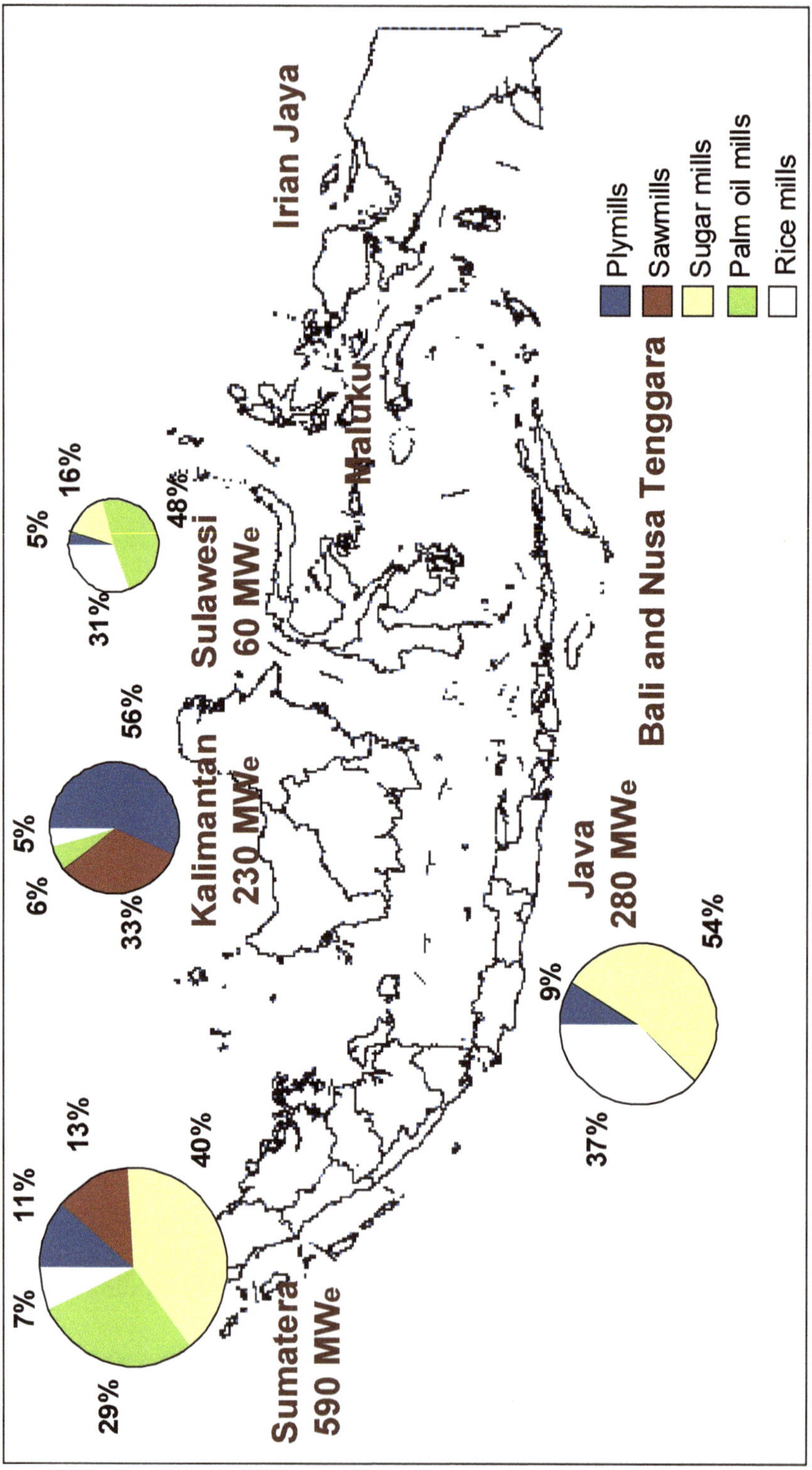


**Figure 11.6: Market Potential for Biomass Power Generation to Electricity in Indonesia**

Ministry of Energy and Mineral Resources. The merit of using bio-diesel is that the exhaust gas emitted 78 per cent less carbon dioxide ($CO_2$) than conventional diesel fuel. Bio-diesel also reduces small carbon particulates, the one that could cause lung-burning endemic, from diesel exhaust by 68 per cent, polycyclic aromatic hydrocarbons, the compound, which could cause cancer, by 75 per cent, carbon monoxide (CO) by 50 per cent, and hydrocarbons (HC) by 37 per cent and zero sulphur emission. Indonesia has more than 50 kinds of plant that produce triglycerides and very feasible for being bio-diesel raw material. Some of them are non-edible fats, which gave bigger advantage to be developed. Based on these facts Indonesian Government gives much support for bio-diesel development and substituting 10 per cent of ADO and IDO consumptions in 2015 as a national goal in energy development.

Other alternative energy source that could be developed in Indonesia are mostly in-situ source, such as waterfall, geothermal, wind, ocean wave and nuclear. Even the investment that would be required are expensive, the operational cost would be very cheap.

Developing this kind of alternative energy source are also give a competitive edge for Indonesian researcher and give very good benefit for human resource development. Other benefit for this choice doesn't have to deal with edible plant competition controversy and forest sustainability. However, detailed planning and supervision are highly required for developing this kind of energy source.

## 7.0 Conclusions

### 7.1 Proposed Action from Research Centre for Chemistry of Indonesian Institute of Science in Bio-energy Research Program in Indonesia

Indonesian Institute of Science (LIPI) is the largest governmental research institutions and manages more than 25 research centres of different fieldwork under five deputies. Together with Ministry of Research and Technology and the other governmental research institution, LIPI arranges Indonesian Research and Technology Policies and all other related policies. At this moment, Indonesian Research and Technology Policies were focused in six major fields which are food supplies stability, medicinal science and pharmaceutical, sustainable and renewable energy, transportations, information and communication technology (ICT) and technology for national defence and security.

Research Centre for Chemistry is one of LIPI research centre. It has a main obligation to give advices and to support LIPI in arranging Research and Technology policies and to help universities in developing education system. RC Chem. LIPI also has obligations to support Indonesian chemical industries in order to strengthening Indonesian economic. RC Chem. LIPI was deeply involved in four of those six major fields and most of research project in Division of Process Technology and Catalytic were concerned to sustainable and renewable energy.

In order to give contribution in solving Indonesian energy problem, RC Chem LIPI will focus on research and development in biomass utilization in Indonesia. For the future outlook, the research will follow the scheme as in Figure 11.7. However, in short time development planning, Indonesian Government will put their attention in

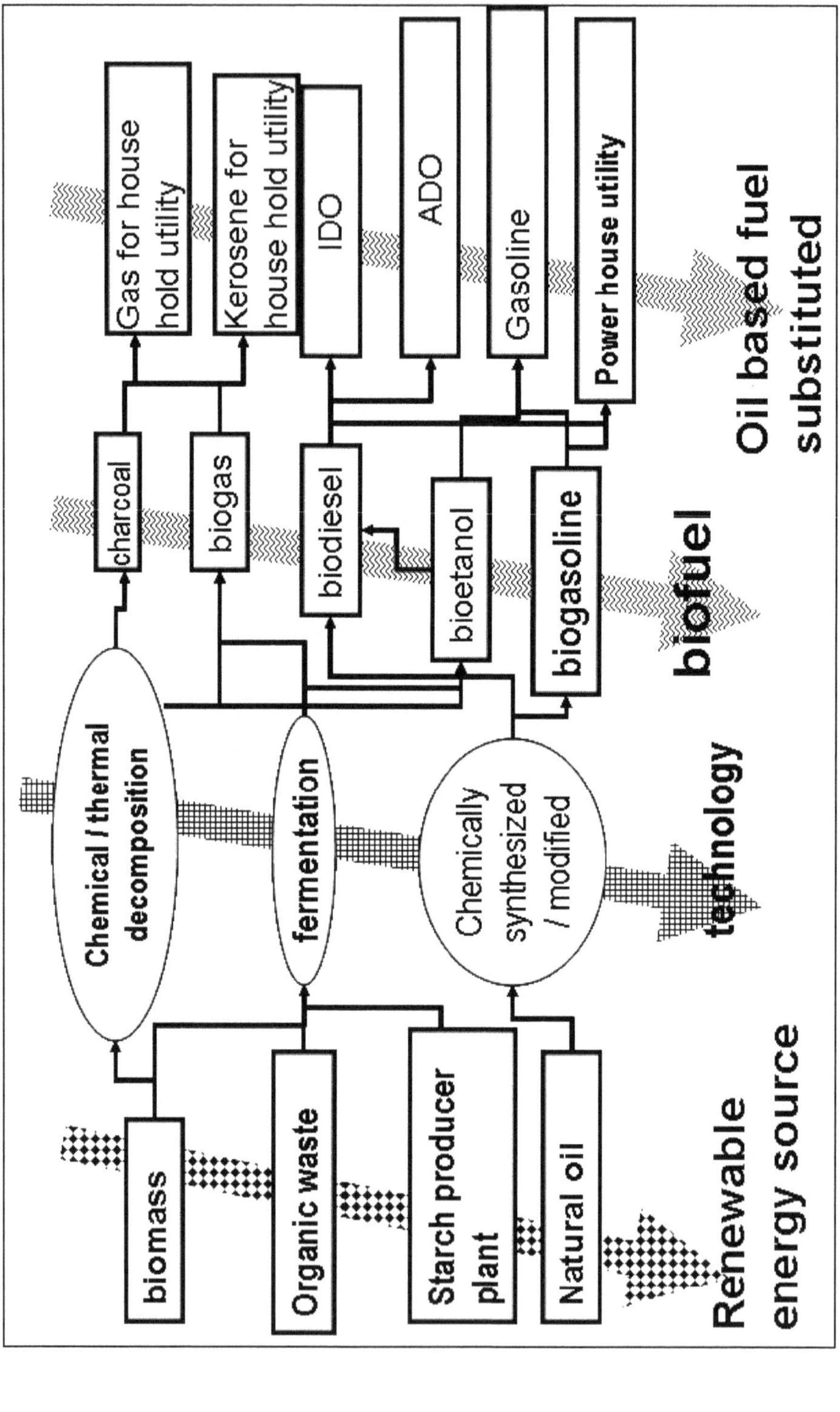


Figure 11.7: Research and Development of Biomass in Indonesia

three research area, biomass utilization for producing electric power house, bio-diesel for substituting Automotive or Industrial diesel oil and bio-ethanol for substituting gasoline.

There are several research areas that become the major concern in Division of Process Technology and Catalysis in biomass utilization, hydrolysis of empty fruit bunch from palm oil plantation using cellulose, designing palm oil incinerator and optimizing the energy efficiency usage for palm oil plant, extracting sodium and potassium from the ash of empty palm fruit bunch, optimizing composting process of domestic organic waste, isolation of composting microbes. Right now, our institution is arranging a joint research project with Chiba University from Japan in developing solid acid catalyst for palm oil biomass hydrolysis and catalyst for producing ethanol from this biomass. Most of the research projects are supporting the national goal in research and technology, developing new or renewable energy source alternatives for reducing fossil based energy source.

Bio-diesel is the most famous bio-energy in Indonesia. All related policies and standard are almost complete. Indonesia had already had Indonesian Bio-diesel Forum and still become the biggest contributor for bio-diesel research and development in Indonesia. This body had already got supported from Ministry of Research and Technology and Ministry of Trade and Industry. Development in bio-diesel is very prospective considering the abundance of raw material in Indonesia, who's recorded as the world's largest palm oil producer, is on target for a record harvest of about 17 million tons in 2007, and a forecast output of 18.4 million tons for 2008. Bio-ethanol lies in the next most popular Bioenergy, which has main goal for substituting gasoline up to 40 per cent of national consumption in 2015.

According to The National Biodiesel Road Map, in 2009 Indonesia is expected to be able to supply 2 per cent of the national diesel fuel supplies, *i.e.* about 720,000 kiloliter methyl ester. At the moment the plant at The Research Centre for Chemistry were able to produce 500 L/day of methyl ester, and will be increased shortly to 1000 L/day.

The specification of methyl ester for biodiesel produced in The Research Centre for Chemistry is as follow: Acid Value is 0.4925, Saponification Value is 199.7113 mg/g, Iodine Value is 200.8535 mg/g, Ester Value is 81.4553 gl/100, Viscosity is 5.9441 cP, Density is 0.8785 g/ml, Caloric Value is 10575.1 Cal/g, Flash Point is 148.0°C, Free glycerol is 0.0433 per cent w, Bonded glycerol is 0.1196 per cent w and Total glycerol is 0.1629 per cent w.

We also have several cooperation projects with essential oil extracting plants in designing and optimizing their incinerator and biogas fermentation process for self utilization. For the incinerator, the basic concept is to design a combustion system for solid fuel with high water content. In biogas fermentation process we are dealing with finding the appropriate microbe for every type of solid waste from essential oil extracting plant.

Our independent division, Technical Service Unit for Chemical Process Technology Development (UPT BPPTK – LIPI), which located in Gunung Kidul Jogjakarta are doing socialization and application consultancy for methane generator for supplying electricity for household utilization. The generator was modified for using methane concentration below 30 per cent volume so it can be connected directly from fermentation tank effluent.

## References


1. BP Migas and Indonesian Petroleum Report 2005-2006 by US Embassy in Jakarta with CIA-The World Book fact, 2005.
2. Director General for Electricity and Energy Utilization Ministry of Energy and Mineral Resources, Republic of Indonesia, Blueprint Pengelolaan Energi Nasional 2005 – 2025.
3. Directorate General of Electricity, website: http://www.djlpe.esdm.go.id/.
4. Indonesian Human Development Report 2004 by Statistic Indonesia, BAPPENAS, UNDP
5. Kamaruddin A.,1998. *Availability of biomass as energy resources.* Proc.National Workshop on Recent Development of Biomas Development in Indonesia, Directorate General of Electric Power and New Energy and Unesco, Jakarta, February.
6. Kamaruddin A., 2000. *Energy Demand for Rural Development.* Proceedings, International Congress and Symposium on South East Asian Agricultural Sciences. ISSAAS, pp. 353-361.
7. Djazuli *et al.*, 2002 Economic feasibility studies of *Mentha arvesis* extraction plant *Final Report. University Research for Graduate Education.* Contract No.032/HTPP-II/URGE/1996. Directorate General of Higher Education, Indonesia.
8. Ministry of Energy and Mineral Resources, website: www.esdm.go.id.
9. PT Pertamina, website: www.pertamina.com.
10. PT PLN, website: www.pln.co.id.

# Chapter 12

# *Biogas Production from Distillery Slops*

**Lat Lat Tun**
*Material Science and Material Engineering Research Department,*
*Ministry of Science and Technology, Myanmar*
*E-mail: sulatt.2008@gmail.com*

## ABSTRACT

Anaerobic digestion is an established technology for the treatment of the distillery slops. Distillery slops (spent wash) was used as the substrate using 4.5 litre digester, 10 litre digester and 26 litre Anaerobic Baffled Reactor (ABR). The performance characteristics of the digesters were investigated on the bases of gas production, COD removal, methane content, and process stability. The average gas production of 15 litre biogas per litre distillery slops was obtained. The maximum methane content was 68 per cent and the COD removal was 67 per cent. The pH of the effluent was stable at 6.9 to 7.2. Therefore, the effluent can be discharged into the sewage with an acceptable quality for the environmental pollution.

*Keywords: Distillery slops, Digester, Anaerobic digestion, COD, Biogas, Environmental pollution.*

## 1.0 Introduction

Everyone has the right to an environment that is not harmful to their health or well being and to have the environment protected for the benefit of present and future generations. To prevent deleterious effects and pollution to the natural environment by the wastes discharged by the industries, the treatment step to the effluent before discharge should be introduced. When the pollution level of the effluent was higher than the legal limits, treatment system must be installed to raise the quality of the effluent. The main objective of the waste water treatment is to remove undesirable impurities from the waste water. It helps to minimize the level of pollution and thus maintains good sanitary conditions in the areas where wastes are disposed off.

In the production of ethanol, the disposal of the concentrated bottom liquid of a distillation column creates serious environmental problems. For every cubic meter of alcohol distilled, there would be 12-15 $m^3$ of slops to be discharged. This is equivalent to pollution in terms of BOD, of sewage produced by 6,000-6,500 people/day.

According to the industrial statistics from the industrial Supervision and Inspection Directorate of the Ministry of Industry (Sack; Buckley, 1998), Myanmar, there are 7 state owned distilleries and 122 private distilleries, producing 25,000 gallons of rectified alcohol per day and release about 500,000 gallons of distillery waste. This implies that huge amount of distillery waste was produced by the alcohol distilleries in the whole country.

The characteristics of spent wash do not allow its discharge into a water body, hence it requires treatment. The treatment processes may be physical, chemical and biological or combination of these depends on the nature of waste. Biological treatment is the most effective method to decompose colloidal or soluble organic matter in the waste water. Anaerobic processes are the most suitable processes for the waste water containing concentration of organic substances and are attracting attention due to its characteristics such as less sludge generation and energy recovery in the form of methane gas.

Biogas is a mixture of methane (55-70 per cent), carbon dioxide (30-45 per cent) and traces of nitrogen, hydrogen, carbon monoxide, hydrogen sulphide and oxygen. It has a net calorific value (n.c.v) of about 16-20 MJ/$m^3$ so that one $m^3$ is equivalent to about 0.6 litre of kerosene or 6 kWhr. The anaerobically digested sludge is useful not only for raising the fertility of soil, increases agricultural production but also be used for feeding fish, pigs and hatching chickens.

In Myanmar, biogas plants were built for rural electrifications, mostly in upper part of the country. All the existing biogas plant used only cowdung.

In this work, distillery slops which are available in considerably large quantities and free of charge are chosen to produce biogas.

## 2.0 Material and Methods

### 2.1 Experimental Analysis of Distillery Slops

The distillery slops obtained from Hlaing Thar Yar Industrial Estates, Yangon, was analyzed by Civil Department, Yangon Technological University and results in details were presented in Table 12.1. The results indicated that distilleries should carryout treatment activities before they discharged the waste water.

### 2.2 Distillery Slops Neutralization Process

The pH of the distillery slops was 2.81. The sodium hydroxide solution was added to the distillery slops until the pH equals 7. After the distillery slops was mixed with sodium hydroxide solution, the distillery slops flowed through the separator for the purpose of separating out excess solids or sludge. After removing the sludge, the neutralized distillery slops flowed to the digester.

**Table 12.1: Analysis of Distillery Slops**

| *Parameter* | *Unit* | *Value* |
|---|---|---|
| BOD | mg/L | 12000 |
| COD | mg/L | 24960 |
| TS | mg/L | 12000 |
| Nitrogen(N) | mg/L | 55 |
| $SO_4$ | mg/L | 150 |
| pH | – | 2.8 |
| Sp.gr at 15.56ºC | – | 0.974 |
| Electrical Conductivity | micro ohm/cm | 3300 |

## 2.3 Sterilization of Neutralized Coconut Water (SNCW)

The coconut water was neutralized with lime. The neutralized coconut water was sterilized to control the fermentation condition. The neutralized coconut water was sterilized in pressure cooker and then cooled down to room temperature.

## 2.4 Starter Preparation

Two types of starters were prepared. Cow dung (CD) has been composed of different variety of bacteria which can produce biogas. Liquid media composition for methanogens contains cystine. Cystine is sulphur containing amino acid, and it can be occurred in coconut water. Therefore, cow dung and coconut water were used in starter preparation.

The amount of used coconut water was less than the maximum allowable concentration of harmful materials in anaerobic digestion. Therefore, the coconut water was not harmful other bacteria and only methanogenic bacteria can consume the cystine from the coconut water.

When cowdung with coconut water was used as starter, the coconut water was sterilized. Initially, the 6.23 pH of coconut water was neutralized with lime. When the enough aging time of slurry for anaerobic digestion was over, the biogas evolved. At that condition, the slurry could be used as the active phase digested sludge starter for another digester.

Similarly, the fermented cow dung starter was prepared according to the above procedure. The cow dung was mixed with distillery slops. Therefore, the slurry was obtained after a few hours. The fermented cow dung starter was used as starter.

## 2.5 Biogas Production

### Biogas Production with 4.5 litre Digester and 10 litre Digester

The neutralized distillery slops and of each starter as mentioned above were filled in the 4.5 litre digester and capped tightly with rubber stopper. The rubber stopper had three holes for NaOH inlet, thermometer pocket and biogas outlet. The drain line with valve was attached at the bottom of the digester to collect the sample for measuring the pH of the slurry. Daily pH adjustment was done and the digester was thoroughly shaken twice a day.

The gas outlet was connected with a plastic pipe which was also joined to the inlet of the gas holder. The gas holder was vertically inverted in water tank. The volume of the gas holder was 5 litres. The volume of the evolved gas was determined by the water displacement method. The volume of the displaced water was indicated as the volume of the evolved gas. The biogas composition was determined by using orsat meter. The reactor temperature was at room temperature and the room temperature was varied from 26°C to 34°C.

Similarly, the neutralized distillery slops and each starter as mentioned above were filled in the 10 litre digester. The digester was capped tightly with rubber stopper. The rubber stopper had biogas outlet. The drain line with valve was attached at the bottom of the digester to collect the sample for measuring the temperature and pH of the slurry. The gas outlet was connected with a plastic pipe which was also joined to the inlet of the gas holder. The volume of gas holder was 14 litres. The gas collection system and the determination of gas composition were the same as the above procedures.

**Construction of Anaerobic Baffled Reactor (ABR)**

Based on the results of lab-scale digesters, mini-pilot scale Anaerobic Baffled Reactor was considered to construct. Afterwards, the pilot-scale digester was constructed with the considerations of cost and the required usage of biogas. If the pilot-scale digester was directly constructed based on the results of lab-scale, many difficulties and errors would be encountered. Therefore, mini-pilot scale digester had to be considered.

The size of the mini-pilot scale ABR was chosen 26 litters. The length, width, and height of the rectangular reactor were obtained from the literature.

**Experimental Set-up for ABR**

The digester has a void volume of 26 litters consisting of four chambers. The first has a volume of 13 litres, while the remaining ones, 4.33 litter each. The shell of the reactor has been constructed out of PVC plate with a thickness of 4 mm. The interior baffles are also PVC plate with a thickness of 4 mm. PVC piping and fitting were used. PVC valves and fittings were used for the sample ports.

The common gas sampling ports were located at the centre of the cover plate of each compartment. A water displacement setup monitors the volume of the gas produced. The gas outlet was connected with a plastic pipe, which was also joined to the inlet of the gas holder.

The volume of gas holder was 14 litres. The gas holder was vertically inverted in water tank. Another open-tube coming out from the gas holder was joined to the orsat meter by using an adapter. All the junctions were sealed with epoxy glue to ensure that there was no leakage within the system. The initial stage of ABR is shown in Figure 12.1.

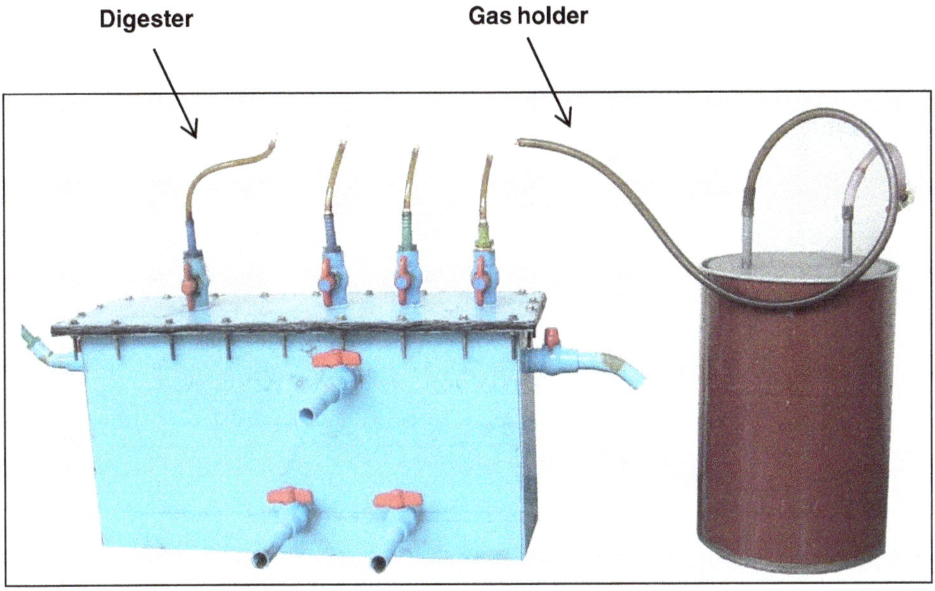


**Figure 12.1: Initial Stage of 26 Litre ABR**

The volume of the evolved gas was determined by the water displacement method. The volume of the displaced water indicated the volume of the evolved gas. The biogas composition was determined using orsat meter.

**Biogas Production with ABR**

The neutralized distillery slops from above pH adjustment was filled in the digester (26 litre ABR) shown in Figure 12.1 and then it was filled with the 20 per cent (CD + CW) starter. The volume of gas production, pH, temperature and total solids were monitored. The COD of the influent and effluent, and the methane content of the gas produced were measured. The reactor temperature was not controlled, with the ambient temperature varying from 27° to 32°C.

## 3.0 Results and Discussion

### 3.1 Study on Anaerobic Digestion for Batch Processes

Anaerobic digestions of distillery slops with various starters gave various results of gas composition and gas production rates. The gas production rates depend on the type of the starter used. In this research two types of starters were used to investigate retention time, pH fluctuation, gas production rate and gas quality.

The anaerobic digestion of using (CD + CW) starter gave more gas production rate than that of using the (CD) starter. Anaerobic digestion processes are dependent on pH. Hence it is very important that the value and stability of the optimal pH in an anaerobic reactor are maintained. Most anaerobic conversion processes operate best near neutral pH.

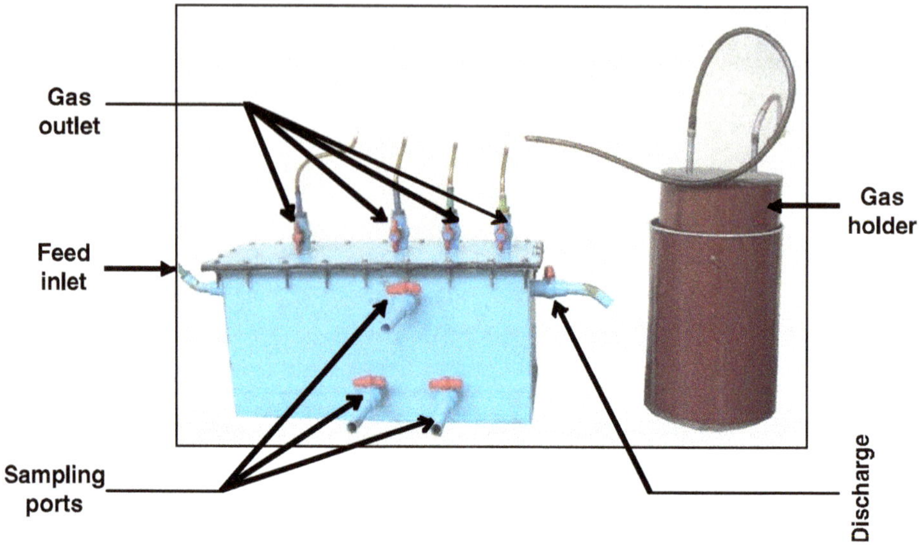


**Figure 12.2: Gas Evolved Stage of 26 Litre ABR**

In 4.5 litre digester and 10 litre digester, it was found that (CD + CW) starter gave the same gas production and the optimum gas volume.

It can be observed that the total slurry was the mixture of culture (*i.e.* not pure methanogenic bacteria), only methanogenic bacteria consumed the cystein from the (CD + CW) starter. The coconut water was not harmful the other bacteria. The amount of used coconut water was less than the maximum allowable concentration of harmful materials in anaerobic digestion.

Fortunately, it can be seen that the small amount of coconut water was used, the gas production rate was significantly high. Therefore, coconut water was the required material in starter preparation.

When using (CD + CW) starter was that initial fermentation stage was rapid and the acids, which produced reduce pH level. But the pH was needed to adjust to neutral in five hours a time. It was found that pH was more stable after six days.

The methane content of using cow dung starter was less than 60 per cent. The maximum methane content of using (CD + CW) starter was 68 per cent after 8$^{th}$ day. Fortunately, it was observed that using (CD + CW) starter improved not only the gas yield but also methane content.

## 3.2 Overall Reactor Performance of ABR

Anaerobic digestion using (CD + CW) starter improved not only the gas yield but also methane content. Therefore, (CD + CW) starter was used as starter in ABR.

In the ABR, the earlier compartments generally have a lower pH as the acidogenesis and acetogenesis phases take place in these compartments. Methanogens are extremely sensitive to pH and work best under neutral conditions (later compartments). The pH in the first compartment was lower than that in last compartment.

The compartmentalized design of the ABR results in a buffering zone between the primary acidification zone and the active methanogenesis zone on the ABR. The methanogens are thus not exposed to low pH during organic shock loads.

During a start-up of a biogas system, the volume of biogas produced will increase daily due to the growth and build up of the microorganisms. It started at a low 30 per cent to a high of 68 per cent at the end of the run. As the microbial population builds up gradually with time, the plug flow configuration of the reaction of hydrogen and carbon dioxide to methane in the later stage of reactor system.

The influent COD and BOD are 24960 mg/L and 12000 mg/L. The COD and BOD of the effluent to the ABR are 8237 mg/L and 3240 mg/L. The COD reduction was 40 per cent at 7$^{th}$ day and the BOD reduction was 48 per cent at 7$^{th}$ day. The COD reduction was increased to 67 per cent when the HRT was increased to 14$^{th}$ day. The BOD reduction was increased to 73 per cent when the HRT was increased to 14$^{th}$ day.

The performance of the reactor showed stable operation without any sign of impending failure. This should indicate that reactor has promise to be high-rate reactor.

## 4.0 Conclusions

The pH of the distillery slops and coconut water are very small. The pH of the reactor is ranging from 7.3 to 7.8. Therefore, neutralization is required to produce biogas. Neutralization affects not only pH adjustment but also dilution to the distillery slops. When the NCW was not sterilized, the pH of NCW drops to 3.21. When NCW was sterilized, the pH of SNCW drops to 4.63. As a result, there was an accumulation of organic acids reducing the pH rapidly. Therefore, it can be seen that the NCW should be sterilized to make the fermentation stable.

In 4.5 litre capacity digester and 10 litre capacity digester, cow dung starter and (CD + CW) starter were used as starters. The amount of gas obtained by using (CD + CW) starter is higher than that of obtained by using cow dung starter. Using (CD + CW) starter has the highest gas volume in the biogas production of distillery slops.

The ABR, with its stagewise configuration, generated 15 litters biogas per litre feed slops with 68 per cent methane content. Both gas production rate and gas quality were distinctly improved when (CD + CW) starter was introduced. It was possible that if CW was present, the methanogenic bacteria were healthier than before and they could work very well and produce more biogas. The pH was stable between 6.9 and 7.2. The ABR was a small reactor but it can treat the large amount of the distillery slops.

The biogas can either be used as fuel for burning lanterns, for cooking in domestic or other uses. The anaerobic digestion would help to treat the waste and would benefit the community in term of value added return in the form of biogas/power and nutrients. If further investigations are carried out to reduce the carbon/nitrogen ratio to about 30, then the distillery slops can serve as a good raw material for yielding biogas.

It is envisaged that the evaluation protocol being developed will assist the municipalities in making informed decisions on whether or not to receive certain high strength organic liquid effluents. Furthermore, the success of this project could provide an alternative sink for liquid wastes generated inland which have no other outlet for disposal other than landfills and it may provide an additional option for the treatment of landfill leachate.

## Acknowledgements

The author would like to express her heartfelt thanks to her professor Dr. Mya Mya Oo, Rector and Head, Department of Chemical Engineering, Yangon Technological University, for her support, suggestions, guidance, understanding, and unfailing kindness throughout her research work. The author's special thanks are also due to Dr. Daw San Myint, Professor (Retired), Department of Chemical Engineering, Yangon Technological University, who supervised closely and systematically, and also provided with valuable guidance, advice, encouragement and understanding. Sincere thanks also go to all teachers from Chemical Engineering Department of Yangon Technological University for their helpful discussions and suggestions.

## References

1. De Sousa, R.A. 2005. "Determination of Metals in Brazilian Coconut Water using An Inductively Coupled Plasma Optical Emission Spectrometer" J.Braz. Chem. Soc. 16, no.3B: 540-544.
2. Dr. Joshi, H.C., 2005. Environmental Sciences Division, Indian Agricultural Research Institute, Bio-energy Potential of Distillery Effluents.
3. Reddy, A.C., 1996. "Biogas from an Industrial Waste" Department of Mechanical Engineering, M.J.C.E.T. India. Energy for Sustainable Development 3, no.1.
4. Sack, J. and Buckley, C.A., 1998. Anaerobic Digestion of High-strength or Toxic Organic Effluents in Available Digester Capacity.
5. Sacks, J., Buckley, C.A., Stuckey, D.C. and Plumb, J.J., 1999. Treatment and Depolarization of Food Dyes in the Anaerobic Baffled Reactor.
6. Stephen Gitonga, 1997. Biogas Promotion in Kenya.
7. Bio-Wise. Biotechnology at work. 2001. "Industrial Wastewater and Effluent Treatment" A Review of Anaerobic Digestion Technology Department of Trade and Industry. UK. May 2005. http://www.dti.gov.uk/biowise.
8. Metcalf and Eddy. 1991. "Treatment, Disposal, and Reuse" Wastewater Engineering 3rd ed. McGraw-Hill, Inc., New York.

# Chapter 13

# *Solar Traffic Lighting System: Use of Multicolour LED Lamps*

**D.K. Shah[1], J.N. Shrestha[2] and S. Poudyal[3]**

*[1]Faculty of Technology, Nepal Academy of Science and Technology (NAST), Khumaltar, Lalitpur, Nepal*

*E-mail: dkshah149@gmail.com, dkshah149@yahoo.com*

*[2]Centre for Energy Studies, IOE, TU, Pulchowk Campus, Lalitpur, Nepal*

*E-mail: jns@heathnet.org.np*

*[3]Traffic Engineering and Safety Unit, Department of Road, Babarmahal, Kathmandu, Nepal*

*E-mail: poudyal_s@hotmail.com*

## ABSTRACT

Energy saving has become the prime consideration for the mankind and is necessary for the preservation of natural resources. It is estimated that by 2050 AD, more than 30 per cent of the electricity worldwide would be generated from Renewable Energy resources.

Energy can be saved by using efficient devices like multi-colour light emitting diodes (LED) lamp instead of traditional incandescent lamp in traffic lighting system.

The power consumption for traffic control system is 24 times less by using multi-colour LEDs lamp instead of currently used incandescent lamp in traffic lighting system. The life of LED lamp is 33 times more than traditional incandescent lamp. The incandescent lamp is replaced about 40 times while LED lamp is replaced only one time during operation of 20 years in the traffic lighting system.

LED lamp greatly reduces the cost of mechanical components as well as operating cost. The total estimated cost for operation in 20 years of system consisting LED lamps and power supply from solar energy with monocrystalline solar cell is 8 times less than system using traditional incandescent lamp and power supply from National grid. The pay back period of system using LED lamp and power supply with solar energy with monocrystalline solar cell is 3.46 years when compared with system using incandescent lamp and power supply from National grid.

***Keywords:*** *Solar energy, Traffic signal, Energy saving, Incandescent lamp, Multi-colour LEDs.*

## 1.0 Introduction

Energy saving is as energy generation. Large amount energy can be saved in transport sector by using multi colour LED lamps in traffic lighting system. By using renewable energy resources (*i.e.* solar energy), consumption of fossil fuel can be reduced. It will be highly reliable for power if solar power is supplied in traffic lighting system.

Energy is an essential tool both for economic development and for an improved quality of life. Life on the Earth, since it's beginning, has been powered by the energy of sun. Every thing what happens in the world is the expression of flow energy in one of its form. Energy is the important requirement for carrying out all human activities.

Now, energy saving has become the prime consideration for the mankind. The rate of consumption of natural resources like fossil fuel, gas etc. if not restricted, soon these will be extinct. So, we should hasten our steps towards using the renewable energy sources. With the availability of petroleum products, the modern man finds life easy, comfortable and smooth going. The consumption of energy is steadily rising year by year because of two reasons:

- Increase in population
- Improvements in the standards of living

## 2.0 Objectives of Work

### 2.1 Main Objective

The main objective of the work is to study on implementation of solar traffic lighting system using multi-colour LED lamps.

### 2.2 Specific Objectives

The specific objectives are:

1. To compare power consumption between the systems using multi-colour LED lamp and traditional incandescent lamp
2. To estimate the array size for solar energy by using multi-colour LED lamps
3. To perform financial analysis of power consumption and compare with cost of National Grid power.

## 3.0 Energy Scenario of Nepal

Total energy consumption is in Nepal for 2002 is 8.205 million toe and per capita energy consumption of about 15 GJ/Year.

The energy resources of Nepal consist of combination of traditional and commercial sources of energy including biomass, hydropower. Biomass energy sources namely fuel wood, agricultural residues, animal dung and biogas meet about only 76 per cent of household energy demand. In spite of its large theoretical hydropower potential of 83,000 MW, of which 42,000 MW is established to be technically feasible, Nepal has developed only 600 MW of hydropower to date (*i.e.* 1.3 per cent of the total commercial potential). It supplies only about 12 per cent of total energy requirement.

Per capita commercial energy consumption is also low as compared to other countries in the sub-region. However, the growth of commercial energy consumption is high about 8.4 per cent per annum.

## 3.1 Use of Solar Energy

Nepal being located in favorable latitude receives ample solar radiation. There are more than 300 sunny days in Nepal with average sunshine of about 6.8 hours. According to World Metrological Organization, the average solar radiation for Nepal is about 4.5kwh/$m^2$/day at optimum tilt. Solar PV sector is a new and growing sector in Nepal.

The solar energy resource in Nepal is abundant, evenly distributed over the country and over the seasons. Nepal being located in favorable latitude has huge potential for solar energy. Solar energy has been the focus of renewable technology in Nepal. Solar photovoltaic technology can play a significant role in rural electrification within a short span of time.

Utilization of solar energy in Nepal is mainly three forms like solar photovoltaic, solar thermal and solar dryer. Solar photovoltaic system is using for lighting, water pumping, entertainment, communication etc, solar thermal for domestic water heating and solar dryers for drying crops. Government and the private sector both are involved in the promotion of solar PV in Nepal.

Taken all system together, total installed capacity of solar PV has reached to 1.99MW. The traffic lighting system of 6975 watt has been installed out of total installed capacity. At different sixteen locations in Kathmandu there are solar PV traffic lighting system.

## 3.2 Solar Energy in Traffic Lighting

The traffic lightning system powered by solar energy already exists in various cities of the world. There are different sixteen locations at which solar traffic lighting system has been installed in Nepal.

Now it is a time to trap solar energy for different devices as well as traffic signals in transport sector in urban area. The prevalent unreliable power situation in the country has demanded the reliable supply for the application in critical areas such as in traffic control system. The traffics in the road are ever increasing without proper and timing management of the traffic, the accident being frequent. The solar energy can be used with storage system to provide the reliable power to the traffic lighting system and the solar power system is very essential if the local power supply is highly unreliable.

### 3.3 LEDs in Traffic Signal

The traditional lamps that are used in traffic lights are relatively powerful, usually rated between 35W (quartz halogen) and 65W (incandescent) for a normal traffic light, and sometimes up to 150W depending on the application. They operate at high temperatures with frequent switching and as such require replacement on a regular basis.

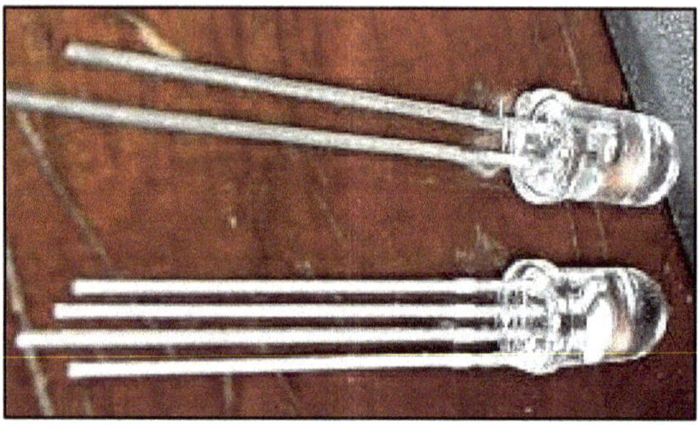

**Figure 13.1: Tri Color (RGB) LED**
**(*Source*: http://ledmuseum.home.att.net/ledgrn.htm)**

A traffic authority that adopted LED technology into its traffic signals would see financial savings begin immediately as maintenance cost is reduced. Along with lower maintenance costs, LEDs use 80-90 per cent less energy than traditional incandescent lamps, providing a relatively quick payback on the investment.

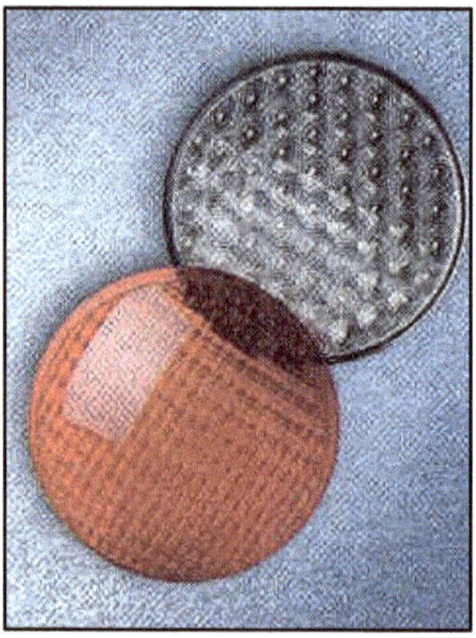

**Figure13.2: White LED Lamp with Red Filter**

Due to the advancement in technology, now very efficient 5.6 candela white and other colour LEDs are available. Replacement of the traditional incandescent bulbs by ultra bright LEDs greatly reduces the operating cost as well as the power consumption. Instead of using the colour filter in incandescent bulbs, the colored LEDs can be used to fulfil the need of three colored lights in traffic control system without any filter incorporated.

## 3.4 Multi-Colour LEDs

Every semiconductor material type has a different energy gap. The wavelength that describes the colour of the radiated light depends upon the magnitude of the energy gap. The relation is:

$$E = hc/\lambda$$

The wavelength of the emitted light, which describes its colour, depends upon the band gap energy, which further depends upon the material of the diode.

The following table shows some of the gallium compounds material and the colour radiated by them.

**Table 13.1: Material of Diode with Band Gap Energy**

| *Sl.No.* | *Material Used* | *Colour* | *Wavelength (nm)* | *Band Gap Energy (ev)* |
|---|---|---|---|---|
| 1. | GaAs:Zn | Infrared | 900 | 1.37 |
| 2. | GaAsP4 | Red | 700 | 1.77 |
| 3. | GaP:N | Green | 546.1 | 2.27 |
| 4. | GaAsP:N | Yellow | 590 | 2.1 |
| 5. | GaAsP5 | Amber | 610 | 2.03 |

## 3.5 Advantages of LEDs

An LED is a semiconductor device that uses solid-state electronics to produce colored light. The LEDs are more efficient than incandescent lamps for different reasons as follows:

1. Energy is not wasted as heat: Unlike incandescent lamps, LED lamps do not produce much heat while producing light..
2. Energy is not wasted by filtering: LEDs create colored light directly
3. Use 90 per cent less energy compared to standard incandescent lamps
4. Last 7-10 years resulting in less labor devoted to re-lamping
5. Produce Colour directly without filter for greater visibility in fog, sun and rain
6. Reduce emergency call-outs due to increased reliability
7. Less susceptible to mechanical failure–Rugged solid state construction
8. Easy to retrofit–the LED module installs directly into the casing of existing traffic signals

## 4.0 Power Consumption

### State of Lamp Signals

Total numbers of signal lamps are 24 if multicolor LED lamps are used for particular location (Gwarko, Lalitpur).

**Table 13.2: States of Lamps**

| *State of Lamps* | *ON* | *OFF* |
|---|---|---|
| Phase I | 18 | 6 |
| Phase II | 16 | 8 |
| Phase III | 16 | 8 |
| Phase IV | 14 | 10 |

Maximum number of lamp glow at a time = 18

Signal working hours in a day = 16 Hrs.

### Traffic Flow at Particular Location

The maximum traffic flow is 2,826 passenger Car unit/hr at 9.00 am to 10.00 am (peak time) (Figure 13.3).

### Maximum Power Consumption by LED Lamp Using

For number of signal lamps which are ON

**Table 13.3: Number of Signal Lamps is on**

| *Phases* | *Number of Green and Red signal* | *Number of Walk and Don't Walk Signal* | *Power Required (Watt)* |
|---|---|---|---|
| Phase I | 6 | 12 | 6 x 7 + 12 x 6 = 114 |
| Phase II | 4 | 12 | 4 x 7 + 12 x 6 = 100 |
| Phase III | 4 | 12 | 4 x 7 + 12 x 6 = 100 |
| Phase IV | 2 | 12 | 2 x 7 + 12 x 6 = 86 |

Maximum energy consumption for 16 hrs in a day = 114 x 16 = 1.824 kWh.

## Findings

### Maximum Power Consumption by Incandescent Lamp

Generally, 150W lamp is used in both pedestrian and vehicles signal in traditional incandescent traffic lighting system. Then power consumption for that system is

Maximum Power required = 18 x 150 W = 2700 W

Maximum energy required for 16 Hrs. in a day = 2700 x 16 Wh = 43.2 kWh

But Maximum energy consumption by multi colored LED lamps for 16 hrs/ day = 114 x 16 = 1.824 kWh.

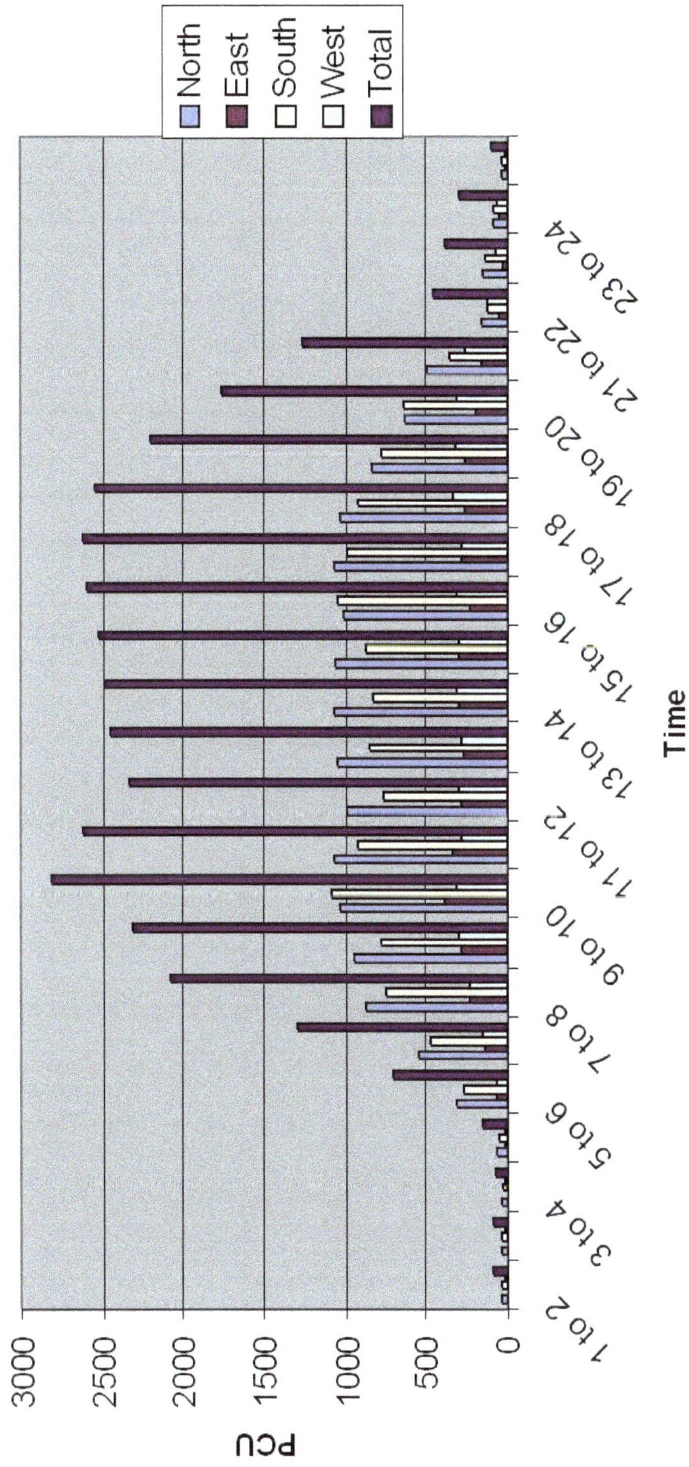


**Figure 13.3: Maximum Traffic Flow**

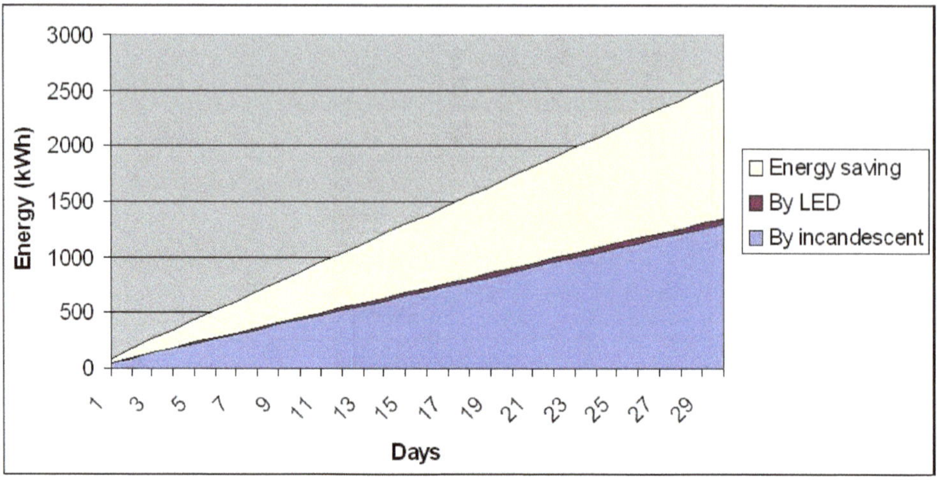


**Figure 13.4: Energy Consumption and Saving in a Month**

## Comparison of Life of LEDs and Incandescent Lamps

Average Life of incandescent lamp = 3000 Hrs

Average life of LED = 100,000 Hrs.

The life of LEDs Lamp is 33.33 times more than incandescent Lamp.

Operation Hour in 20 Years = (16 x 365 x 20) = 116800 Hrs.

The replacement times of LEDs lamp during 20 years: 1 time

The replacement times of incandescent lamp during 20 years: 40 times

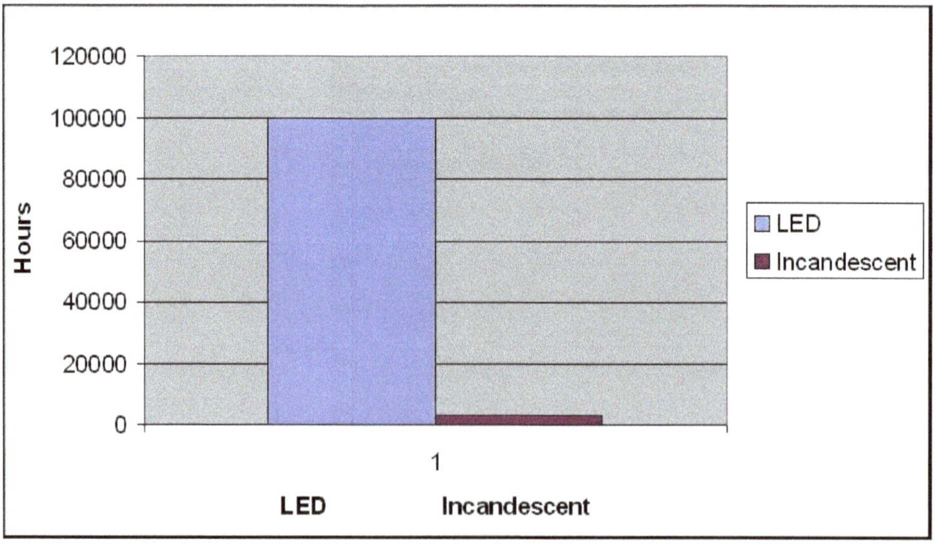


**Figure 13.5: Comparison of Life of Lamps**

## Cost Comparison

The estimation for different situations is as follows:

Using LED lamp and solar energy (monocrystalline solar cell), the total estimated cost for operation 20 years is $ 5163.1 (*i.e.* NRs. 387232.5) (Since 1 $ @ NRs.75).

- Using incandescent lamp and National Grid, the total estimated cost for operation 20 years is $ 39143.13 (*i.e.* NRs. 2935735.75).
- Using LED lamp and National Grid, the total estimated cost for operation 20 years is $ 2957.61 (*i.e.* NRs. 221820.75).

The repair and maintenance cost of system in the course of operation in 20 years is not included in these figures.

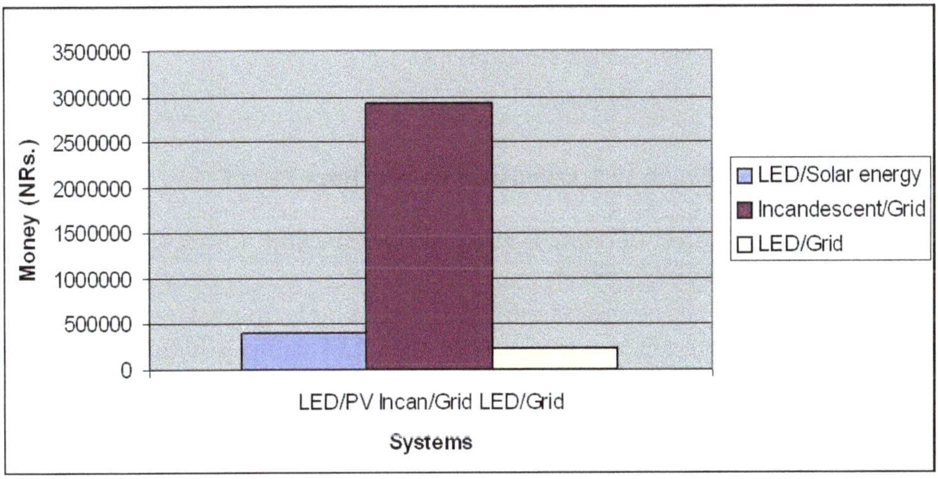


**Figure 13.6: Cost Comparison of Different Systems**

## Comparison of Pay Back Period

The pay back period of system using LED lamp and solar energy having monocrystalline solar cell is 3.46 years and system using LED lamp and power supply by National grid is 2.11 years when compared with incandescent lamp (Figure 13.7).

## 5.0 Conclusions and Recommendations

Solar energy for power generation is one of the promising renewable energy resources, which is available through the world. Because of its availability and environmentally benign the use of solar PV power for various purposes are increasing day by day.

The power consumption for traffic control system is 23.69 times less (*i.e.* 95.78 per cent efficient) by using multi-colour LEDs lamp instead of traditional incandescent lamp in traffic lighting system. The life of LED lamp is 33.34 times more (*i.e.* 97 per cent efficient) than traditional incandescent lamp. The incandescent lamp is replaced

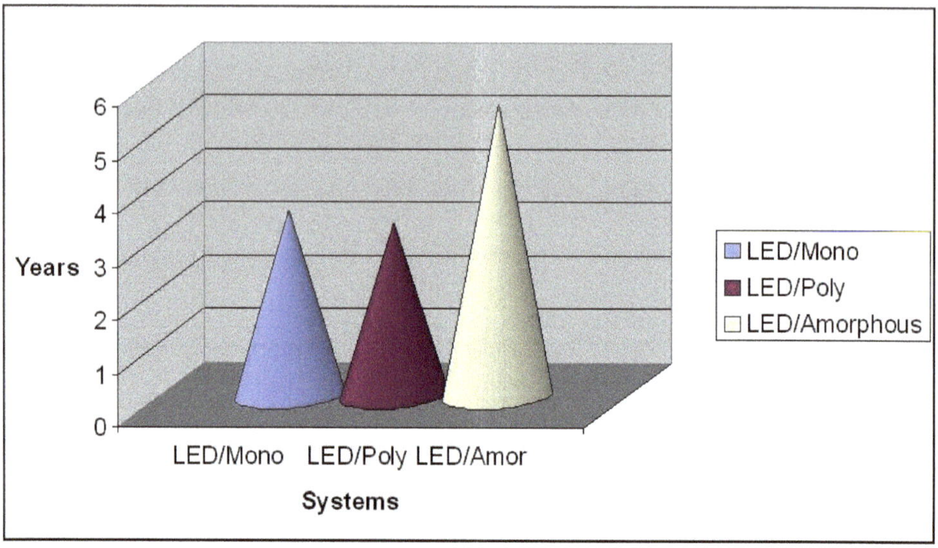


**Figure 13.7: Comparison of Pay Back Period**

about 40 times where as LED lamp is replaced only one time during operation of 20 years in the traffic lighting system.

The economic analysis indicates that the power supply from solar energy having monocrystalline solar cell by using multi-colour LED lamp is highly feasible as comparison to system using traditional incandescent lamp and National grid. The total estimated cost for operation in 20 years of system using LED lamp and power supply by solar energy having monocrystalline solar cell is 7.58 times less (*i.e.* 86.80 per cent efficient) than system using traditional incandescent lamp and National grid.

The pay back period of system using LED lamp and solar energy having monocrystalline solar cell is 3.46 years and system using LED lamp and power supply by National grid is 2.11 years when compared with incandescent lamp.

In existing solar powered system, only single colour white LED with filter is used which consume more power. It is recommended to use multi colour LED lamp that does not need filter in traffic lighting system. For reliability of power supply for traffic control system, solar energy should be used, which is economically feasible, environment friendly as well as available abundantly.

It is recommended to replace all existing traffic lighting system using incandescent lamp by system using multi-colour LED lamp with solar energy for power supply.

## References

1. Avinash Tripathi, 2000. "Energy Audit and PV Sizing for back up power at International Centre for Genetic Engineering and Biotechnology, New Delhi" M.Sc Thesis, CES, IIT Delhi.

2. Deependra Pokharel, Sharad Kumar Ghimire, 2000. "Efficient Light Emitting Diode based low powered traffic light control system" Project Report, BE Electronics, Department of Electronics, Pulchowk campus, IOE.
3. Ljubisav stamenic and George W. Ingham; "Solar photovoltaic Revolution", a Canadian Handbook for engineers, electricians, inspectors, builders.
4. Satish Chandra, Upendra Kumar, 2003. "Effect of lane width on capacity under mixed Traffic Condition in India" Journal of Transport Engineering, ASCE/ March/April pp. 155-160.
5. Shah, D. K., 2004. "Study on implementation of solar photovoltaic traffic lighting system for Gwarko Lalitpur, Nepal" Department of Mechanical Engineering, IOE, TU.
6. Shingh Gurcharan, 1999. Highway Engineering Third Edition, standard Publisher Distributors, 1705-B, Nai Sarak Delhi-110006.
7. CES, 2000. Renewable Energy Perspective Plan of Nepal, 2000-2020: An Approach, *Report*, Center for Energy Studies, Institute of Engineering, Tribhuvan University, Kathmandu, Nepal.
8. Overseas Road Notes 13; 1996. Transport Research Laboratory, Crowthorne, Berkshire, United Kingdom.
9. Siemens Solar Photovoltaic Training program "System Design and Engineering Course Training Manual ', IREDA /World Bank / SIEMENS, Siemens solar Industries, California, 1996.
10. Internet surfing.

# Chapter 14

# *Sourcing of Local Raw Material for Sustainable Manufacturing of Solar Energy Powered Components in Nigeria*

**G. Ladan[1], D.B. Ayo[2] and P.A. Onwualu[3]**
*Raw Materials Research and Development Council (RMRDC)*
*Abuja, Nigeria*
*E-mail: [1]gladanrmrdc@yahoo.com; [2]danielayo@hotmail.com; [3]ceo@rmrdc.org*

## ABSTRACT

This paper presents efforts and challenges of the Raw Materials Research and Development Council in promoting local raw materials for production of solar powered components in Nigeria. Availability of local raw materials, indigenous efforts in the development of renewable technologies and their investment opportunities in manufacturing of solar powered components/devices are highlighted. An overview of environmental consequences of energy derived from fossil fuels and the prospects of renewable energy resources in Nigeria are examined. The different roles that can be played by academia, industry, and government towards sustainable application of solar energy as clean technology are suggested. The paper concludes that for sustainable production of competitive solar components, a deliberate plan of action that will provide an enabling environment for sufficient investment in the area of research and development, training, marketing, and public awareness program on solar technologies should be embarked upon.

***Keywords:*** *Raw materials research and development council, Raw materials information system, Solar energy, Renewable energy, Clean technology and Small-scale industry.*

## 1.0 Introduction

The Raw Materials Research and Development Council (RMRDC) was established in 1987 as a parastatal under the supervision of Federal Ministry of Science and Technology. The purpose of establishing the Council is to support and expedite industrial development and self-reliance through optimal utilization of local raw materials as inputs for Nigerian industries.

While pursuing this objective, the Council was mandated to carry out the following functions:

1. Promotion of raw materials development and utilization.
2. Publicity of research findings and other information relevant to local sourcing of raw materials.
3. Technology development and its popularization.
4. Advise on adaptation of machinery and processes for raw materials utilization.
5. Encourage growth of implant research.
6. Advice on and device awards or systems for industries that achieve any break through or make innovations and inventions.
7. Draw up policy guidelines and action programmes on raw materials acquisition, exploitation and development.
8. Organize workshops, symposia, and seminar designed to enlighten people on the new developments and solutions discovered from time to time.
9. Review from time to time raw materials availability and utilization with a view to advising the Federal Government on the strategic implications of depletion conservation or stockpiling of such resources.
10. To consider and advise on any issue capable of enhancing the objectives of the Council.

It is suffice to state that right from its inception, the Council evolved several programmes, and strategies aimed at promoting the exploitation, development and utilization of local raw materials. These include the following:

1. Techno-Economic survey of Nigeria's Resources and Technologies.
2. Raw Materials Information system (RMIS).
3. Promoting Small and Medium Scale Local Resource-Based Investments.
4. Sponsoring Research and Development (R&D) activities for raw materials utilization and new products.
5. Upgrading of Research laboratories and Promoting Science Education in Nigeria.
6. Promoting the Development of Indigenous Technology/Capacity Building in Process Technology Design and Fabrication.
7. Technology exposition.
8. Supporting Agro-Industrial Raw Materials Production.

9. Providing Industrial Support Services (Consultancy).
10. Publication of technical reports/studies on raw materials development and utilization.
11. International Collaboration on Raw Materials Development and
12. Policy Advisory.

With array of these programmes, the Council is today the focal point for the exploitation, development and utilization of Nigeria's vast industrial raw materials, their locations, reserves and investment potentials as well as a linkage to the outside world. This feat also placed on the Council the onus of providing credible information on raw materials to investors, industrialists, policy makers, researchers, and other stakeholders. However, this presentation is limited to the promotion of local raw materials, which could be utilized in manufacturing of solar energy powered components for sustainable economic growth in Nigeria.

Energy is an essential ingredient for socio-economic and industrial development of any nation. It is for this reason that the present Government of Nigeria placed energy as one of the seven point agenda with the sole aim of moving the country forward industrially and putting it among the largest economies of the world by the year 2020.

Nigeria is blessed with abundant fossil fuels as well as renewable resources. Some of the renewable energy resources that have been identified are solar, hydro, wind and biomass. Among these renewable, solar energy has a great potential as a clean and reliable energy for the future. A major constraint is however; inadequate process technology, which will transforms local raw materials as industrial inputs for making solar powered devices on commercial scale. At present, most of the essential components are imported. Although some of the raw materials for making these components are available in their natural forms, they still have to be transformed to make them available for making solar technology devices. The objective of this paper is to highlight the challenges of sourcing local raw materials for sustainable manufacturing of solar energy powered components for small-scale industry in Nigeria.

## 2.0 Raw Materials for Manufacturing of Solar Energy Powered Components

About 90 per cent of raw materials required for manufacturing of solar powered components can be locally sourced in Nigeria. However, the technology in which they can be effectively processed for the production of standard solar powered components on commercial basis is not locally available. This constitutes a major problem for meaningful industrial development. Raw materials which are locally available include silicon, steel, metal sheets, aluminium, metal mirror, glass mirror, copper, foils, plastic sheet, wood and galvanized reinforced plastic (GRP). Others are pipes, hoses, glazing materials (glass, acrylic and fibreglass) and insulation materials, (wool, and natural materials residue from processing of coconut, rice, corn, peanut, etc.). Increased R&D activities on raw materials for manufacturing of solar energy powered components and appliances will also bring out other materials with higher quality and efficiency for industrial applications.

## 3.0 Indigenous Efforts in the Development of Renewable Energy Technologies

In Nigeria, as a result of research and development activities of the Universities, Research Institutes and Energy Research Centres supported by such organizations as Raw Materials research and Development Council (RMRDC) and Energy Commission of Nigeria (ECN), a large number of renewable energy devices have been developed. Although some of these systems were even installed for community utilization on pilot scale, they are however yet to be commercialised for meaningful integration into economic activities of the country. Some of the renewable indigenous technologies developed are as follows:

### Solar Dryers

Various types of solar dryers developed for different applications include Triple pass solar dryer, dehumidified air solar heated tray dryer and passive solar heated poultry chick brooder.

#### Triple Pass Solar Dryer

This type of dryer is used in drying variety of agricultural produce such as cereal grains, tuber crops and vegetables. It consists of a flat plate collector, a portable drying cabinet and dehumidification chamber. These three units incorporate unique design features which result in higher heat gain through the collector, and sustained a uniform drying for day and night time operation.

#### Dehumidified Air Solar Heated Dryer

This consists of three functional units, namely flat plate collector, dehumidification chamber and dryer cabinet. The dryer is suitable for fish drying and also vegetables such as okra, pepper, etc. One major advantage of this dryer is that its design was carried out in such a way that a number of modules can be added in-situ without undue structural modification. Similarly, all major components are detachable and thus making it easy for mass production, transportation and assembly of the dryer.

#### Passive Solar Heated Poultry Chick Brooder

In brooding of chicks, environmental conditions such as temperature, relative humidity, etc. are necessary during the first six weeks of chicks' life. Usually the brooder is made of a wooden box roofed with solar collector that stores solar radiation for day and night operation.

### Solar Stills

These are designed to produce distilled water from brackish water that can be used in hospitals, laboratories and industries, etc. The stills can also be used in production of distilled water for wet batteries.

## Solar Pumps

Photovoltaic solar modules for pumping water from boreholes have been developed to provide water requirements, especially for rural communities. The pumps can also be employed for irrigation purposes.

## Solar Powered Refrigerators

Photovoltaic power components have been produced to derive electricity from the sun for generators and deep freezers in which vaccines, drugs, and medical instruments can be safely stored without loosing their potencies.

## Solar Cookers

These cookers of box-type arrangements have been developed to cook local dishes under average sunshine conditions.

## Solar Water Heaters

Solar heaters have been locally produced to provide hot water for both domestic and industrial purposes in hospitals, hotels, private residences, cottage and small-scale industries. These types of heaters are also capable of significant reduction of electricity bills.

## Solar Street Lights and Traffic Controls

Through local efforts, photovoltaic modules have been produced to provide uninterrupted power supply during day and night for traffic control in some city centres. With the use of power storage device they have also powered streets light continuously without the power outages commonly associated with grid system.

## Biogas Digesters

Digesters are usually constructed from sheet metals or empty drums and fed with slurries of animal dung to produce biogas. The gas produced by these digesters has a significant content of methane, which is combustible and suitable for domestic cooking. The gas can also be used in empowering internal combustion engines for electricity generation in rural areas.

Other solar systems, which have also been locally developed, and waiting for commercial production are devices that use wind and micro hydro in the generation of electricity. Although most of these appliances have more than 90 per cent local content of raw materials in their production, they are still limited by unavailability of energy storage devises, and thus their usage is mostly limited to sunny days.

As it is today, the development of cheap and efficient indigenous solar appliances is still at the elementary stage. A lot of work needs to be done to establish standard codes and certification, which are essential for consumer protection and acceptance of these appliances in the competitive energy market. Systematic plan of action that will ensure integration of efficiently proven appliances into our energy mix is also very important. Research efforts by the various Universities, Energy Research Centres and other Organizations should be harmonized and be more responsive to specific economic ventures, especially for rural industrialization. In general, solar powered devices required in rural and semi-urban areas in Nigeria can be categorized as follows:

### Domestic/Community Needs

Cooking, house lightening, refrigeration, water heating, traffic control, television and radio powering, schools, hospitals and clinics.

### Agricultural Production

Water pumping and distribution for irrigation, operation of various agricultural implements, processing, drying and storage of agricultural products.

### Industrial/Commercial Needs

Cottage, small and medium scale Industries, and business establishments (restaurants, bakeries, grocery shops, petrol filling stations etc.).

**Table 14.1: Solar Energy Requirements of some Economic Activities in the Rural Areas**

| *Sl.No.* | *Economic Activity* | *Use* | *Energy Requirements (KW)* |
|---|---|---|---|
| 1 | Cottage and Small Scale Industries | i. Groundnut oil processing | 5–30 |
| | | ii. Garri and cotton wool processing | 3–25 |
| | | iii. Stone crushing | 10–25 |
| | | iv. Fish smoking | 2–20 |
| | | v. Saw milling | 10–30 |
| 2 | Agro–processing | i. Crop drying | 2–5 |
| | | ii. Threshing | 2–5 |
| | | iii. Grinding/milling | 1–2 |
| 3 | Household and other domestic activities | i. Cooking | 0.4 |
| | | ii. Lighting | 0.2 |
| | | iii. Refrigeration | 0.3 |
| | | iv. Ironing | 0.5 |
| | | v. Water pumping | 0.5–1 |
| | | vi. Radio/TV | 0.1–3 |

*Source*: Global Climate Change: Impact on Energy Development (1995).

## 4.0 Renewable Energy Resources and their Prospects in Nigeria

Most of the energy used today in Nigeria is derived mainly from fossil fuels. These are fuels that were created several millions of years ago by the decay of plants and animals buried between layers of earth and rocks. Consumption rate of these fuels is much higher than their creation. In no distant future these sources of energy will be depleted. It is because of this attribute they are considered as non-renewable. Daily usage of these fossil fuels will also lead to environmental pollution and sustainability problems. Therefore, it becomes imperative to look for other sources of energy that will last forever without pollution and degradation of our environment. These resources of energy, which have been discovered, are called renewable.

Nigeria is blessed with abundant fossil fuels as well as renewable resources. Renewable energy is a domestic resource, which has the potential to contribute to or provide complete security of reliable and limitless energy supply. Renewable system use resources that are constantly replaced without polluting the environment. All renewable energy sources have their origin in the activity of the sun. Solar Energy, Wind Energy, Hydropower and Biomass are some of the renewable energy sources waiting to be effectively utilized for sustainable economic growth in Nigeria.

## 4.1 Solar Energy

The geographical location of Nigeria in the tropics coupled with its large expanse of land area of about 923,770 sq km make her receives abundant solar radiation of about 16.7KJ in almost every day. It has been estimated that using 1 per cent of the available land area in the country for 180 sunny days and operating at 5 per cent technical efficiency, an equivalent of about 4.2 x 10 GWh of electricity can be generated.

Solar energy can enhance life of the poor in many ways. The escape from poverty by both rural and urban poor requires such things as good lighting, cheap solar cookers, adequate sanitation and health services, clean water, good educational system and communication network. Electricity from solar energy extends the day by providing extra hours for reading, work and even leisure. Solar cookers can save women and children from daily exposure to noxious cooking fumes from burning wood and other sources. Solar refrigerator allows rural clinics to safely keep medicines and sterilising of basic medical instruments.

For long term and wide range of applications, solar energy can be technically feasible and economically viable as no moving parts are required in its operations. Also the energy can easily be utilized at the point of use without environmental problems. As sunlight is widely available in all the states of the federation, Nigeria can build solar electric infrastructures that are geographically diverse and less vulnerable to international politics and volatile markets of fossil fuels. Furthermore effective utilization of solar technology can directly reduce poverty by raising a nation's productivity and enhancing the quality of goods and services, and thus improving the living standard of the poor.

## 4.2 Wind Energy

Wind energy is a form of solar energy produced by uneven heating of the earth's surface. The sun radiates billions of kilowatt-hours of energy to the earth per hour. About 1 per cent of the energy coming from the sun is converted into wind energy.

In the high wind speed areas of Nigeria, it was estimated that at the height of 30 meters, a 25-meter diameter turbine with 30 per cent efficiency could generate electricity up to 97 megawatt per annum. Nigeria has great potential for harnessing wind energy. However, its development and contribution to the national energy mix is yet to be fully ascertained in terms of economic and technical feasibility. Winds as the primary source of energy costs nothing and can be decentralized without extensive infrastructures as required in the operations of conventional power supply.

### 4.3 Hydropower

Nigeria has great potential for generating hydroelectric power. This is because of the existence of several river basins that have been identified as economically viable for small-scale hydropower sources capable of generating a total of about 1000 megawatts, which could be supplied to rural dwellers and industries. Currently, the installed capacity of hydro power plants at Jebba, Kainji and Shiroro is totaled to about 1900 megawatts. Estimate of future capacity has been put at 8000 megawatts with an annual energy generation potential of 36,000 GWh. This revealed a great prospect for further harnessing of hydropower in Nigeria.

In all accounts, renewable energy resources will be increasingly important parts of the power generation mix in the coming decades. Usage of renewable technologies will not only reduce global carbon emissions, but they would also add the much needed flexibility to the energy mix, and thus decreasing dependence on limited reserve of fossil fuels.

## 5.0 Investment Opportunities in the Manufacturing of Solar Energy Powered Components/Appliances

Global economic and population growth, technological advances, power quality, reliable sources of clean energy, environmental challenges and utility restructuring underscore the need for re-engineering energy generation and delivery system, especially in the developing countries. According to a study conducted recently by the International Energy Agency (IEA), about 3000 GWh of electricity will be required worldwide by the year 2020. This study also revealed that the fastest growing sources of energy would be supplied by renewable out of which solar energy has comparative advantage. With this picture in mind, a market with substantial economic returns is waiting to be tapped by the entrepreneurs and industrialists that will invest in the solar operational activities such as silicon processing, crystalline and thin film production, PV cells/modulus production and manufacturing of electronic devices for interface and other balance of system mechanism.

Spin-off effects of investment in the solar energy components manufacturing industry will also significantly contribute to the business activities of such profession as Architects and Builders, who will utilize skilled laborers such as roofers, electrical and metal workers to integrate photovoltaic into commercial structures industries and residences. The Glass Industry, Plastics and Polymer Industry, Wood Industry, Wire and Cable Makers, Steel, Aluminium and other Metal Industry will also derived significant investment from activities of a pragmatic solar energy industry.

Investment in the nation's intellectual and research resource at the Energy Research Centres, Universities and other Industrial Research Organizations will also guarantee the needed improvement in the existing technologies, and development of better ones for manufacturing of efficient solar energy devices and appliances.

## 6.0 Strategy to Enhance Indigenous Efforts in the Manufacturing of Solar Energy Powered Components

The availability of renewable energy resources in all parts of Nigeria and the existing level of development of solar systems and devices necessesitated the need for adoption of practical measures that will systematically introduce various renewable energy technologies into the national economy. A proactive relationship between academia, industry and government is therefore important as integral approach for sustainable manufacturing of solar energy powered components. However, the following roles are envisaged for these key actors.

### Academia Role

1. The Universities, Research Institutes and Energy Research Centres should be more dedicated in producing applied R&D that can be translated into commercial solar products.
2. The academia should also increase their R&D emphasis on manufacturing.
3. Improvements of indigenously produced solar components and appliances.
4. R&D on new materials and devises with potentials of higher efficiency and low cost production of solar products should be encouraged.
5. Integrate solar energy R&D activities in all the research outfits, academia and the industry.

### Industry Role

1. Production of standardized solar products.
2. Improve throughput of products in manufacturing processes.
3. Create manufacturing partnership among the fabricators of solar energy components and appliances.
4. Enhance capital investment on the manufacturing of solar components.
5. Encourage establishment of manufacturing centres of excellence for solar products.
6. Increase understanding and public awareness for business executives, federal, state and local government policy makers and consumer about solar products.

### Government Role

The government should provide sound policies and institutional initiatives through:

1. Development of regulatory and policy frame work that will encourage the use of solar electricity and solar products in the country.
2. Appropriate funding of basic research on materials for both present and next generation of improved technologies for manufacturing of solar components and appliances.

3. Provision of tax incentives structure that would encourage development of clean energy.
4. Creation of incentives for solar energy products manufacturers.
5. Investment on solar energy infrastructural network, especially in rural areas.
6. Create an enabling environment to attract investments in the renewable energy industry.
7. The government should support higher-risk long-term R&D that can leapfrog manufacturing of efficient and competitive solar energy components and appliances.

## 7.0 Conclusion

Nigeria is blessed with abundant agricultural and mineral raw materials as well as renewable energy resources. However, there is need to put in place an integrated approach for the development and adoption of appropriate technology for utilization of these resources towards meaningful contribution to the national economy. Even though tremendous progress has been made by the Raw Materials Research and Development Council, Universities, Research Outfits and Energy Research Centres in developing solar energy powered components with higher percentage of local raw materials, a lot of effort is still required to galvanize active participation of private sector for commercialization and subsequent mass production of these devices. Affordable and accessible solar energy components have potentials of eradicating/ reducing poverty, especially in the rural areas.

In general, sustainable manufacturing of solar energy powered components entails embarking on proactive measures that will provide enabling environment for sufficient investment on research and development, technology, manufacturing, training, marketing, and public awareness program on solar technologies and applications for cleaner production and protection of our environment.

## Acknowledgements

I wish to express my profound appreciation to the Centre for Science and Technology of the Non-aligned and other Developing Countries (NAM and ST Centre) and Society of Engineers and Managers (SEEM), India for accepting me to participate in the International Workshop on Cleaner Production and Energy Conservation for Sustainability, Cochin, India. I would also like to acknowledge and appreciate the efforts of Prof. Arun P. Kulshretha, Mr. C. Jayaraman, M. Bandyopadyay, Mr. A.K.Ramadas, and Mr. R. Harikumar towards hosting of a successful Workshop.

## References

1. Achieving Sustainable Development in Nigeria; National Report for UNCED, Rio de Janeiro, Brazil, 1 – 12 June, 1992.
2. Federal Environmental Protection Agency, 2000. National Policy on Environment, Nigeria.

3. Nicolas Lensen, 1993. "Providing Energy in Developing Countries." In the state of the world. A worldwatch Institute Report on Progress towards a sustainable society. Ed Linda Starke.

4. Onwualu, P.A., 2007. "Promoting local content in Manufacturing of Solar Energy Technology Components" National Solar Energy Forum (NASEF'2007) and Awareness Programme on Solar Energy, Abuja, Nigeria.

5. Pauchauri, R.K., 1995. "Energy, Environment and Sustainable Development" Global Climate Change: Impact on Energy Development, pp 317-331. Jos, Plateau State, Nigeria.

6. Report of the Techno – Economic Survey/Industrial Study Review Committee, 2008. Raw Materials Research and Development Council (RMRDC), Abuja.

7. Renewables, 2007. Global Status Report: Renewable Energy Policy Network for the 21st Century: www.ren21.net.

8. Sambo, A. S., 1991. "Renewable Energy Resources in Nigeria" proceedings of the conference on Energy issues in Nigeria": Today and Tomorrow held on 22 – 24 May, published by Gilspar Ltd, Lagos, Nigeria.

9. Wikipedia, Sustainable Industries: http:// en.wikipedia.org/wik/ sustainable_industries (accessed 2008).

# Part IV

# *Energy Efficiency*

# Chapter 15

# *Global Overview of Energy Efficiency in Buildings*

***Binu Parthan[1] and Udo Bachhiesl[2]***
*[1]REEEP, VIC D1732, Wagramerstrasse 5, Vienna, A-1400, Austria*
*E-mail: bp@reeep.org*
*[2]IEE, Graz University of Technology, Inffeldgasse, 18, Graz, A-8010 – Austria*
*E-mail: bachhiesl@TUGraz.at*

## ABSTRACT

The paper attempts to highlight the importance of energy efficiency in buildings and lists the key barriers to development of this opportunity. It also lists the key approaches to promote the development of this market and lists some of the key initiatives in this space. Thereafter conclusions are drawn based on the overview provided.

***Keywords:*** *Awareness, Buildings, Barriers, Energy efficiency and Energy pricing.*

## 1.0 Energy Efficiency in Buildings

The construction and buildings sector is a significant economic sector globally. On a global basis the buildings and construction sector typically generates 5 to 15 per cent of the GDP (UNEP, 2007).

Buildings need energy for space heating and cooling, appliances, lighting, building services equipment, water heating, cooking etc. Buildings and the appliances and equipment used in the buildings are responsible for almost 35 per cent of the energy use globally (REEEP, 2008). The absolute figures of energy consumption in buildings are rising fast as construction booms especially in emerging countries such as China and India. There is a wide variation in terms of the per capita energy use in buildings globally. The per capita use is high and in the range of 35-40 GJ per capita/year in North America and in the range of 20-35 GJ per capita/year in Europe. The per capita energy use in buildings is considerably low in developing countries and is in the range of 4 to 10 GJ per capita/year (REEEP, 2008).

Energy Efficiency in buildings has also been receiving considerable attention from a policy perspective also. The IPCC fourth Assessment report recognizes that it is often more cost effective to invest in energy efficiency improvements than in increasing energy supply to satisfy the demand for energy services. The IPCC Fourth Assessment Report also shows (Figure 15.1) that energy efficiency in buildings offer the maximum potential amongst all the other sectors and is in the range of 4.7 – 6.7 $GtCO_2eq$/year (IPCC, 2007).

The McKinsey Global Institute study of global GHG abatement opportunities also show that several mitigation opportunities in the buildings sector such as insulation, air-conditioning and lighting constitute some of the most attractive options for emission reductions globally. These key studies as well as several political developments around energy and climate change issues have raised the political profile of the issues surrounding energy efficiency in buildings. The continued rise of crude oil prices in during 2007 and 2008 has ensured that energy efficiency and energy efficiency in buildings has continued to remain as an issue in the public domain.

## 2.0 Barriers to Energy Efficiency in Buildings

While the potential for savings in the buildings sector appear very high the achievements in building energy efficiency have been rather modest. There are several barriers, which prevent higher levels of energy efficiency in buildings, some of which are:

### Investment Barriers

More energy efficient buildings and appliances imply higher upfront investment costs and the general objective of the construction companies and the homeowners is to reduce the initial investments in a house and the appliances. The effect of the higher investment costs is more pronounced in the residential buildings than commercial buildings.

### Energy Pricing

Energy pricing and especially electricity pricing decisions across the world do not often reflect the true cost of the electricity service. In many countries, especially developing countries and in economies in transition, energy supply to households and agricultural applications are highly subsidised. These low tariffs encourage inefficient use of energy and significantly reduce the incentives for energy efficiency in residential buildings.

### Financial Risk

Banking and financial institutions consider the financial model for energy efficiency projects as risky. This is because they are based on energy savings and

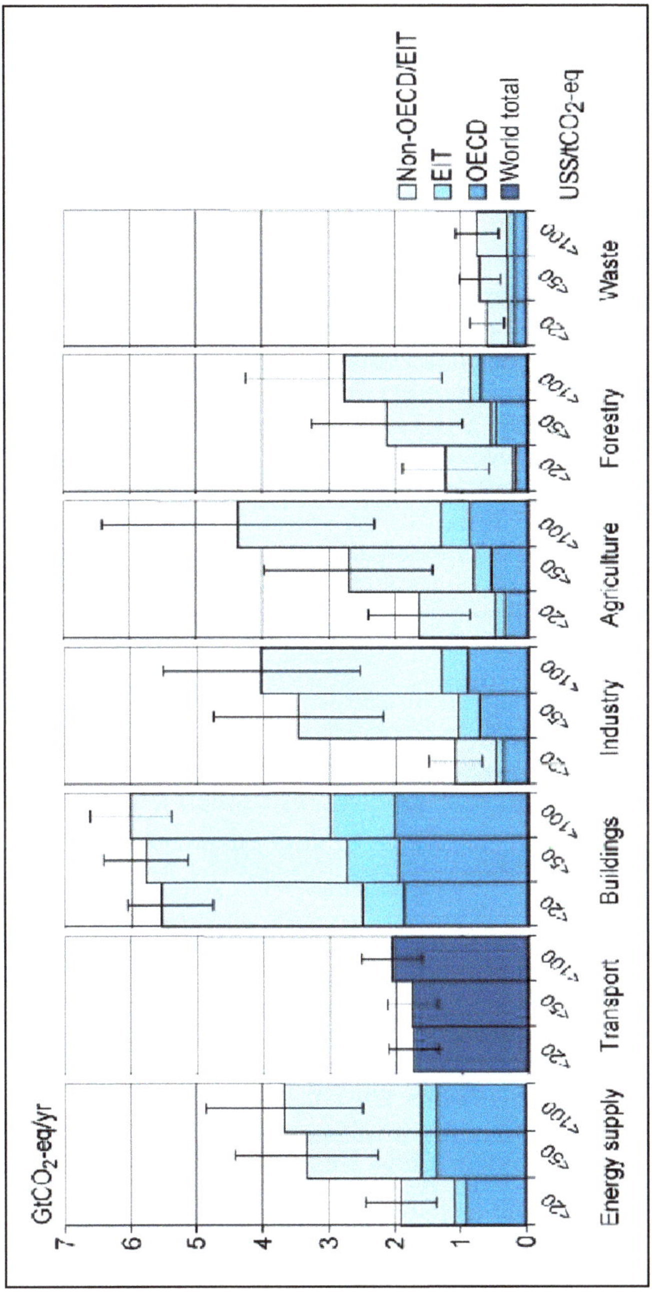


**Figure 15.1: Economic Mitigation Potentials by Sector in 2030**
**(*Source:* Intergovernmental Panel on Climate Change, 2007)**

without an additional revenue stream. The business model that is based on a shared savings contract[1] is also considered to be riskier than normal business models. Uncertainties relating to the baseline, incremental cost of energy efficiency investments, protocols for monitoring and verification etc. increase the risk perception of the financial sector. These high risk perceptions result either in the building energy efficiency projects not being financed or being financed at a high risk premium. When building energy efficiency projects are financed at a high risk premium, it increases the cost of funds thereby adversely affecting the financial profitability of the projects.

## Structural Issues

There is a structural misalignment in the investment and returns of energy efficiency investments in that the ones who make the decision about energy efficient building envelope, energy consuming equipment and devices are not the ones who pay the energy bills. This misalignment means that mostly only price considerations go into building construction and purchase of appliances. This is particularly a major barrier in building energy efficiency where the investments in the building, insulation, equipment and electrical fixtures are made by the builder or the building owner and the energy bills are paid by the tenant. As Figure 15.2 illustrates, more than 80 per cent of energy consumption of buildings occurs during the operational phase of the buildings.

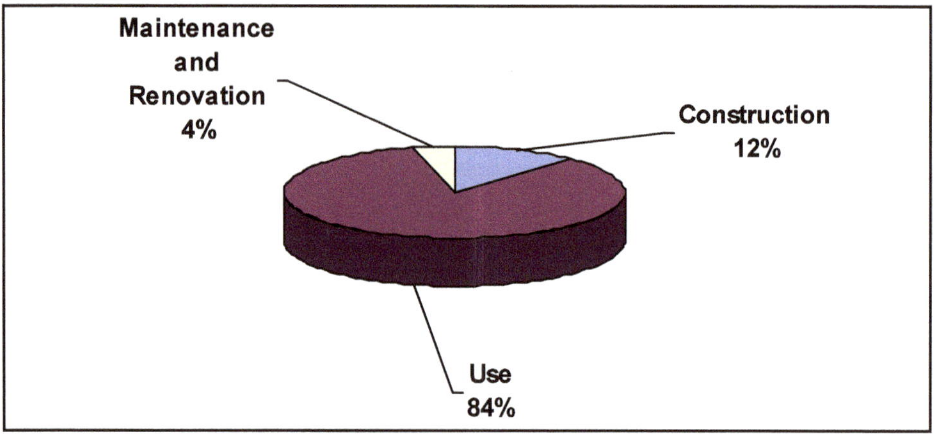


**Figure 15.2: Life Cycle Energy Use in Buildings**
**(*Source*: WBCSD, 2007)**

## Absence of Policy Framework and Regulatory Instruments

In most parts of the world, there is an absence of a policy framework to promote energy efficiency in buildings and regulatory instruments to enforce the policy. It is possible to regulate the thermal and electrical energy requirements of building

1 Normal contracts used by Energy Service Companies (ESCOs).

components, service systems and equipment. Absence of the policy framework results in increasing number of buildings being constructed and appliances purchased which then use energy at lower efficiencies for a longer period of time.

### Awareness

There is limited awareness of the opportunities available for energy efficiency in the building sector and in appliances. Many households and commercial enterprises do not consider expenses on energy important and do not keep track of energy use and related expenditure. When appliances used in buildings are purchased emphasis is placed on the features offered by the product and its purchase price. Operating costs of appliances is given limited consideration during purchase decisions. Apart from end users, awareness about energy efficiency products, projects and their economics is limited in the finance and banking sector and energy efficient products and energy efficiency projects are considered to be riskier. Similarly policy makers are also unfamiliar with characteristics of energy efficiency business and policies and regulatory instruments are often designed with a limited appreciation of issues surrounding energy efficiency in general and specifically in buildings.

## 3.0 Promotion of Building Energy Efficiency

The buildings and appliances sector is a complex system involving a large number of stakeholders, longer lifecycles and a range of user preferences. A number of approaches and programmes involving policy, regulation, financing, institutional models and technologies is required to promote energy efficiency in the building sector. Some of the most effective approaches are listed below:

### Building Codes and Performance Standards

Building codes and standards regulate the physical, thermal and electrical requirements of building components, service systems and equipment. Increasingly building codes include energy performance standards, which specifically limit the amount of energy the buildings can consume. One of the major challenges with building codes is that they can easily be applied to new buildings but not to existing building stock. The EU Energy Performance in Buildings Directive has resulted in significant savings already and the scope of this directive will be expanded substantially in 2009. Canada has a model national energy code for buildings and a voluntary standard (R 2000), which requires about 40 per cent higher performance than the standard.

### Appliance Standards and Labelling

Energy-efficiency labels and standards for appliances, equipment, and lighting offer an opportunity to improve energy efficiency in residential and commercial buildings. Energy-efficiency standards are specifications and regulations that prescribe the technical specifications or energy performance of manufactured products. Energy-efficiency standards can transform markets by removing inefficient products and encouraging efficient energy services.

Energy-efficiency labels are informative labels affixed to manufactured products to describe the product's energy performance (usually in the form of energy consumption, efficiency or energy cost). These labels give prospective consumers the data necessary of their future energy performance. Energy labels empower consumers to make informed choices about the products they buy on the basis of their future energy performance and to manage their energy bills. In India the Bureau of Energy Efficiency (BEE) launched the Energy Labels programme in 2006. The Indian labelling scheme is voluntary and currently includes the following appliances: refrigerators, tubular lights, air-conditioners, distribution transformers and motors. Schemes for fans, gas burners and standby power are also under preparation.

### Building Rating Systems

Building rating systems are based on building performance standards and can either be mandatory government programmes or voluntary programmes. The mandatory programmes are prevalent in Japan, Taiwan and Singapore and one of the largest voluntary rating programmes is the Leadership in Energy and Environmental Design (LEED) programme by the World Green Building Council. LEED has a rating system for buildings based on the incorporation of building techniques and design. Based on the rating the buildings are classified as silver, gold or platinum.

Both the rating systems provide an indication to the prospective buyers of a property of the buildings future energy performance and effect of energy costs. The voluntary schemes like LEED are popular with commercial and corporate buildings in US, Canada and India.

### Kyoto Protocol Mechanisms

The market based mechanisms of the Kyoto Protocol, which allows for emissions trading between governments that are signatories to the Kyoto Protocol are the Clean Development Mechanism (CDM) and Joint Implementation (JI). CDM and JI allow the energy efficiency gains from building energy efficiency projects and programme to be quantified as avoided $CO_2$ emissions which can then be traded. Since 2007 with the advent of the programmatic CDM (pCDM), it is possible to combine several housing units into a single CDM Programme of Activity to gain scale advantages. 9 projects are currently[2] in the CDM pipeline with a potential to generate over 1.3 million CERs by 2012. The revenues from the trade in emission reductions provide added incentives to the building energy efficiency projects and programmes.

## Awareness and Capacity Building

Information and education campaigns are of considerable importance in improving building energy efficiency. The information dissemination programmes and campaigns need to be carried out in the local languages and with specific target groups such as builders, architects, homeowners, tenants etc. using the appropriate industry/trade media. Apart from the information dissemination programmes, there is a need for capacity building of financiers and policy makers in the features, characteristics, opportunities and economics of energy efficiency opportunities in buildings.

---

2 In early June 2008.

## 4.0 Initiatives in Building Energy Efficiency

While there are several institutions such as the UN, IEA and World Bank and the industry, which do have projects or publications that target energy efficiency in buildings, the following organisations have energy efficiency in buildings as a core area of operation.

### SBCI

Sustainable Buildings and Construction Initiative (SBCI)[3], was launched by United Nations Environment Programme (UNEP) in 2006 and brings together a number of companies and initiatives globally from construction, metals, consulting and rating organisations. The objective of SBCI is worldwide adoption of sustainable buildings and construction practices and to provide a common platform on related issues. The initiative will focus on energy efficiency and greenhouse gas emissions and will establish baselines, develop tools and strategies and implement pilot projects.

### CLASP

The Collaborative Labeling and Appliance Standards Programme (CLASP)[4] was a partnership launched in 2002 at the Johannesburg Renewable Energy Summit and focuses on energy efficiency standards and labels for appliances, equipment and lighting. CLASP originated from an initiative started in Lawrence Berkeley National Labs (LBNL) in 1996 and a subsequent project in 2000 by United Nations on standards and labeling. CLASP is registered as a non-profit corporation and headquarter is located in Washington DC, US. CLASP works with governments, industry, inter-governmental organizations, NGOs and technical support groups. CLASP received project-funding support from UNDP/GEF, USAID, USEPA, Renewable Energy and Energy Efficiency Partnership (REEEP), APEC etc.

### REEEP

Renewable Energy and Energy Efficiency Partnership (REEEP)[5] was launched by the Government of United Kingdom in 2002 at the World Summit on Sustainable Development (WSSD) in Johannesburg. REEEP currently consists of over 245 partners including 38 governments. REEEP supports renewable energy and energy efficiency projects in its priority countries of Brazil, China, India, Mexico and South Africa as well as African and Pacific LDCs. The key REEEP donors are Australia, Austria, New Zealand, Norway, Netherlands, Ireland, Italy, Spain, UK and US. REEEP has so far supported 84 projects with over 6 million which has been able to leverage over 23 million in co-financing. Of these about 2.7 million of REEEP resources have supported energy efficiency projects. REEEP projects are also implemented in 50 countries.

---

3 For more details refer to www.unepsbci.org.

4 For more details visit www.clasponline.org.

5 Refer to www.reeep.org for information on REEEP activities.

REEEP in late 2007 launched the Energy Efficiency Coalition (EEC) consisting of a number of global stakeholders from industry, government, financiers and NGOs active in energy efficiency. The initial focus of this coalition will be building energy efficiency. REEEP was also mandated by the G8 in its Gleneagles Plan of Action to target energy efficiency improvements in buildings. REEEP also accords priority funding to projects targeting energy efficiency in buildings and is currently supporting such projects in China, Poland, India and Eastern Europe.

## 5.0 Conclusions

The conclusions that can be drawn from the brief overview of this rather complex sector are:

- There is significant potential for energy efficiency gains in the building sector globally. There is an increasing political support for realizing this potential.
- There are several barriers, which prevent the realization of potential in energy efficiency in buildings. The key barriers include investments, pricing, risk perceptions, structural issues, policy, regulation, capacity and awareness.
- Some of the effective approaches to address these barriers and develop this market segment are building codes and performance standards, standards and labeling of appliances, building rating systems, financial gains through CDM/JI and building capacity and creating awareness.
- Initiative such as SBCI, CLASP and REEEP all target development of energy efficiency in buildings as a core area of activity.
- However a great deal needs to be done and will involve significant efforts on parts of the key stakeholders – homeowners, policy makers, financiers and other development and facilitating agencies–to make significant achievements in this important energy efficiency sector.

## References

1. Bachhiesl, Udo, 2004. Measures and Barriers towards a Sustainable Energy System, 19th World Energy Congress.
2. BEE, 2006. Energy Efficiency Labels: Details of Scheme for Energy Efficiency Labeling.
3. Commission of the European Communities, 2006. Action Plan for Energy Efficiency: Realizing the Potential.
4. De T'serclaes, Philippine, 2007. Financing Energy Efficient Homes: Existing policy responses to financial barriers, IEA..
5. European Commission, 2005. Doing More with Less: Green paper on energy efficiency.
6. IPCC, 2007. Summary for Policymakers. In: Climate Change 2007: Mitigation. Contribution of Working Group III to the Fourth Assessment Report of the

Intergovernmental Panel on Climate Change [B. Metz, O.R. Davidson, P.R. Bosch, R. Dave, L.A. Meyer (eds)], Cambridge University Press, Cambridge, United Kingdom and New York, NY, USA.

7. Mathur, Ajay, 2007. Promoting Energy Efficiency through Regulatory Framework and Financing Options-Experience from India, BEE;
8. Parthan and Bachhiesl, 2007. Barriers to Energy Efficiency under the Clean Development Mechanism, Proceedings of the Global Forum for Sustainable Energy-7, Vienna, Austria.
9. REEEP, 2007. Global Status Report on Energy Efficiency 2008.
10. United Nations Environment Programme, 2007. Buildings and Climate Change: Status, Challenges and Opportunities.
11. United Nations Environment Programme, 2007. Assessment of Policy Instruments for Reducing Greenhouse Gas Emissions from Buildings.
12. United Nations Foundation, 2007. Realizing the potential of energy efficiency: targets, Policies and Measures for G8 countries.
13. Vattenfall AB, 2007. Global Mapping of Greenhouse Gas Abatement Opportunities;
14. World Business Council for Sustainable Development, 2007. Energy Efficiency in Buildings: Business realities and opportunities.

# Chapter 16

# *Technology of Energy Efficient Motors: Some Latest Developments*

**S. Mahadevan**
*Consultant, Energy and Energy Efficiency,*
*Tiruvanmyur, Chennai, India*
*E-mail: smdevan_2000@yahoo.com*

## ABSTRACT

With the rapid rise and advancement of the production and usage of Rotors and Motors technologies in Europe and USA, India is not far behind. India is keeping up pace with world around in terms of standards and technology too. Efforts are also being put on to cover Agricultural and Submersible Rating motors. The technology used is easily applicable for small medium ratings of Motors. This paper discusses the advancements and technological improvements of Rotors and Motors that convey the potential in store for energy Efficiency, Energy Conservation and cost effectiveness. The author examines the concept of usage of various types of Rotors that has proved to be the beginning of a great revolution towards technological efficiency. Applications of Die-casting technology for the Rotors have also been discussed for efficiency and reliability improvements.

***Keywords**: India, Energy efficiency, Energy conservation, Energy efficient motors and rotors and Die-cast technology.*

## 1.0 Introduction

In the context of Application and usage at the present times, the very term Motors refers to AC Squirrel Cage Induction Motors, which comprises of more than 90 per cent of the total usage. With the advent of cost effective and versatile AC Motors Drives and controls, this percentage is actually on the increase. The very concept of 'Squirrel Cage' construction of the Rotor is an interesting big leap forward, in the technology, as it made the 'Motors' simple, usable and affordable.

This in fact made the Electrical Prime Movers so popular that helped speeding up the Industrial revolution in the last century. At present Motors consume almost 75 per cent of the total Electrical Energy, which means literally that one of the prime reasons for increased demand for Electricity is increased need for usage of Motors.

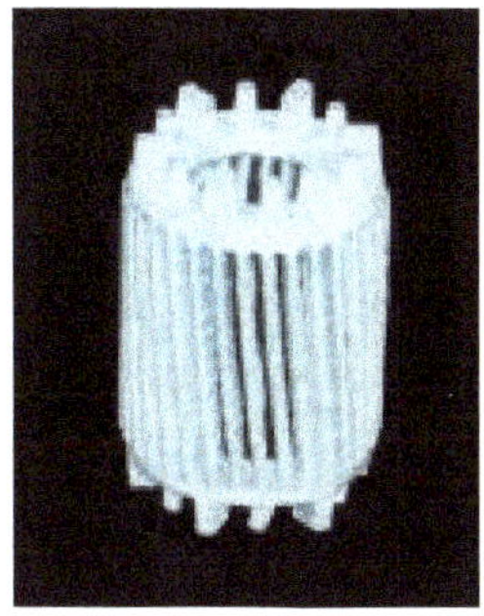

**Figure 16.1: The Basic Constructional Aspect of the Rotor**

Apart from concern for Reliability, very rightly, there is equal concern today for Efficiency as it addresses the issue of Energy conservation. Conserving Energy through cutting down usage was alright so long as we were an 'under developed' country, but today with advancements, affordability and increased usage, addressing 'Efficiency' is very important to ensure best usage of the Energy with least consumption.

## Energy Efficient Motors

Efficiency Standards of Motors have been reviewed and revised upwards from time to time and the First BIS Standard for Energy Efficient Motors (IS 12615) was brought out in 1989 around the time when Energy Efficient Motors were catching great attention in Europe and USA. With the 'Energy Conservation Act' of 2001 by the Govt. of India and the formation of Bureau of Energy Efficiency, there have been positive moves with regard to Energy Efficient Motors. BIS Standard for Energy Efficient Motors has been revised now matching with the Energy Efficient Motors Standards of Europe and USA.

Basically the efficiency improvements are achieved through reduction of various losses and the Fig 1below will show the typical apportionment of different categories of losses in Induction Motors.

From Figure 16.2, we can see the dominant roles played by both Core or Iron Losses and the Stator and Rotor losses or Copper Losses. Stray load losses are another area of concern today, which is dealt later along with Rotor Losses in the context of 'Recent Technological Advancements'.

**Is this too Fundamental?**

$$\text{Efficiency} = \frac{\text{Output Power}}{\text{Input Power}} \times 100\% = \frac{\text{Input Power - Losses}}{\text{Input Power}} \times 100\%$$

## 2.0 Some Recent Technology Developments and the Scope for the Future

As detailed in the Introduction, invention of 'Squirrel Cage' Rotor is the beginning of a great revolution, which is continuing with 'Leaps' forward towards betterments. Application of Die-casting for the Rotors and then the metallurgy of the Rotor

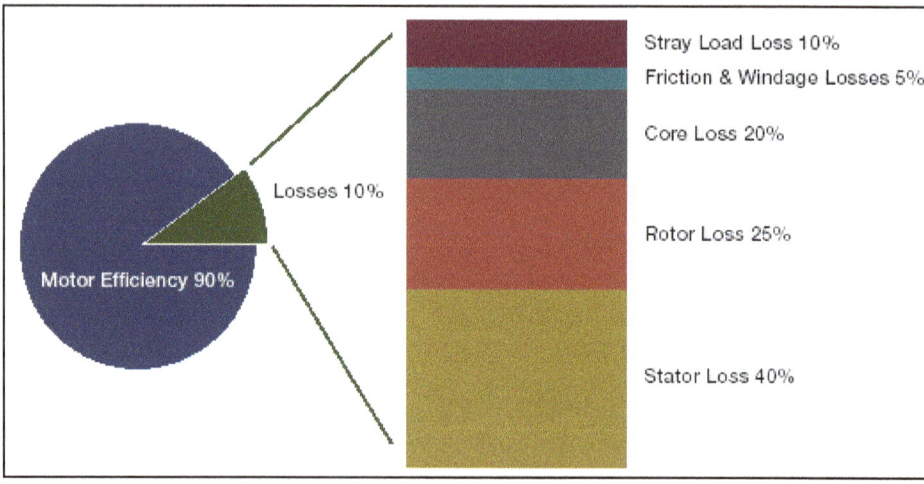


**Figure 16.2: Typical Motor Losses**

conductors formed main areas of attention resulting in Efficiency and reliability improvements.

## Introduction of Die-cast Technology with Aluminum as the Rotor Conductors

Almost after 60 years after invention of Squirrel Cage Motors, the Innovation appeared in the form of Aluminum Die-cast Rotors. This eliminated most of the Rotor Problems resulting in Motors with much higher reliability.

## Experiments with Die-cast Technology with Copper as the Rotor Conductors

Towards the end of last Century in 1998 or so the idea of trying out Die-casting of Rotors with Copper commenced, Problems addressed and the Research and Development started in many places.

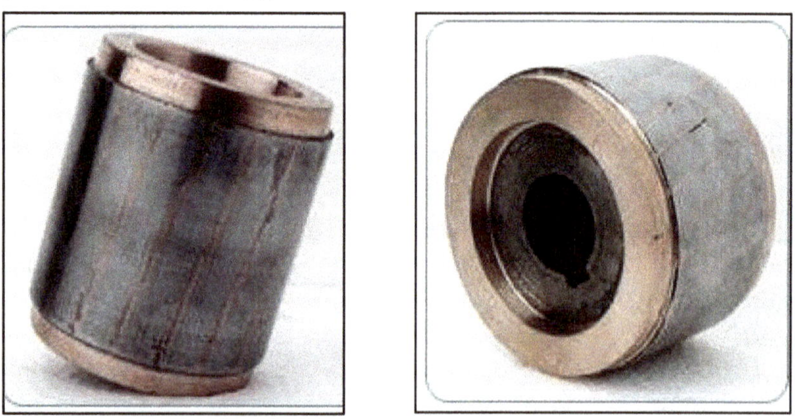

**Figure 16.3: Some of the DCRs Manufactured in India**

Given below are the details of evolution of Efficiency Values for a sample rating of Motor to illustrate how improvements are achieved and the role DCR or Die-cast Copper Rotors can play in taking the Efficiency levels forward by another 3 to 5 per cent.

Copper being a Better conductor and most preferred for carrying Electricity, in course of time, with cost effective and streamlined Production, DCRs can replace all the Die-cast Al. Applications as there is even scope for overall economy.

Lot of experiments have been carried oust in India as well in all varieties of Motors including Industrial Motors, Agricultural Motors including Submersibles, Three Phase and Single Phase Motors etc. Comparative Alpha and Beta Tests have been performed between DAR and DCR as well as between Fabricated bar Copper versus Die-cast Copper Rotor Motors. The results have been very encouraging in all cases confirming the need for change over.

**Results of Experiments and Tests Conducted and Scope for the Future**

Given below are comparative tests information in brief based on some of the Indian Experiments. Experiments have actually been carried out in many varieties and large samples.

## 3.0 A Case of Comparative Test Results with Die-cast Al. vs Die-cast Cu Rotors

MOTOR RATING: 5 HP/3.7kW, 3ϕ, 415 V

NO LOAD:

Current increases by 2.29 per cent

Input power decreases by 16.66 per cent

FULL LOAD:

Efficiency increases by 2.96 per cent

Current decreases by 1.31 per cent

Slip decreases by 2.65 per cent

Power factor decreases by 4.7 per cent

TEMPERATURE RISE:

Decreases by 7.15°C

## 4.0 Conclusion

In Europe, USA and in India, this has been commercialised with regular Production of Rotors and Motors. In India the sizes covered at present are up to 15 hp Industrial TEFC and IEC Frame Motors. Efforts are on to cover Agricultural and Submersible Ratings. It is interesting to note that this technology is easily applicable for small and medium ratings of Motors where the Efficiencies are Low and Good improvements can be achieved. Proposals are also underway for trying out this Technology in fans and other mass Induction Motors applications for improving Efficiencies.

It can be concluded that Die-Cast Copper Rotors for Squirrel Cage Induction Motors is another leap forward to make these Machines more Energy Efficient, as the present Efficiencies and Premium Efficiencies Levels can be surpassed with Die-Cast Copper Rotors and Dedicated Designs.

AC Squirrel Cage Induction Motors will continue to play a very important role in all spheres of activities of using Electricity, and the advancements and technological improvements discussed briefly in this presentation convey the potential in store for Energy Efficiency and Energy Conservation clubbed with the potential of cost savings.

# Part V
# *Case Studies*

# Chapter 17

# *Energy Conservation Measures at Visakh Refinery*

***B.J. Baruah*[1], *G. Bhagavan*[2] *and P.S. Rao*[3]**
*Technical Services Department, HPCL Visakh Refinery, P.O. Malkapuram*
*Visakhapatnam – 530 011, Andhra Pradesh, India*
*E-mail:* [1]*bjbaruah@hpcl.co.in;* [2]*gbhagavan@hpcl.co.in;* [3]*Psrao@hpcl.co.in*

## ABSTRACT

Energy cost constitute the major part of manufacturing cost of an oil refinery and the annual energy cost of a medium sized oil refinery runs into Billions of Rupees. Various energy conservations measures are implemented in Visakh Refinery of Hindustan Petroleum Corporation Limited, India, and fuel worth hundreds of millions of Rupees are saved annually. The energy performance of the Refinery, in terms of fuel and loss and specific energy consumption has improved significantly over the years due to energy conservation measures. The initiatives include close monitoring, periodic surveys, audits, development of new energy saving schemes and implementation of ideas related to energy saving.

This paper addresses methodology adopted for monitoring energy performance in the Refinery, details of improvements made in energy front and future plans. The paper is meant to share the experience of Visakh Refinery with the industry on the benefits derived from various energy conservation measures taken in the Refinery.

***Keywords:*** *Fuel and loss, Online cleaning, Specific fuel consumption, Steam and air leak survey and Theoretical fuel.*

## 1.0 Introduction

Visakh Refinery of Hindustan Petroleum Corporation Limited was established in 1957 with an installed crude processing capacity of 0.65 Million Metric Ton per Annum (MMTPA) and was the first Refinery on east coast of India. The current installed crude processing capacity of the Refinery is 7.5 MMTPA. In addition to the three number of crude and vacuum distillation units, the Refinery has two fluidized catalytic cracking (FCC) units for cracking vacuum gas oil to liquefied petroleum gas and cracked run naphtha (Gasoline). The Refinery also has one Visbreaker unit for reducing viscosity of Vacuum residue, hydro desulphurization unit for treating diesel, naphtha fired cogeneration power plant with a capacity of 68 MWH and other facilities such as product and effluent treatment plants, bitumen blowing, propylene recovery, utilities etc. The present capacity utilization of the Refinery is at around 125 per cent. The Refinery produces wide range of fuel products, from liquefied petroleum gas to aviation turbine fuel for aircraft and bitumen for road laying.

Fuel is used in the Refinery in various processes such as hydrocarbon heating and vaporization in furnaces, steam and power generation and FCC regenerator. The main fuels used are low sulphur heavy stock (LSHS), naphtha, fuel gas, coke (in FCC units). The fuels used are internally generated and constitute around 5 wt per cent of the total crude thruput. Around 70 per cent of the total steam requirement of the Refinery is met by waste heat generation. The annual energy cost of the Refinery was ~~is~~ around Rs. 11.11 billion during 2007-08. As Refinery is an energy intensive industry, utmost importance given in improving energy performance of the Refinery.

This paper addresses various issues such as details of energy monitoring in the Refinery, details of improvements made in energy front and future plans.

## 2.0 Monitoring Energy Performance

In Visakh Refinery, Energy Conservation is a continuous thrust area at the top most level.

The month wise target specific energy consumption for the Refinery is set at the end of the previous year. The actual performance is evaluated for each month and compared with the set target. Theoretical fuel consumption and loss for the Refinery is calculated for every fifteen days period and also for every month and compared with actual fuel consumptions. (Theoretical fuel consumption is calculated basis furnace duty, efficiency, service factors of units, coke yield, fuel gas generation, power inputs etc.). The energy performance of the Refinery for every quarter is also compared with target performance and past best performance and detailed analysis is done for deviations.

The Refinery has an Energy Conservation Cell, which is part of Technical Services Department of the Refinery. There is a system of day to day monitoring and report preparation covering furnace/boiler efficiency monitoring, steam balance, monitoring of service factors of equipment related to energy conservation such as air preheaters, fuel consumption details, specific fuel consumption in captive power plant and all other parameters which have a direct bearing on fuel and energy consumption. Preheat monitoring and optimization of steam loads is also done. The abnormalities in the plant on the day-to-day basis, which may cause the energy loss, are also monitored. The abnormalities are evaluated to identify opportunities for improvement.

All possible sources of energy saving in the refinery are also identified and schemes are developed to reduce energy consumption. Periodic surveys in the field on steam leaks, compressed air leaks, identification of passing safety valves connected to flare etc. are conducted as per a predetermined schedule. Energy conservation studies/ audits are also conducted periodically in the Refinery.

As active participation of workman is very important for energy conservation, periodic energy conservation training is being organized for the Refinery technicians with external faculty. The Refinery has also introduced the concept of energy champions, where one officer from each plant is designated as energy champion and takes up issues related with energy conservation and follows on energy conservation activities in the plant

## 3.0 Energy Conservation Measures

Various energy conservation measures are taken up in the Refinery to improve energy performance. Some of the measures taken are periodic survey of steam leak and arresting, periodic compressed air leak survey and arresting, online cleaning of furnaces thereby increasing efficiency, periodic cleaning of exchangers, reducing operating pressure in Crude Distillation Columns, optimization of steam usages, reduction in operation of steam pressure reducing stations, reducing specific fuel consumption in captive power plant, increasing service factor of CO boilers, installation of traps for copper tubing etc. The details of some of the measures taken are given below:

### Periodic Steam Loss Survey

Periodic surveys are conducted at an interval of four months with external party to identify and quantify steam loss from various sources inclusive of copper tubing, direct leaks, passing/ malfunctioning steam traps and any other sources. Ultrasonic leak detector and visual methods are used for identification and quantification of leaks. The identified losses are arrested (both in-house and through external agencies).

The total steam loss of the Refinery has been brought down drastically over a period of last four years (Figure 17.1). The cumulative monetary gain over past during the period of 4 years due to reduction in steam loss is estimated at Rs 162 Million. The target steam loss is being set at 0.5 MT/Hr for the entire Refinery.

In Refinery, flow transmitters, small diameter pipelines, sample points etc., which are on heavy oil applications needs a heat tracing to prevent congealing of the fluid. In the original design, copper tracing line is provided as a heat tracer on the lines and the exhaust of the copper tubing are kept open to atmosphere, resulting in steam loss. Hundreds of ¼" new generation of steam traps are being provided for arresting steam leak through copper tubing (Figure 17.2). The reduction in steam loss achieved due to provision of ¼" traps on copper tubing is estimated at around 2.5MT/Hr.

### Compressed Air Leak Survey

Compressed air is used in the Refinery as instrument air and plant air. Plant air is required for cleaning, blowing, operating pneumatic tools etc. Instrument air is required for operating control valves. The compressed air system consists of air compressors, instrument air receivers and instrument dryers. Compressed air is an expensive utility. At Visakh Refinery, the cost of one $NM_3$/Hr of compressed air is

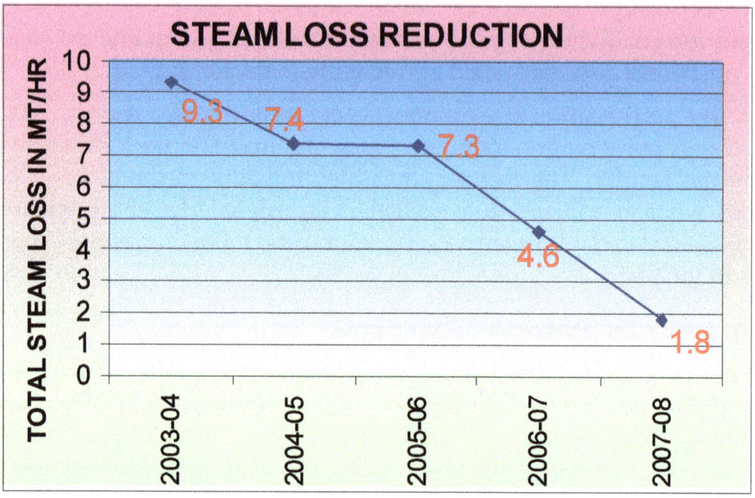


**Figure 17.1: Steam Loss Reduction**

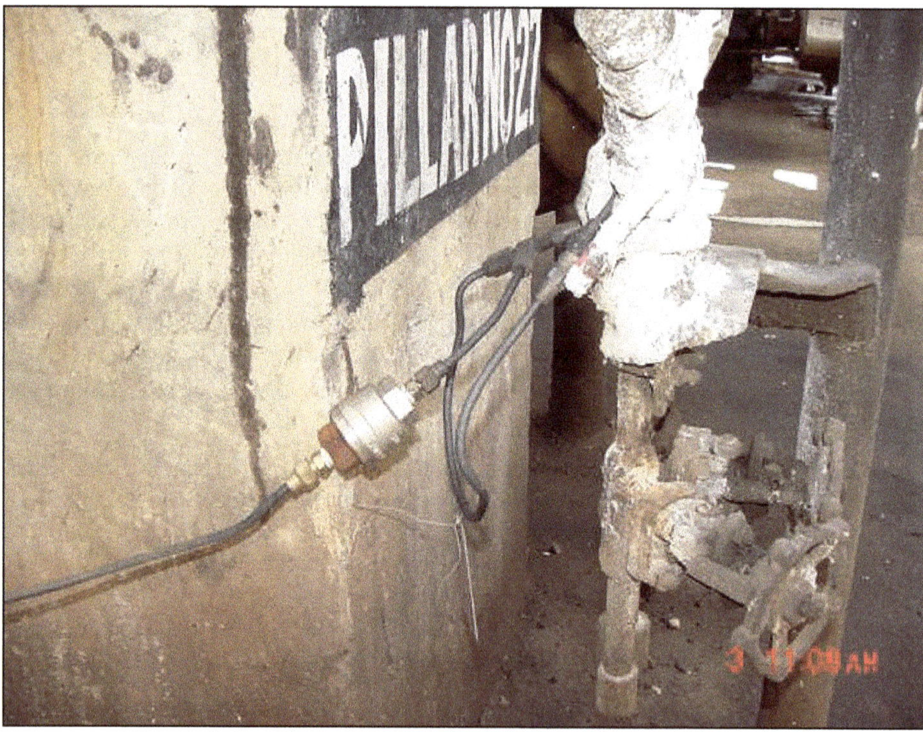


**Figure 17.2: Steam Trap on ¼" Copper Tubing**

estimated at Rs 0.94/-(2007-08), which means, loss due to per NM3/Hr of compressed air leak over a period of one year is Rs 7520/-(Considering 8000 operating hours). Periodic compressed air leak surveys are being carried out at an interval of every six months through external agencies to identify leaks. The leaks are identified using ultrasonic leak detector and are arrested in house. Around 93 per cent reduction in compressed air leak is achieved over a period of last three years and the cumulative saving due to arresting of compressed air leak during the period is estimated at around Rs.54 Million.

## Online Cleaning of Furnaces

The fuels fired in the Refinery heaters are LSHS and fuel gas. These fuels produce metallic, organic and inorganic foulants inside the furnaces, which gets coated on the furnace tubes (Figure 17.3). This reduces the heat transfer across the tubes, necessitating an increase in fuel firing to get the same absorbed heat duty and capacity. Increase in heater arch temperatures, fire box temperatures follows and affects the heater duty and capacity. Normally the heaters tubes are manually cleaned during shutdown. At Visakh Refinery Online Chemical cleaning of heaters, a proprietary technology available with very few experts in the world, is being used to clean the tubes, periodically, without shutting down the heater or affecting unit operation. These results in cleaner tubes (Figure 17.4), substantial reduction in heater arch temperatures (by 80-120 deg.c.), stack temperature (10-15 deg.c.) and fire box temperatures. Unit throughput could be maximized and fuel firing could be brought down significantly. The present schedule of cleaning for Atmospheric Furnace is three times per year and vacuum furnace twice per year.

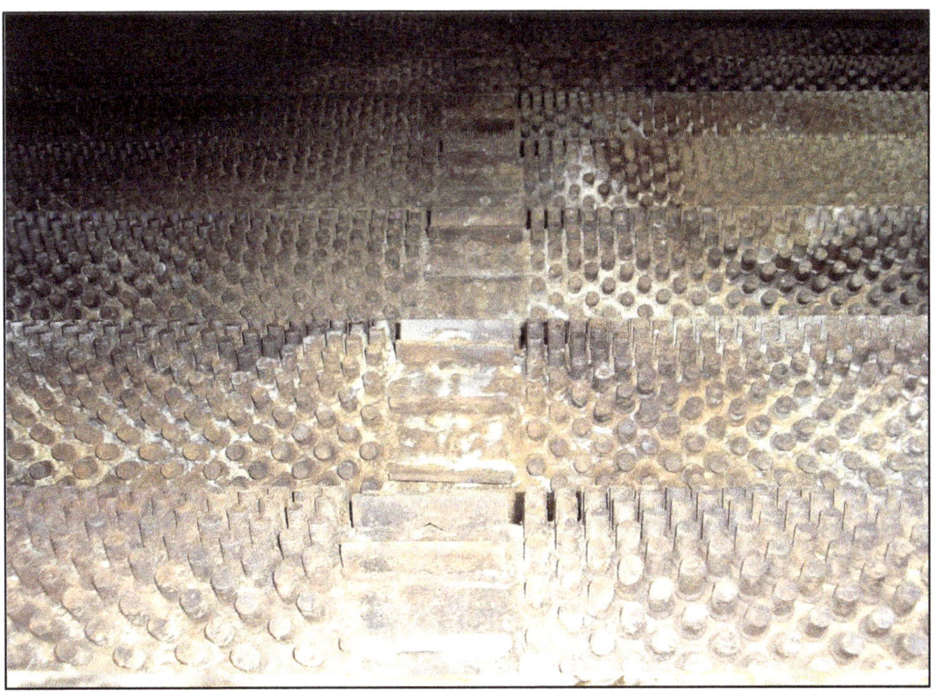

**Figure 17.3: Convection Tubes before Cleaning**

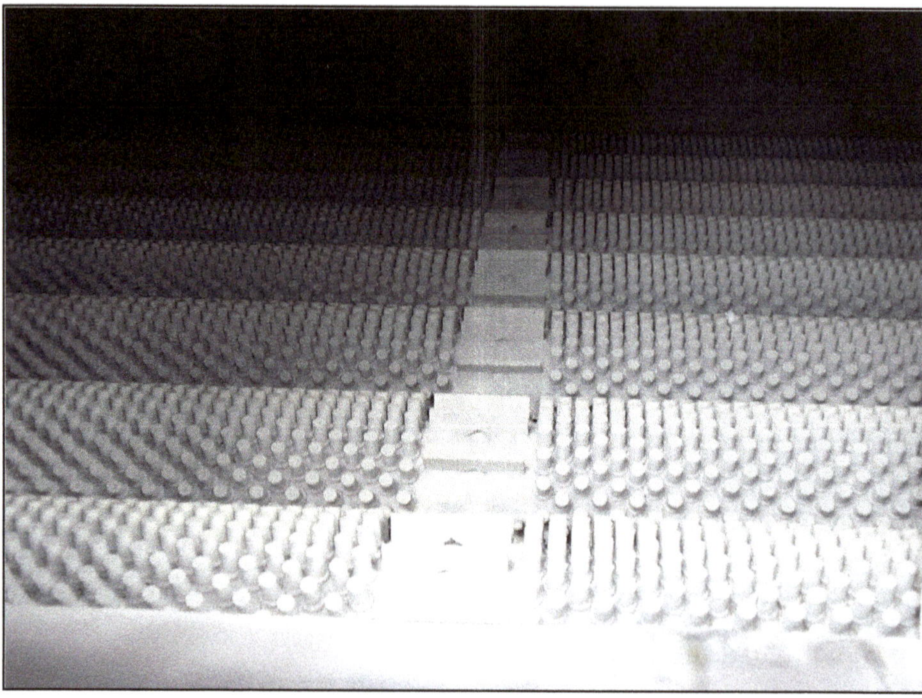

Figure 17.4: Convection Tubes after Cleaning

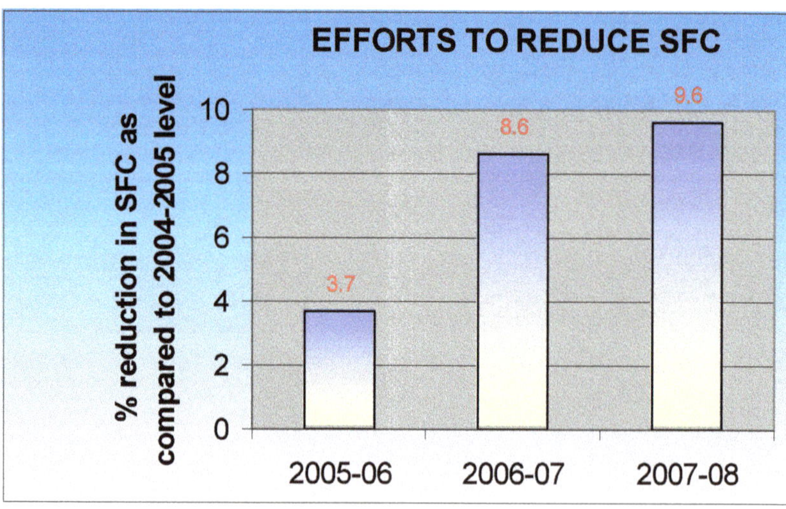


Figure 17.5: Improvement in SFC in CPP

## Minimizing PRDS Operation

In HPCL-Visakh Refinery, Steam pressure reducing stations are available for reducing Very High Pressure, High Pressure and Medium Pressure steam to lower pressure steam. However, by placing turbines in place of motors for pumps, both steam reduction as well as saving of electric power to motor can be achieved. Accordingly, in place of reducing stations, turbines are used for driving pumps and at the same time reducing steam pressure. This helped in saving power and total monetary savings was estimated at Rs19.9 Million per year due to implementation of this idea.

## Improvement in Specific Fuel Consumption in Captive Power Plant

In cogeneration captive power plant at HPCL Visakh Refinery, there are total four numbers of gas turbines. Depending on power demand and utility supply logistics, the number of turbines in service is minimized and less number of turbines are run at higher load. As running of less number of turbines at higher load is beneficial than running more number of turbines at lower load, specific fuel consumption (SFC) of Captive Power Plant has improved by around 9.6 per cent over a period of three years (Figure 17.5).

## Other Energy Conservation Measures Taken

The other energy conservation measures include routing of hot well off gases to Atmospheric furnace in Crude / Vacuum Distillation Unit-II (Off gas was routed to Atmosphere as per original design), *in-situ* processing of high oily sludge inside tanks using chemical, ex-situ processing of high oily sludge by external agencies, optimization of stripping steam in Crude, Vacuum columns and side strippers, increasing service factor of CO boiler located at down stream of FCC Regenerator, reducing operating pressure in Crude Distillation Columns and in preflash drum, monitoring preheat exchangers fouling and cleaning, condensate recovery from steam trap outlets, maximization of hot feeding to FCC unit which minimizes cooling of VGO in vacuum unit and subsequent heating in FCC unit before routing to Reactor, hydrokinetic decoking of furnace coils in place of steam – air decoking there by avoiding consumption of large amount of steam during decoking, insulation survey for steam lines and implementation of recommendations, implementation of energy audit recommendations, hydrocarbon loss survey etc.

## Future Energy Conservation Plans

Several energy conservation measures are in progress for further improving the energy performance of the Refinery. The stack dampers of old furnaces are planned to be replaced with new multiple leaf type dampers to avoid leakage of flue gases through dampers. Steam generators are planned for slop cut stream in vacuum distillation units, which goes directly to coolers along with vacuum residue steam as per design. Insulation survey of hot hydrocarbon lines is being planned. Natural gas conversion is being planned which will reduce fouling and improve service factor and efficiency of furnaces and waste heat recovery equipment. Supplementary firing is being planned in Heat Recovery Steam Generators in Captive Power Plant to generate additional quantity of steam with near 100 per cent efficiency. Automatic air / fuel ratio control is being planned for furnaces. Flare gas recovery, antifoulant for preheat train, provision of low pressure steam generator in product rundown upstream of cooler (basis feasibility) are some of the ongoing projects.

## 4.0 Conclusions

Due to various energy conservation measures, substantial energy savings and monetary gain could be achieved in the Refinery. The fuel consumption of the Refinery for last financial year is the best ever (Fuel and Loss of the Refinery is improved from 6.13 per cent in 2004-05 to 5.58 per cent in 2007-08). The specific energy consumption of the Refinery (MBTU/Bbl/NRGF) has improved by 14 points over past five years. As energy conservation is a continuous activity, there are several plans in the pipeline for further improving the energy performance of the Refinery. Sustenance of the improvements made and future improvements are the main focus area.

# Chapter 18

# *Energy and Emission Study in Ceramic Industry*

**M.T. Sambandam**
*Mechanical Engineering Department, Kalasalingam University*
*Krishnankoil, Tamil Nadu, India*
*E-mail: mt_sambandam@yahoo.com*

## ABSTRACT

The NSIC-Technical Service Centre Rajkot has carried out a comprehensive study to identify areas for improving energy efficiency and emission levels in a Ceramic Industry. The unit is engaged in the manufacture of Ceramic tableware. This project was carried out under the sponsorship of Naroda Industrial Association as a part of GEF-SGP Project. In this study emphasis is given to identify areas for reduction of energy loss, level of emission and ways and means to improve energy efficiency, reducing air pollution and optimizing process parameters. It is observed through the study that the potential for primary energy (oil) conservation lays in furnace area, which consumes a major chunk (98 per cent) of the total Energy input in the Manufacturing of Ceramic Tableware. This study was also focused on the consumption of electrical energy though it is mere 2 per cent of the total energy input. With all the above analysis done for energy conservation the present level of emission as well as possible reduction in emission levels were quantified. The present levels of emissions were compared with the standards recommended for ceramic industries and found to be fairly good.

The present specific energy consumption was calculated as:

Electrical: 41 kWh/Ton (35, 26 kcal/kg)

Fuel (GK): 188 Lit /Ton (188*9800*0'95)= (1750 kcal/kg)

Fuel (BK): 88 Lit/Ton (88*9800*0.95) = (819 kcal/kg)

The overall plant specific energy consumption in terms of total energy (E + T) value was 2604 kcal/kg.

The estimated fuel saving is in the tune of 22 Ton/annum which will reduce the $CO_2$ emission to a tune of 62.56Ton/annum and SOx to a tune of 0.7 ton/annum. The projected electrical energy saving is 4.68 MWh/annum which will result in a reduction of 6.56 Ton of $CO_2$/annum at generating point. The total estimated reduction in Green house gas is about 10 per cent of the present value including thermal and electrical pertaining to this industry.

***Keywords:*** *Kiln, Specific energy consumption, Glazing emissivity and Pre-heating.*

## 1.0 Energy and Emission Study in Ceramic Industry

The NSIC-Technical Service Centre Rajkot has carried out a comprehensive study to identify areas for improving energy efficiency and emission levels in a Ceramic Industry. A detailed Energy and Emission audits were carried out in Dipali Ceramics to ascertain energy conservation potential and the potential for Emission reduction by way of improving energy efficiency. Energy audit was carried out for both thermal and electrical energy consuming equipments. The plant produces glazed ceramic tablewares and annual production is 64000 bundles of weighing each 13 kg. (828 ton/annum.) The plant uses mainly fuel and electricity in its manufacturing process. The major energy consuming equipment are:

1. Glaze Kiln
2. Biscuit Kiln
3. Electrical Motors in process equipments
4. Electrical Distribution network

## 2.0 Manufacturing Process

The manufacturing process involved is given in the process flow chart Figure 18.1. The major energy consuming processes involved are:

1. Biscuiting
2. Firing

A brief of various manufacturing process are as under:

### 2.1 Raw material

The plant uses raw material for manufacturing as given in Table 18.1. To maintain the quality of ceramic product the raw material should not contain iron and Titanium. The components of the raw material vary depending on the lot to be produced which is purely market demand based.

### 2.2 Mixing and Grinding

In this process one Ton of raw material of above composition is mixed with 60 per cent water and grind in the Grinding mill for about 8 hrs till the grind size becomes 50 micron standard. At this stage the material gets good forming and sintering properties.

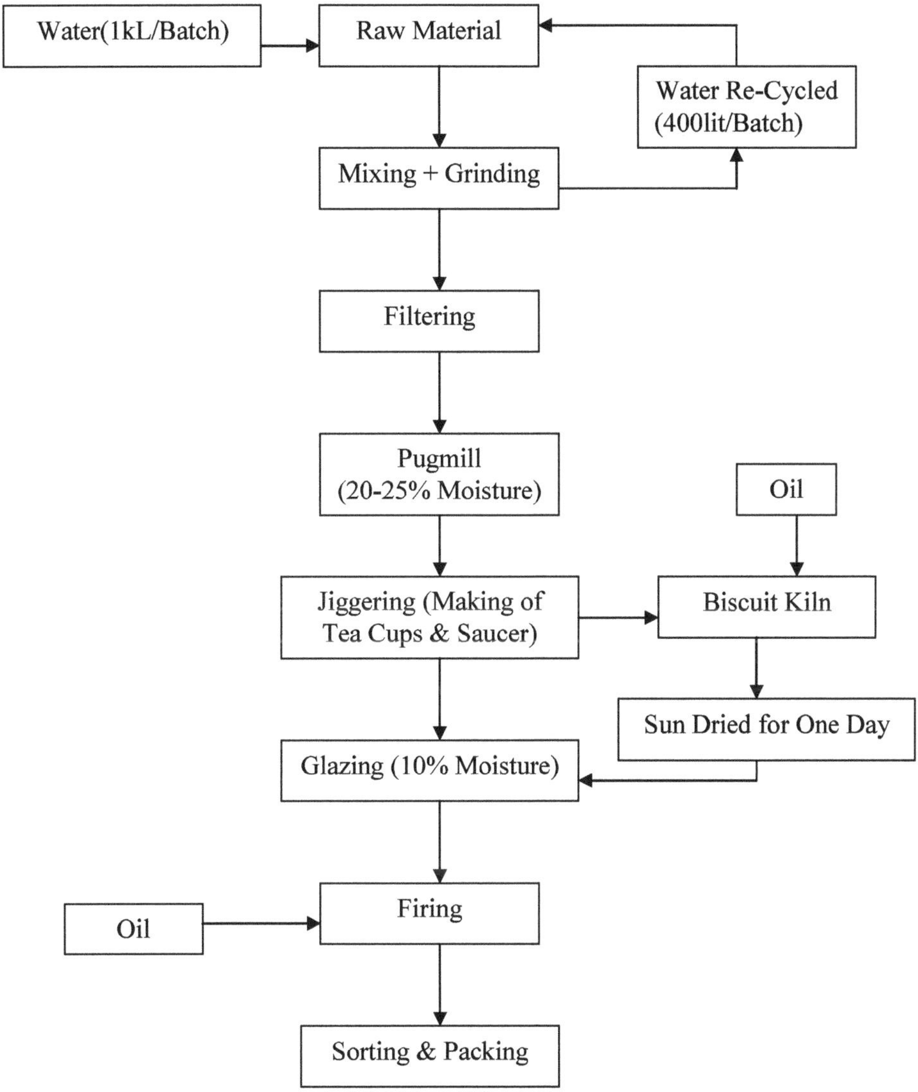


**Figure 18.1: Process Flow for Ceramic Table Ware Manufacturing**

## 2.3 Blending and Kneading

The raw material from grinding mill is filtered in the filter press and the water is recycled in the mixing process. The filtered raw material is put into Pug mill and Blended for about 2-3 hrs to get a homogenous mixture. The moisture content is about 20-25 per cent. This process is important to get maximum yield, quality and workability for the consecutive process.

**Table 18.1: Raw Material**

| *Components* | *Per cent* |
|---|---|
| **Basic Body** | |
| Quartz | 26 |
| Flourspur | 22 |
| China clay | 20 |
| Ball Clay | 10 |
| Fire Clay | 22 |
| | 100 |
| **Glaze** | |
| Quartz | 19 |
| Flourspur | 40 |
| Calcite | 15 |
| China clay | 5 |
| Zirconium | 14 |
| Zinc | 2 |
| Barium Carbonite | 2 |
| Cobalt Chloride | 2 |
| Talc | 1 |
| | 100 |

## 2.4 Jiggering

The homogenous raw material from blending is formed into required tableware, in this case Teacup and saucer using moulds. The process is done manually by placing the raw material in the rotating moulds.

## 2.5 Drying (Biscuiting)

The saucers from the above process are kept for a day in solar drying. The initial moisture of 60 per cent is reduced to about 40-50 per cent. Then these saucers are dried in Biscuit Kiln. The drying temperature is about 800-850 °C. During drying, sintering also takes place, which will increase the hardness of the product. Drying cycle time is 30-45 minutes.

## 2.6 Glazing

The sintered saucer and cup is given glazed coating and kept for a day for drying. The exact coating thickness is maintained by practice and no scientific measurement is made. However the plant maintains a uniform coating for getting good quality product.

## 2.7 Firing

The firing process is done at 1240 °C in glaze kiln. The firing cycle is 22 hrs during the process crystals combine with one another to improve the hardness of the product. The firing is terminated before the material gets melted.

## 3.0 Status of Energy Consumption

The plant consumes both thermal and electrical energy in the manufacture of ceramic ware. The average monthly energy consumption is as follows

| | | |
|---|---|---|
| Electricity | 3890 | kWh/Month |
| Glaze Kiln Fuel | 13 | kl/Month |
| Biscuit kiln Fuel | 2.2 | kl/Month |
| Production in | | |
| Glaze Kiln | 69 | Ton/month |
| Production in | | |
| Biscuit Kiln | 25 Ton/Month | |

The specific energy consumption is:

Electrical: 41 kWh/Ton (35.26 kcal/kg)

Fuel (Glass Kiln): 188 Lit /Ton (188*9800*0'95) = 1750 kcal/kg)

Fuel (Biscuit Kiln): 88 Lit/Ton (88*9800*0.95) = 819 kcal/kg)

The overall plant specific energy consumption in terms of energy (E + T) value is 2604 kcal/kg.

The plant specific energy consumption is at comparatively at good level due to the continuous improving operational parameters as a member of waste minimization circle formed by Central Glass and Ceramic Research Institute, Naroda, Ahmedabad.

In the total energy consumption 98 per cent of the energy input is consumed in the form of fuel and used in kiln and remaining in the form of electricity used in other manufacturing equipments. The energy share is shown in Figure 18.2 and Table 18.2.

**Table 18.2: Present Energy Consumption**

| *Energy Consumption* | *Consumption/ Day* | *Heat Value/ Unit* | *Total Heat Input Kcal/day* | *Percentage* |
|---|---|---|---|---|
| Electrical | 130 kWh | 860 kcal/kWh | 111,800 | 2 |
| Fuel Oil | 560 litres | 9800 kcal/kg | 5,213,600 | 98 |
| Total | | | 5,325,400 | |

## 4.0 Present Emission Level

The SPM, SOx and NOx Emission levels were actually measured at the following location in the plant.

1. Process Area
2. Kiln Area

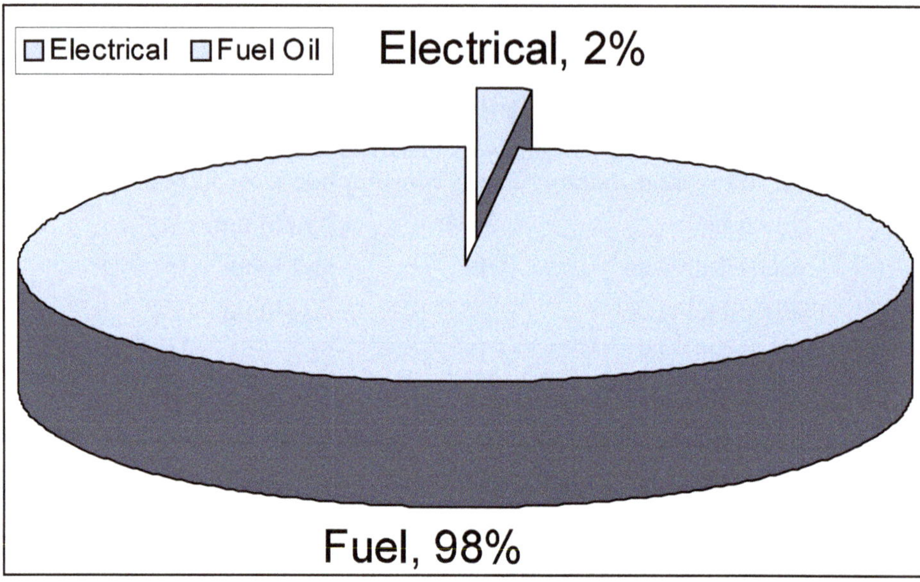


**Figure 18.2: Energy Share of the Plant**

Measured values and recommended values are given in Table 18.3.

**Table 18.3: Emission Levels**

| *Location* | *SPM (mg/Nm³)* | | *SOx (μg/Nm³)* | | *NOx (μg/Nm³)* | |
|---|---|---|---|---|---|---|
| | *Actual* | *Recommended* | *Actual* | *Recommended* | *Actual* | *Recommended* |
| Ball Mill (Process Area) | 600 | 150 | 16.40 | ** | 6.26 | ** |
| Tunnel Furnace | 546 | 150 | 14.50 | ** | 4.30 | ** |

SPM: Suspended Particulate Matter.

*Source* : EPA(US) Notification GRE 475(E) dt.5.5.1992.

The Emission Level in the form of green house gas and air pollutants was calculated as follows:

## 4.1 Carbon Dioxide Emission

The plant uses mixed fuel, which has carbon f 87 per cent by weight. The $CO_2$ generated per kg of carbon burnt is 3.67 kg [Tables 18.4(a) and (b)]. Therefore total $CO_2$ generated by complete combustion of 182.4 kL/annum of fuel having density 0.95 and 87 per cent carbon by wt with complete combustion of 95 per cent of carbon, will be 525.6 Ton of $CO_2$ per annum.

$(182.4 \times 0.95 \times 0.87 \times 0.95 \times 3.67 = 525.60)$

$CO_2$ generated at power generating point for supplying 3890 kWh/ month to the plant with 20 per cent T&D loss will be as follows:

Fuel (lignite) equivalent

$= (3890 \times 12 \times 1.2 \times 0.56 \times 0.95 \times 3.67/1.67)$

$= 65.48$ Ton/Annum

Total $CO_2 = 591$ Ton/annum (525.60+65.48) T

**Table 18.4(a): Weights of Products of Combustion**

| *1 kg of Combustible* | *Burnt to* | *Kg. required* | | *Kg. Products of Combustion in Air* | | | | | |
|---|---|---|---|---|---|---|---|---|---|
| | | *$O_2$* | *Air* | *$CO_2$* | *$H_2O$* | *$N_2$* | *CO* | *$SO_2$* | *Total* |
| C | $CO_2$ | 2.67 | 11.52 | 3.67 | - | 8.85 | - | - | 12.52 |
| S | $SO_2$ | 1.00 | 4.32 | - | - | 3.32 | - | 2.00 | 5.32 |
| $H_2S$ | $H_2O + SO_2$ | 1.41 | 6.10 | - | 0.53 | 4.69 | - | 1.88 | 7.10 |
| $H_2$ | $H_2O$ | 8.00 | 34.56 | - | 9.00 | 26.56 | - | - | 35.56 |
| $CH_4$ | $CO_2 + H_2O$ | 4.00 | 17.28 | 2.75 | 2.25 | 13.28 | - | - | 18.28 |

**Table 18.4(b): Fuel Properties**

| *Sl.No.* | *Property* | *Mixed Oil* |
|---|---|---|
| | **Name of Fuel: Mixed Oil** | |
| 1. | Density (kg/lit) | 0.95 |
| 2. | Viscosity, Cst | 9 @ RT |
| 3. | Total Sulphur, wt per cent | 2 |
| 4. | Pour Point Deg C | < 20 |
| 5. | Mineral Matter Na K | 10ppm, 2ppm resp. |
| 6. | Ash content per cent wt. | 0.06 per cent |
| 7. | Carbon content | 87 per cent |
| 8. | Hydrogen content | 8.50 per cent |
| 9. | Oxygen, Nitrogen | traces |
| 10 | Gross Calorific value, Kcal/kg | 9.8 |
| 11. | Flash point, Deg C | > 66 |
| 12. | Storage Classification | Class C |
| 13. | Acidity | Nill |
| 14. | Free water | 0.80 per cent |

## 4.2 Sulphur Dioxide Emission

The Sulphur content in the fuel is 2 per cent by wt. and SOx generated by burning 1 kg of sulphur is 2 kg. Therefore total SOx generated by complete combustion of 182.4 kL/annum of fuel having density of 0.95 with 2 per cent sulphur and having

100 per cent combustion of sulphur will be 6.58 Ton of SOx per annum.. (182.4 × 0.95 × 0.02 × 0.95 × 2 = 6.58).

Other emissions in the plant at different locations are discussed below:

It is observed that the particulate matter emission is higher than recommended. The SOx and NOx levels observed are within the permissible limit for any ambient air quality. Since the plant uses raw material containing chloride to very minimum 1 per cent no traces of chloride was observed in the furnace area. No fluoride-based raw material is used by the plant; hence no fluoride emission was traces. Suitable suggestions are given for reduction of particulate matter in this report under *Cleaner Production Options.*

**Table 18.5: Fluoride and Chloride Emission**

| *Location* | *Fluoride (mg/Nm³)* | | *Chloride (mg/Nm³)* | |
|---|---|---|---|---|
| | *Actual* | *Recommended* | *Actual* | *Recommended* |
| Ball Mill (Process Area) | ** | ** | ** | ** |
| Tunnel Furnace | _ | 10 | _ | 100 |

Source for Recommendation: EPA Notification GRE 475(E) dt.5.5.1992

## 5.0 Energy Conservation Measures Identified

Energy conservation measures were identified in the following categories:

1. Improving Kiln energy efficiency
2. Adopting energy efficient technology
3. Reduction in supply voltage for electrical energy saving

Apart from the above areas good house keeping, selection of good raw material was also considered for reducing energy consumption, which is discussed in the detail below:

### 5.1 Biscuit Kiln

The plant has installed Biscuit Kiln for drying and sintering of saucers. The furnace size is L55′ x W3.5′x H2′. The furnace is operated for 8 hrs/day and the production is 8000 pieces/8 hrs. (8000 * 130 gm = 1040 Kg) and 25 ton/Month. Air curtains are provided for maintaining uniform temp in the furnace. The firing cycle is 35 minutes and Car Pushing speed is 35 minutes per Car. The total nos. of car in the furnace is 14. The operating temp. of the furnace is 820° C. The fuel consumption is 2.2 KL/Month. The specific energy consumption of this furnace is 88 lit/kg (819 Kcal/kg). This value is 25 per cent higher side compared to the energy consumption in developed countries for Biscuit Firing as mentioned in Table 18.6.

The detailed energy Balance calculation shows that the furnace efficiency is about 40 per cent. The major energy loss is found in Radiation and conduction due to poor insulation of the furnace. The average temp. of the furnace wall is about 80° C which is to be maintained at about 60° C by improving the furnace insulation. It is observed from the gas profile that the oxygen percentage in the furnace is at 11 per cent, which is at higher side due to air infiltration into the furnace from both open end and through bottom. Air curtains are provided at both ends of the furnace to get uniform temperature in the furnace. Because of the furnace draft at both ends oxygen is also sucked inside the furnace, which enhances the excess air inside the furnace.

**Table 18.6: Specific Energy Consumption**

| *Product Name* | *Process* | *Specific Energy Consumption Kcal/Kg* |
|---|---|---|
| Table ware | Biscuit firing | 2500–4000 |
| | Glaze firing | 4000–5000 |

*Source*: Handy manual, Ceramic Industry, UNIDO-1994.

## 5.2 Glaze Kiln

The plant has installed Glaze Kiln for drying and sintering of saucers. The furnace size is L 104′ x W 4″x H 4.75′. The furnace is operated for 24 hrs/day and the production is 69 Ton/Month. The furnace is fitted with two Burners, which has separate entry for atomizing and combustion air. Cold air supplied for atomization and combustion. Air curtains are provided for maintaining uniform temp in the furnace. The firing cycle is 22 hrs and Car Pushing speed is 60 minutes per Car. The total nos. of car in the furnace is 22. The operating temp. of the furnace is 1240° C. The fuel consumption is 13 KL/month. The specific energy consumption of this furnace is 188.4 lit/ton (1750 Kcal/kg). This value is at very much lower level compared to the reported value as mentioned in Table 18.6, due to the continuous improvement in the operating parameters of the furnace as well as good insulation. The furnace is very recently modified with improved insulation, increased length and increased height of the furnace. These modifications had helped to produce more volume of products. The present Width to Height ration is maintained at 0.84 which is at optimum level.

The detailed energy balance calculation shows that the furnace efficiency is about 43 per cent. The major energy loss is found in Radiation and conduction due to the size of the furnace. The average out side temp. of the furnace wall is about 60° C which is at satisfactory level. The plant maintains inside temp of the furnace at par with the recommended temperature for effective Pyro-chemical reaction to take place to get better quality. It is observed form the gas profile that the oxygen percentage in the furnace is at 11 per cent, which is at higher side due to air infiltration into the furnace from both open end and through bottom. It is also suggested that the plant should use high quality Fluorspar and china clay with high calcium content which will reduce the firing time. Further to get uniform temp in the furnace air curtains are provided which further increases the oxygen percentage in the furnace.

Furnace inside temp, Skin temperature and hot gas composition profiles are shown in Figures 18.3–18.5 and the Energy balance is shown in Figures 18.6 and 18.7.

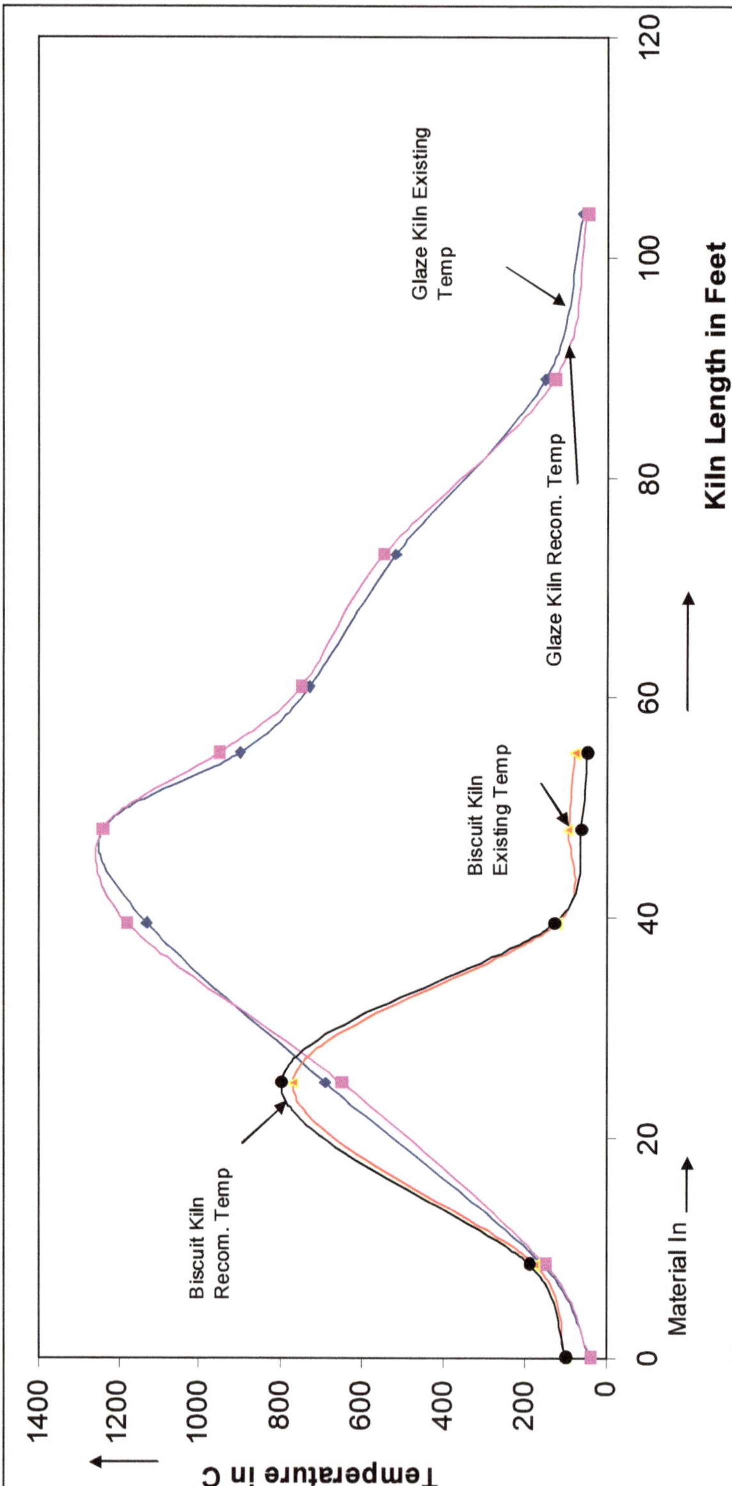


**Figure 18.3: Kiln Inside Temperature Profile**

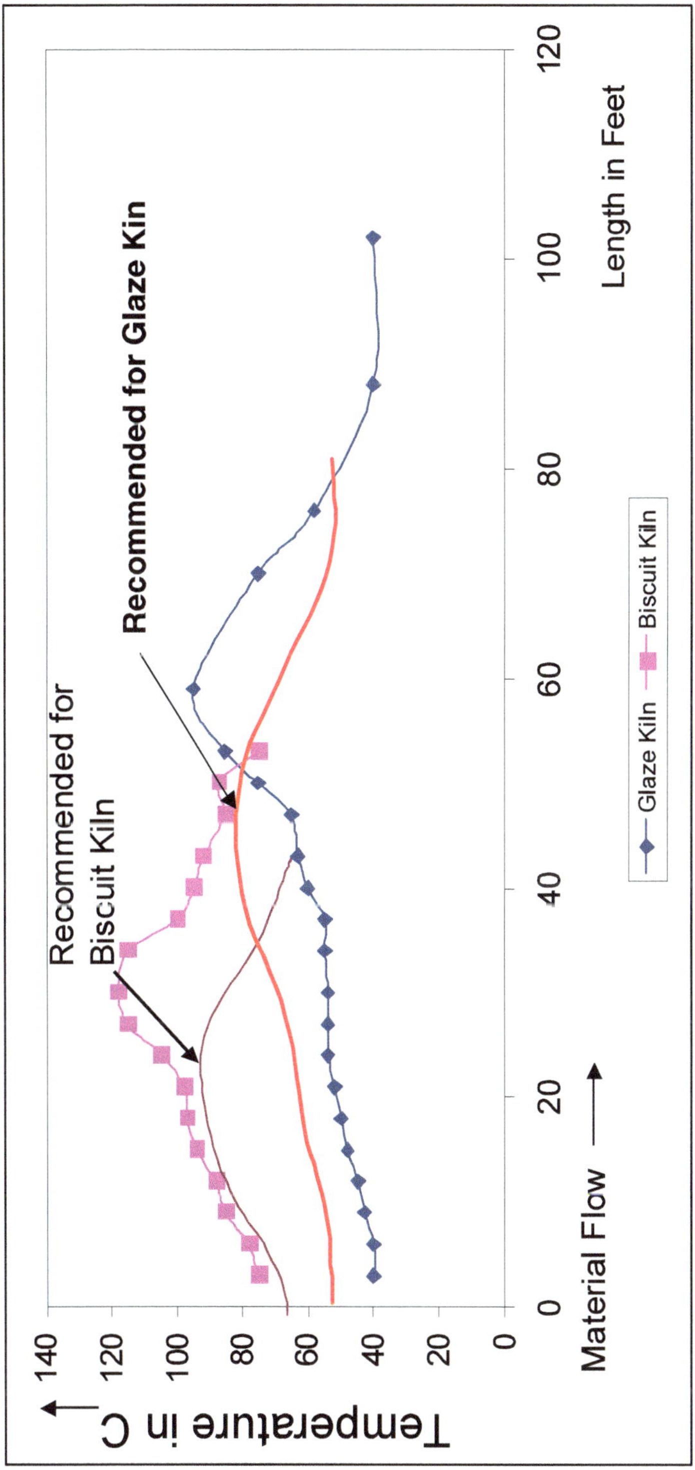


**Figure 18.4: Skin Temperature Profile in Kilns**

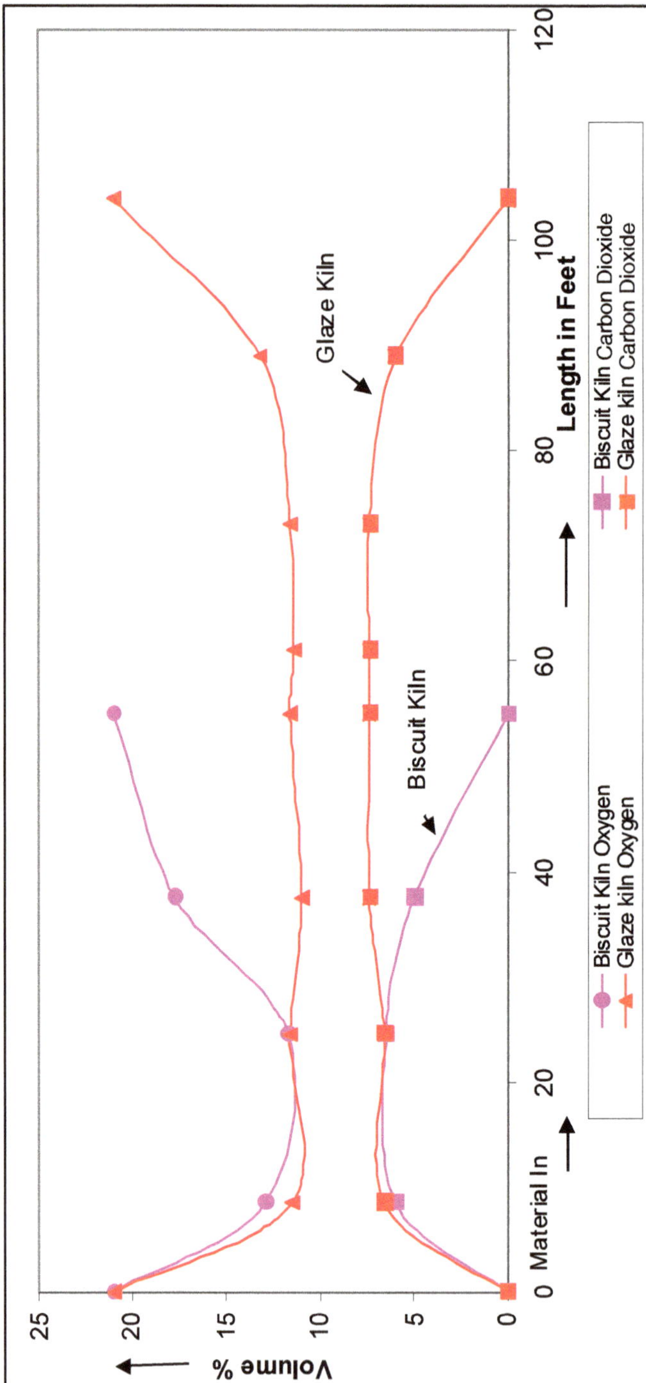


**Figure 18.5: Gas Profile inside Kiln**

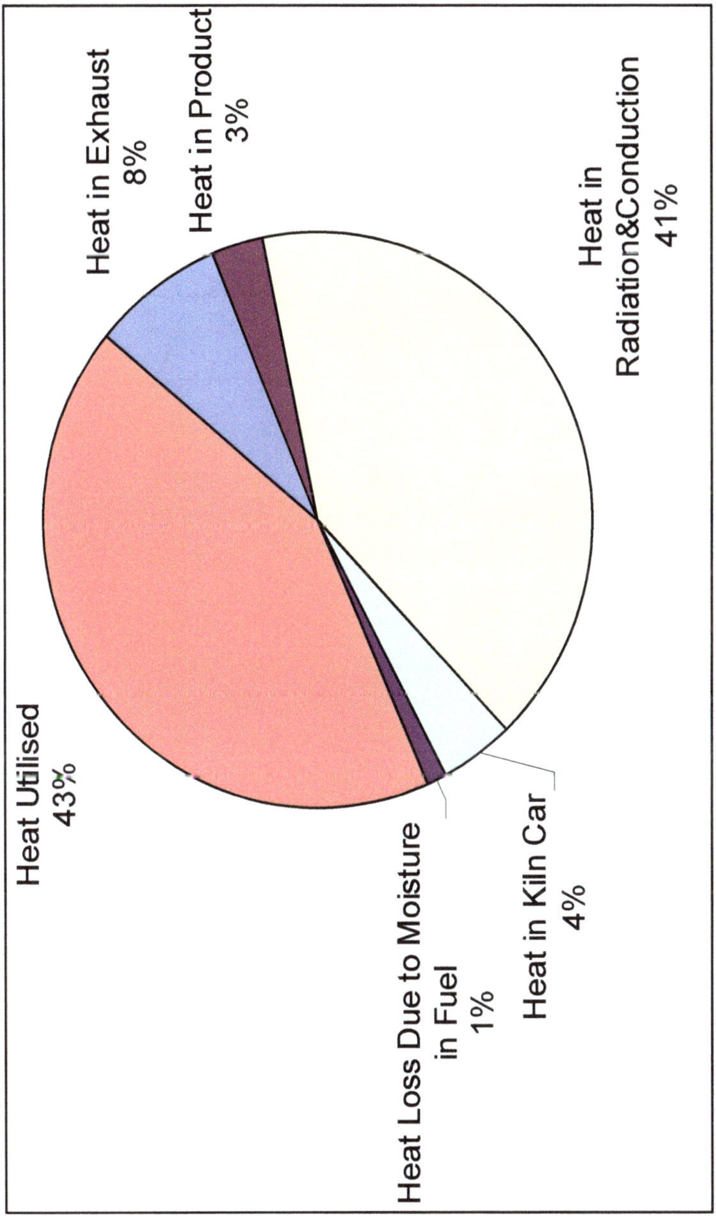


**Figure 18.6: Energy Balance Biscuit Kiln Dipali Ceramics**

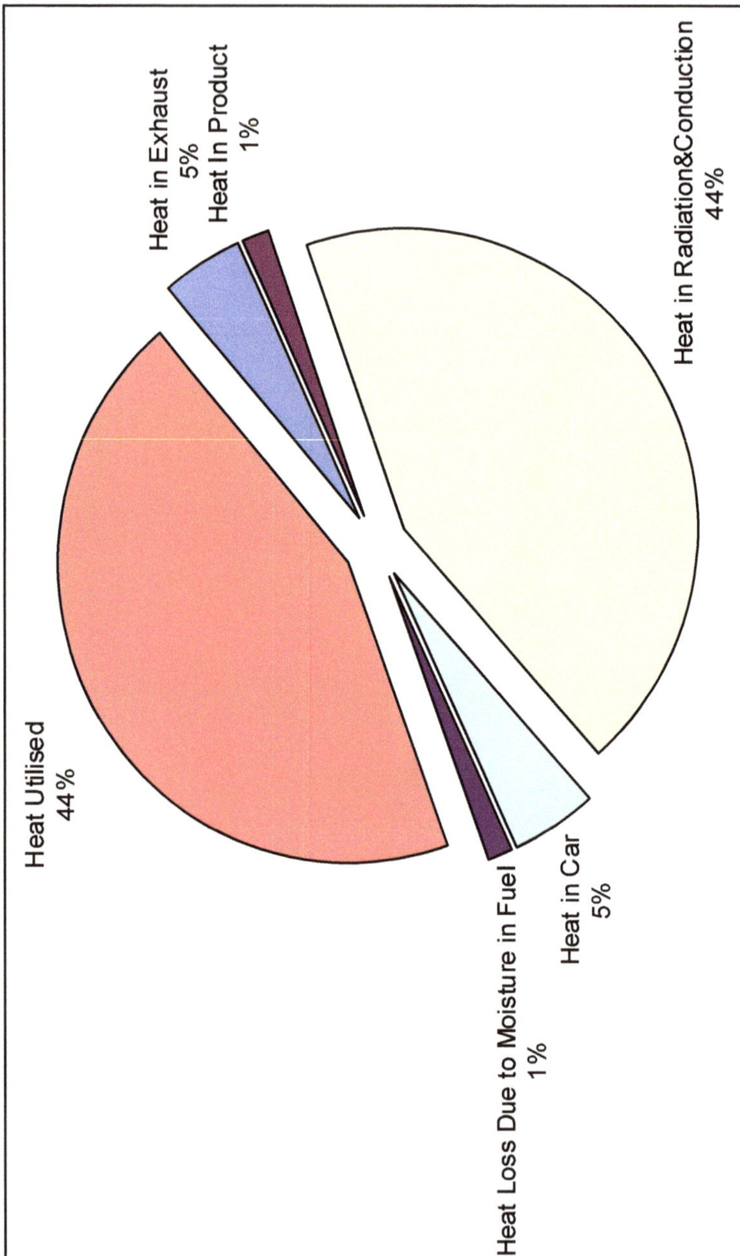


**Figure 18.7: Energy Balance in Glaz Kiln Dipali Ceramics**

# 6.0 Energy Conservation Opportunities

## 6.1 Thermal Energy Conservation

### Preheating of Combustion Air in Glaze Kiln

The secondary combustion air is supplied at ambient temp. It is recommended to preheat the secondary combustion air by drawing from cooling zone of the furnace by installing a separate blower. The present and the proposed combustion air supply scheme shown in Figures 18.8 and 18.9. By this arrangement secondary combustion air can be preheated up to 250 ° C. For every 50° C raise in combustion air temp. the fuel saving of 1.5-2 per cent can be achieved. The fuel saving calculation is as follows:

| | |
|---|---|
| Present fuel consumption | : 18 litres/hr |
| Anticipated saving @ 7 per cent | : 1.26 litres/hr |
| Say | : 1.3 litre/hr |
| Annual Fuel saving | |
| | : 1.3 * 24hrs * 30 days * 12 month. |
| | : 11.232 KL ~ 11 KL |
| | : 10450 kg/anm.(11000*0.95=10450) |

### Energy Efficient Coating

Both furnaces can be given energy efficient coating on the sidewalls inside the furnace to improve the emmissivity. This coating should be applied only in firing zone and preheating zone. By using this coating the fuel consumption can be reduced by about 5 per cent.(BK-11 lit/h and GK-18lit/h, Avg working hrs~16hr/day and 27 days/month)

| | |
|---|---|
| Present fuel Consumption | : 29 litres/hr (11+18) |
| Anticipated saving@ 5 per cent | : 1.45 litres/hr |
| Annual Fuel Saving | : 1.45 * 16 hrs * 27 days * 12 month = 7.5 KL |
| | = 7125 kg/ annum. |

### Improving Insulation in Biscuit Kiln

The present average surface temp of Biscuit kiln was measured as follows:

Cooling zone: 87°C

Firing zone: 112°C

Preheating zone: 82°C

These temp. are at higher side which are to be brought down as follows:

Cooling zone: 60°C

Firing zone: 80°C

Preheating zone: 60°C

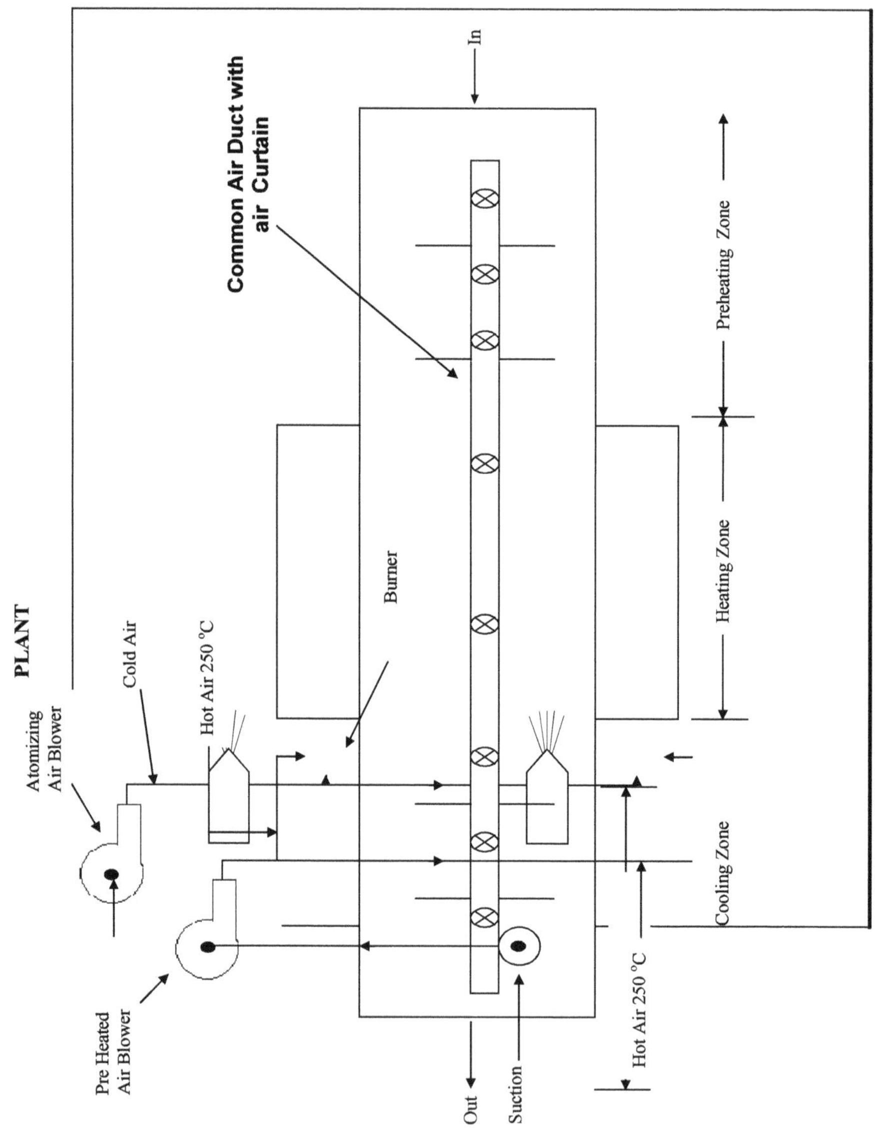


**Figure 18.8: Proposed Burner Air Line For Glaze Kiln Dipali Ceramics**

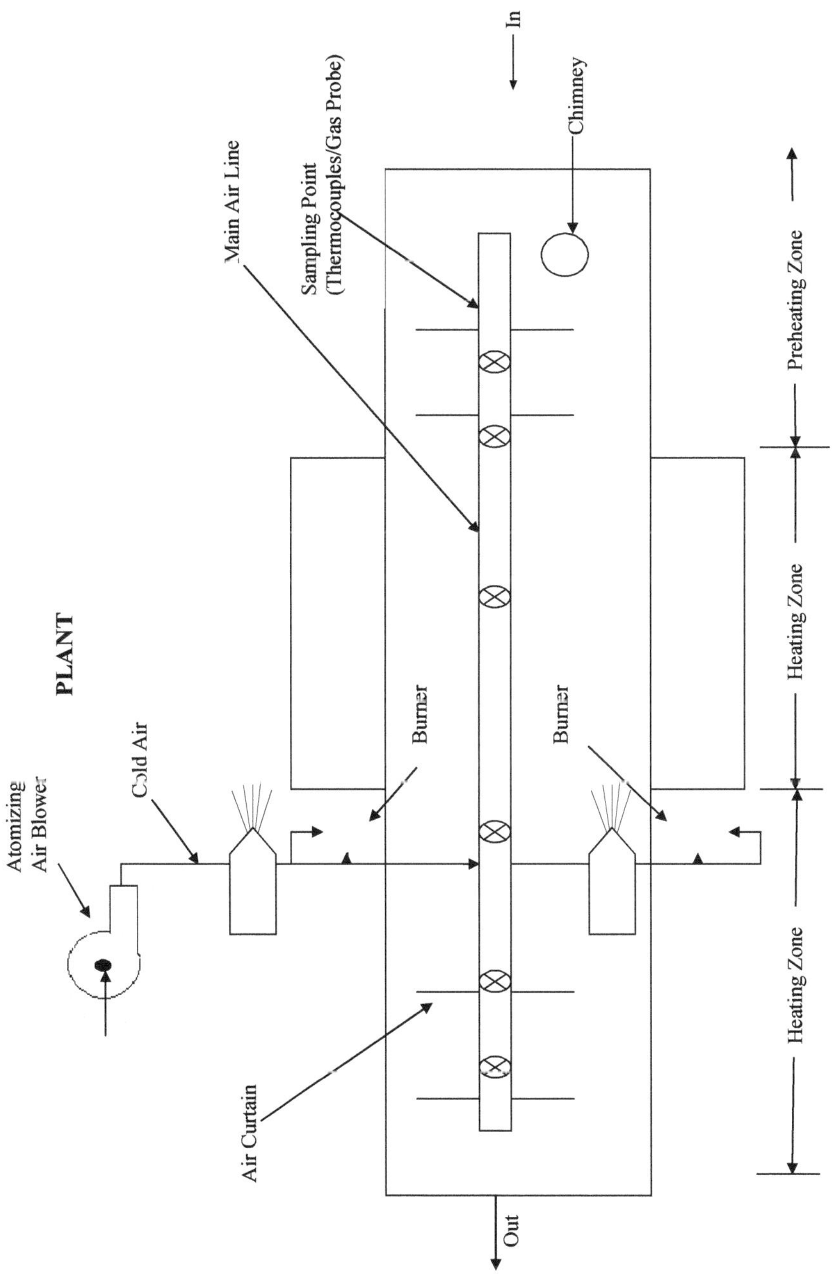


**Figure 18.9: Existing Burner Air Line For Glaze Kiln Dipali Ceramics**

It is recommended to improve the insulation by replacing the old insulation with the new insulation material of ceramic fibre and calcium silicate blocks. The anticipated energy saving is appx. 10-15 per cent.

Present energy loss by radiation and conduction: 40406 Kcal/hr

| | |
|---|---|
| Saving @ 10 per cent | : 4040 Kcal/hr |
| Fuel equivalent | : 0.41 kg/hr |
| | : 0.43 litre/hr |
| Saving per month | : 0.43*25days*8hrs |
| | : 86 litres/ month |
| Annual fuel saving | : 86 * 12 months |
| | : 1032 litres/annum |
| | : 980 kg/annum |

## 6.2 Electrical Energy Conservation

The plant has installed motors ranging from 0.3 Kw to 7 Kw(0.5 HP to 10 HP). These motors are installed in grinding machine, pug mill, press, jigger machine and air blowers. The lighting load is very minimum about 10 fluorescent tube lights of 40 W capacity. The plant has LT connection from AEC (Ahmedabad Electricity Company Ltd.). The average monthly power consumption is only 3900 kWh which is only 2 per cent of the total energy input in the plant. Therefore the scope for energy saving is very limited however an attempt has been made to identified areas for electrical energy conservation. During the study it is observed that all motors are working under loaded with load percentage varies 17 per cent to 15 per cent. These motors though they are under loaded the initial torque required is very high in Ball mill, Pug mill, and Press. The voltage variation was observed 440 to 420 Volts in the plant. The study shows that there is no appreciable potential for electrical energy saving.

## 6.3 Equivalent Emission Reduction

### Thermal

| | |
|---|---|
| Anticipated total fuel saving | : 10450kg+7125+980 kg |
| | : 18555 kg/annum |
| Equivalent $CO_2$ Reduction | : 18555* 87 per cent *0.95* 3.67 |
| | : 56 Ton of $CO_2$/annum |
| Equivalent SOx reduction | : 18555 * 2 per cent *0.95* 2 |
| | : 0.705 Ton of SOx/annum |

### Electrical

There is no significant scope for electrical power saving hence there is no saving of green house gas emission due to power saving:

Total Reduction in $CO_2$ = 58 Ton/Annum

The reduction level are shown in Figures 18.10 and 18.11.

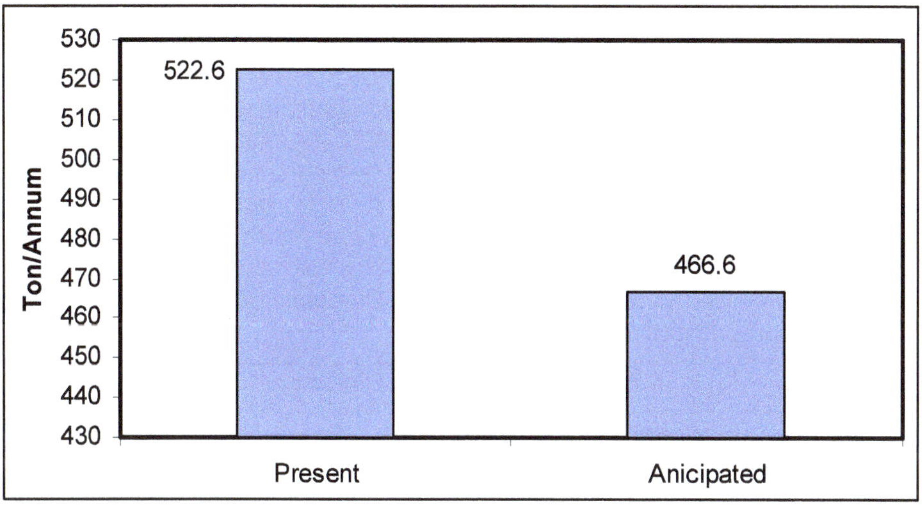


**Figure 18.10: Carbondioxide Emission**

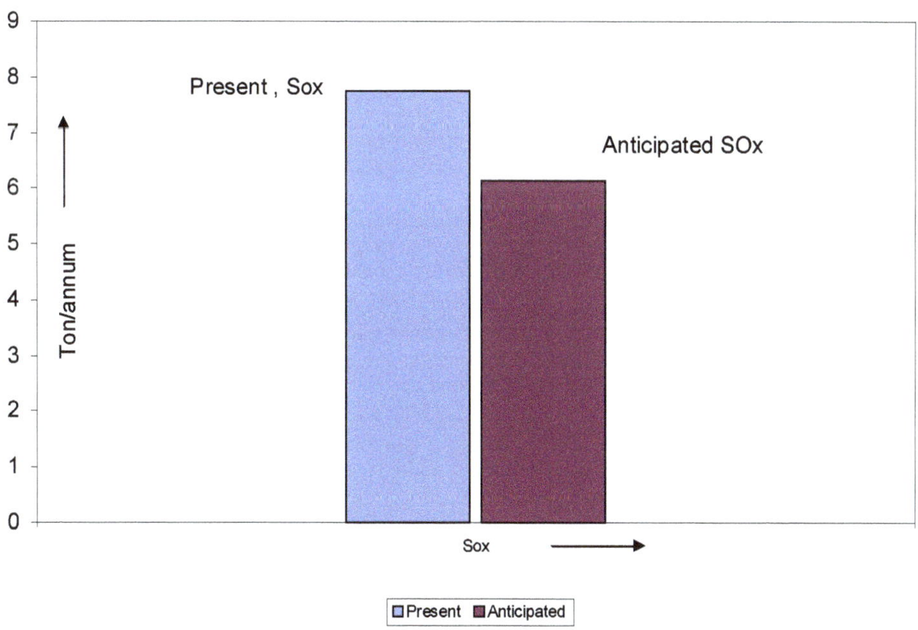


**Figure 18.11: SOx Emission**

# 7.0 Cleaner Production Options

## 7.1 Waste Minimization

The plant's present production is 45.5 Ton/month. The rejection rate is 5 per cent, which is due to breakage. The quantum of rejection is about 2.275 Ton/month. The plant has already taken several measures to optimize process parameters by taking advices from Central Glass and Ceramics Research Institute, Ahmedabad. Recently the plant has started recycling the waste by mixing the grind waste material with raw material about 2 per cent. Through this study it is recommended to mix the recycled ceramic waste in the making of saggar since the chemical analysis of the waste shows that 22 per cent $Al_2O_3$, Table 18.7). This has been done experimentally during our audit and found to be a better option. The plant is advised to continue this with various proportions in the future.

**Table 18.7: Chemical Analysis of Rejected Material**

| *Components* | *Per cent* |
|---|---|
| $SiO_2$ | 64.98 |
| $Al_2O_3$ | 22.9 |
| $Fe_2O_3$ | 0.4 |
| $TiO_2$ | 0.01 |
| CaO | 10.9 |
| MgO | 0.81 |
| Total | 100.00 |

The other various low-cost solutions to increase Efficiency and Reduced waste are listed below:

| | |
|---|---|
| Avoid or reduce the use of heavy metals in mixes and glazes | Glazes that do not contain lead or other heavy metals are commonly available. Metals in ceramic mixes are used for their engineering properties. Product engineers should research non-toxic ingredients. |
| Dry cleanup | Vacuums and damp brooms can be used to clean up dust from mixes and grinding, reducing the amount of waste that goes into the wastewater from final clean up. |
| Water recirculation | Wastewater can be stored so the suspended particles settle to the bottom, and then the water can be reused for cleaning. Sludge can be collected, dried and made into inexpensive bricks rather than disposing it. |
| Improve product drying before firing | Extended drying time reduces fuel requirements. Even drying throughout product stacks reduces defective firing. |
| Improve air flow control | Stopping all air leaks and controlling the kiln opening size allows better control of airflow speed and direction to improve combustion. |
| Switch to LPG or natural gas fuel | If available and competitively priced, these fuels have significantly less emissions and can increase production quality and speed. |
| New kiln design | Vertical shaft brick kilns allow increased production rates and significantly decreased emissions through improved combustion air flow efficiency. Several other kiln designs have also proven to be relatively low-cost and much more efficient than traditional ovens or kilns. |

## 7.2 Reduction of Particulate Matter in Process and Furnace Area

It is observed from the study that the particulate matter in the process and furnace area is very much higher (600 mg/Nm$^3$ and 546 mg/Nm$^3$) than the recommended value (150 mg/Nm$^3$). The following suggestions are made for reducing the particulate matter in the plant:

1. Vacuums and damp brooms can be used to clean up dust. It is also observed that the raw materials are fed manually with out any protective hood.
2. The Raw material should be stored separately with a covered area so that dust is contained in the stored area only.
3. Raw material bags should be covered so that flying dusts could be reduced.
4. All processing machineries like ball mill, grinding m/cs should be isolated from the store area.
5. Curtains should be provided between process sections like, furnace and sager molding, ball mill and sager molding. These curtains should be sprayed with water at regular intervals.
6. To avoid the dust due to vehicle movement and wind water should be sprayed on the open area at regular intervals.
7. Ventilation should be improved by installing exhaust fan on roofs with suitable provision for rain protection.
8. Training should be imparted to the working personnel on cleaner production.

The target for Emission reduction of particulate matter is shown in Figure 18.12.

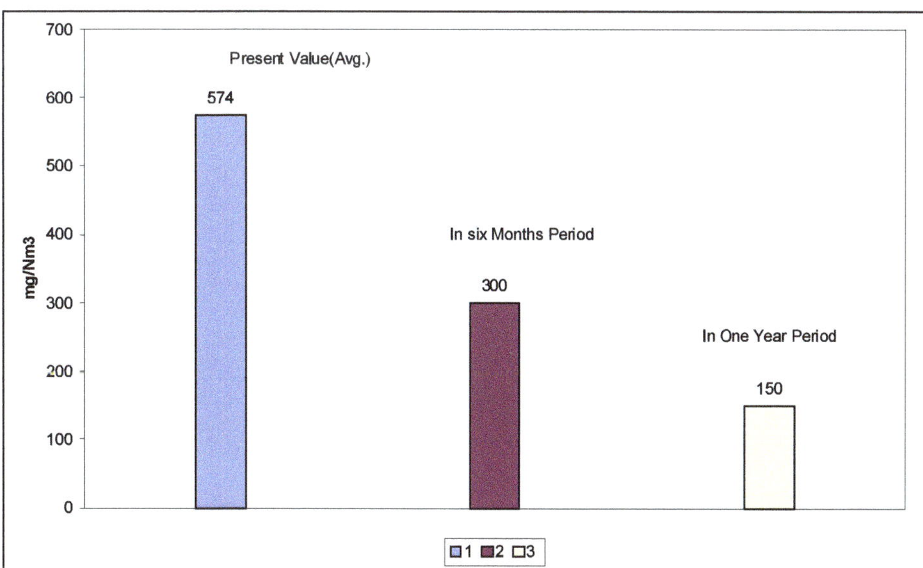


**Figure 18.12: Reduction Particulate Matter**

## 8.0 Summary

As a concluding remark it is observed that the plant has energy saving potential of about 10 per cent in primary energy fuel. The anticipated reduction in the emission level particularly in the green house gas is about 10 per cent. The summary of anticipated Savings and emission reduction are Annexed at I and II. Therefore a systematic and an integrated approach combining energy and emission study will certainly improve the efficiency will result considerable reduction in green house gas emission in ceramic industries.

## Acknowledgement

Author sincerely thanks Naroda Industrial Association and M/s. Dipali Ceramics for entrusting the project to NSIC-Technical Services Centre, Rajkot-India.

## References

1. "Energy performance assessment for equipment and utility systems", 2004. Published by Bureau of Energy Efficiency, Govt. of India.
2. "Energy efficiency in electrical utilities", 2004. Published by Bureau of Energy Efficiency, Government of India.
3. "Handy manual for Ceramic Industries", 1994. Published by UNIDO.

### Annexure I: Summary of Emission Reduction

| Type of Energy | Present Consumption/ Annum | Emission Present Ton/Annum | | Anticipated Saving in Energy Ton/annum | Anticipated Emission Reduction Ton/annum | | Percentage Reduction in Emission | |
|---|---|---|---|---|---|---|---|---|
| | | $CO_2$ | SOx | | $CO_2$ | SOx | $CO_2$ | SOx |
| Fuel | 173 Ton | 522.6 | 6.53 | 18.55 Ton | 56 | 0.705 | 10.72 | 10.71 |
| Total | | 522.6 | 6.53 | | 56 | 0.705 | 10.72 | 10.71 |

### Annexure II: Summary of Anticipated Energy Savings

| Sl.No. | Details of Proposal | Saving Energy | Saving in INR | Invest. in INR | Payback in Months |
|---|---|---|---|---|---|
| **CATEGORY I : 0 < PAYBACK < 1 Month** | | | | | |
| | NIL | | | | |
| **CATEGORY II : 1 < PAYBACK < 12 Months** | | | | | |
| 1 | Preheating of combustion Air in Glaze Kiln | 11.00 kL | 130,000 | 25,000 | 2 |
| | Total | 11.0 kL | 130,000 | 25,000 | 2 |
| **CATEGORY III : 12 < PAYBACK < 36 Months** | | | | | |
| 2 | Energy Efficient coating in Biscuit and Glaze Kiln | 7.5kL | 90,000 | 100,000 | 13 |
| 3 | Improving Insulation in Biscuit Kiln | 1.03kL | 13,000 | 25,000 | 23 |
| | Total | | 103000 | 125000 | 15 |
| | Grand Total | | 233000 | 150000 | 8.0 |

## Chapter 19

# *Case Study of Energy Saving Armec Jet Towers: Without Fans and Fills*

**P.M. Kudva**

*Vice President (Marketing), Armec Group,*
*Armec Chambers, 307, 3rd floor, Decision Tower, Pune-Satara Road,*
*Pune, Maharashtra, India*
*E-mail: armecmktg@airtelmail.in*

## ABSTRACT

Central Air-conditioning is one of the most simple and most common applications all over the World. However, in this there is immense scope for Energy saving at this application level. For the interest of readers the following is a genuine case study based on the Minutes of Meeting signed by the users representative of a cooling tower after about 15 years with the cooling tower Manufacturer. Only people ought to have a heart, faith in alternate technology and courage to accept this technology at the project stage, as it is costlier by about 10-20 per cent.

***Keywords:*** *India, Central air-conditioning, Energy saving, Energy efficiency, Cooling tower, Alternate technology.*

## 1.0 Introduction

*Project is where cost is the prime concern....,*

*.....Working is where maintenance is of grave concern!!!*

The case study given here below is an eye-opener.

The place of the case study is the hottest place in central India: Nagpur, Maharashtra. The institution of case study is the zonal office of the premier bank of India-Reserve Bank of India (RBI), Dr. Raghavendra Rao Road, Civil Lines, Nagpur–440 001. The use of cooling tower is for central Air-conditioning application.

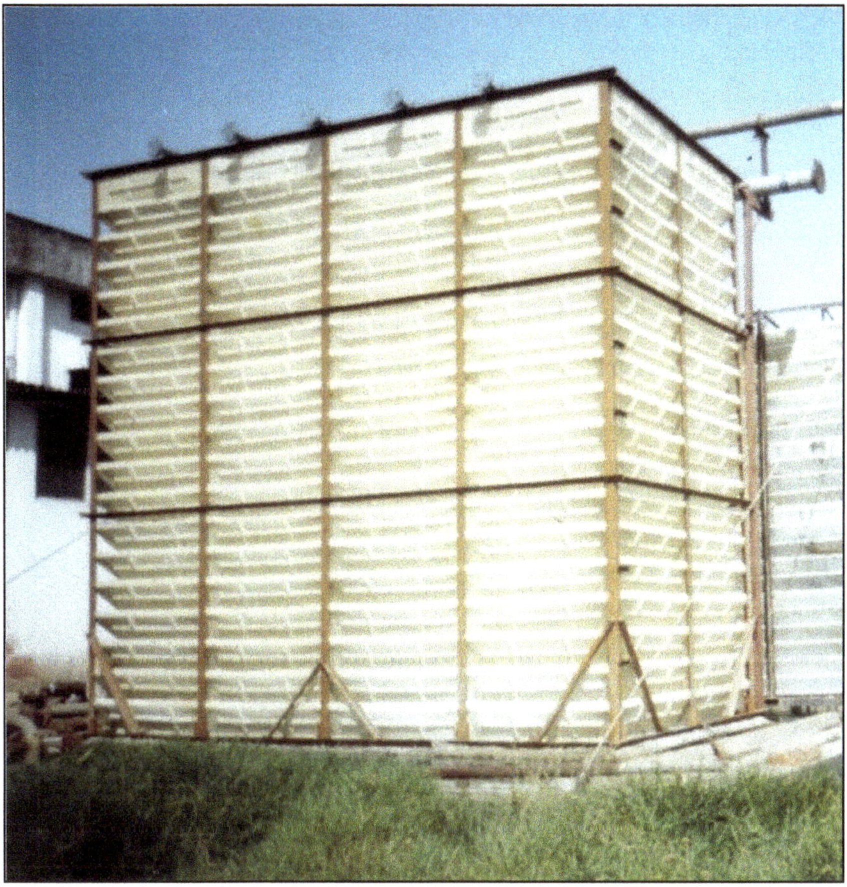

**Figure 19.1: Cooling Tower Used for Air-Conditioning Application**

The year of installation was the end of 1993, and the study was conducted on 11.04.2008.

Persons who did the survey work were Mr.V.K.Daithankar for ARMEC JET TOWERS and Mr. G.Y. Mahashabde for RBI.

Some of the facts for consideration are that, from our records we found that the RBI has asked for replacement of the cooling tower at the same basin of the existing fan type tower and even that they wanted the pump also to remain the same. Replacement work was carried on these terms and conditions.

Rated capacity of cooling towers = 300 TR each

Rise in the system for design = 8°C

Therefore, rated capacity in LPS = 77 LPS

$$\text{Therefore, capacity in Mtr}^3/\text{Hr} = \frac{77 \times 3600}{1000} = 280\ \text{Mtr}^3/\text{Hr.}$$

Rated capacity of the Fan motor required if it was working on induced draught, cross-flow fan and fill type cooling tower = 25 KW

The cooling towers were working for the last 15 years without consuming this amount of energy for each of them. The savings only in electrical energy are as under:

1. For 5 years the tariff per unit is taken as Rs.3.50 per KWH
   Working hours are taken as 12 hours and 2 hours for starting and closing of system. 300 will be working days.
   Therefore, savings in electricity = 25 x 14 x 300 x 3.50
   = Rs.3,67,500.00 per year.
   For 5 years = 3,67,500.00 x 5
   = Rs.18,37,500.00
2. For next 5 years the tariff is taken as Rs.4.50 per KWH
   All other things remain the same.
   Therefore, savings in electricity = 25 x 14 x 300 x 4.50
   = Rs. 4,72,500.00 per year.
   For 5 years = 4,72,500.00 x 5
   = Rs. 23,62,500.00
3. For next 5 years the tariff is taken as Rs.5.50 per KWH
   All other things remain the same.
   Therefore, savings in electricity = 25 x 14 x 300 x 5.50
   = Rs. 5,77,500.00 per year
   For 5 years = 5,77,500.00 x 5
   = Rs. 28,87,500.00
   Total savings in electricity alone per cooling tower Rs. 70,87,500.00
   For 2 cooling towers = Rs.70,87,500.00 x 2
   T O T A L = Rs. 1,41,75,000.00 = 328,429.10 USD
   = 218,733.78 Euro (*i*)

This is the savings due to not having a motor in ARMEC JET TOWER. In addition to this there are other savings due to labor charges incurred for cleaning of fills on regular basis, oil topping of the gearbox of fan motor, replacement of fills, replacement of blades, revamping of cooling tower if the make is of wooden cross-flow etc.,

Savings in labor charges for regular cleaning of fills are as under:

| Persons | Rate Per day (in Rs.) | Reason |
|---|---|---|
| 1 | 200.00 | Electrician for disconnection and connection of motor. |
| 1 | 200.00 | Fitter for removing assembly. |
| 1 | 250.00 | Mukadam/ supervisor for work. |
| 5 | 200.00 | Workers to carry out the work. |
| Total | 1,650.00 | |

This will be the minimum labor bill for cleaning of fills every fortnight of each of the cooling tower after providing all the material handling equipments like ropes, tackles and chains. If the work is outsourced then, the profit of the contractor also gets added. Since, there are two cooling towers, every week one cooling tower will be on maintenance.

Therefore, this cost may be taken as minimum per week cost.

Labor cost = Rs.1,650.00 x 52 weeks in a year

= Rs. 85,800.00

Labor cost for 15 years = 85,800.00 X 15

= Rs. 12,87,000.00 =29,819.28 USD

= 19,859.64 Euro *(ii)*

Replacements cost for replacement of various moving and stationery parts of the Fan and Fill type conventional cooling towers for 15 years.

1. Fills replacement cost (Min. once in 5 years) = Rs. 1,50,000.00

   For 15 years = Rs.1,00,000.00 For first 5 years

   Rs.1,25,000.00 For next 5 years

   Rs.1,50,000.00 For last 5 years

   Total for 15 years = Rs. 3,75,000.00
2. Motor replacement (once in 15 years) = Rs. 25,000.00
3. Blades replacement (Twice in 15 years) = Rs. 30,000.00
4. Gear box replacement (once in 15 years) = Rs. 40,000.00
5. Revamping expenses (Once in 15 years) = Rs. 3,00,000.00

   Total replacement expenses = Rs.7,70,000.00 x 2

   = Rs.15,40,000.00 =35,681.19 USD

   = 23,767.67 Euros *(iii)*

Grand Total = *(i)* + *(ii)* + *(iii)*

= Rs. 1,70,02,000.00 = 393,929.56 USD

= 262,357.09 Euro

This is the extent of monetary benefit accrued to the client in 15 years on account of replacing the existing cooling towers by ARMEC JET TOWERS. You will agree that, the rates quoted herein are on the lower side and only a regular energy auditor can compile a proper study.

Even based on these you will realize that, the saving per year is

= Rs. 11,33,470.00 = 26,262.05 USD

= 17,490.52 Euros

This in itself is much more than the cost of the cooling tower of the same capacity even to day. A decision taken in 1993-94 has benefited this organization to such a large extent through saving valuable energy.

## 2.0 Analysis

The savings is not only in Power usage, but man-hours and spares are also integral parts of savings that is achieved. Man-hours saved can only be realized in huge cooling towers of 1,000Mtr3/Hr or more water flow, as one has to hire internal or contract gangs to remove the fan motor, clean the fills, change the blades etc. In the current case the following are the percentage of expenses that would have incurred on various counts in percentages if the client has used the conventional cooling tower.

Savings due to electrical energy savings: 83.38 per cent

Savings due to labor for regular maintenance: 7.56 per cent

Savings due to spares and breakdown maintenance labor: 9.06 per cent

## 3.0 Synthesis

It is practically difficult to measure any thing other than electricity and water, which are normally metered. Hence, measuring the labor expenditure on cooling tower is very difficult. Further, it is still more difficult to measure the inventory maintained and expenses incurred on account of a cooling tower as it is normally charged to the machine or process in any industry and no inventory is maintained only for cooling towers. This case study is just a guideline to understand the reality, which can be used for future case studies as an approximate assumption for analysis. As more number of case studies is made on this subject, the realistic assumptions can be assumed. Till then, this is just a beginning.

### Effect on Global Warming

This is the most environment friendly cooling tower which works without Fan on the top and therefore, there is no heat dissipation into atmosphere due to moving of motor for suction of hot water and discharge it into the atmosphere in the form of drift. Further, the company has developed and is under patent the ZERO-DRIFT model of the cooling tower, which to a large extent eliminates drift losses.

## 4.0 Conclusions

There are immense possibilities of Energy saving around us and one has to be vigilant to take stock of these and go-ahead to make the World in general benefit from this GREEN technology. Invariably we see a cooling tower on top of every plush and Hi-end building used for Central Air-Conditioning purpose. The conventional cooling towers shall have a fan motor of 5 to 50KW running continuously. Imagine the extent of Energy that can be saved if these are eliminated. In addition the drift that is blown continuously in to the wind containing high temperature water vapors shall heat the surrounding area and the temperature around such cooling towers shall be about 1-2°C higher than the temperature at other places. Imagine the increase in the atmospheric heat in the long run if 45 to 50°C water vapors are constantly blown to the atmosphere. Hence, it is time people at appropriate level realize, at least for domestic buildings and areas, the cooling towers with out FANS and FILLS to be used as it will benefit more ways than one as cited above.

## Acknowledgements

We thanks the officials of Reserve Bank of India, Nagpur, Mumbai and New Delhi for the assistance given to write this article.

# Chapter 20

# *Gasifier Based Crematorium: An Analysis*

***M. Venkata Ramanan*, V. Muthu, R. Velraj and R. Sethumadhavan***

*Institute for Energy Studies, Anna University,*
*Chennai, Tamil Nadu, India*
**E-mail: venkat@annauniv.edu, madhavan@annauniv.edu*

## ABSTRACT

Indian religions, such as Hinduism, Jainism and Buddhism mandate cremation. In these religions the body is seen as an instrument to carry the soul. As an example the Bhagavad-Gita quotes *"Just as old clothes are cast off and new ones taken, the soul leaves the body after the death to take a new one"*. Hence, the dead body is not considered sacred since the soul has left the body and the cremation is regarded as ethical. Cremation–in earlier days–and now also to a greater extent is performed by direct wood burning in an open space. According to 2001 Census, Hindus constitute 80.5 per cent of the total Indian population of 1,028,610,328. With a mortality rate of 8 per 1000 lives (2006 data), about 66.25 lakhs of Hindus die every year in India. This amounts to a cremation of about 18 000 corpses daily. It is estimated that the annual consumption of fuel wood for cremation alone is about 1.75 to 2 million tons with a per body fuel wood consumption of about 250 kg and above at the minimum. This paper enunciates on the concept of Gasifier based Crematorium as an alterative to cheaper and pollusive open burning and costlier electrical cremation.

***Keywords:*** *Electrical cremation, Incineration, Open burning, Gasifier crematorium, AAQ norms, energy conservation and Cleaner production.*

## 1.0 Introduction

Cremation is the act of reducing a corpse by burning, generally in a crematorium furnace or crematory fire. The former is widely adopted in developed nations while the latter is followed in developing countries. Cremation may serve as a funeral or post-funeral rite, which is alternative to the interment of an intact body in a casket. The remains (often called cremains) after cremation, which are not a health risk, may be buried or immured in memorial sites or cemeteries or they may be legally retained by relatives or dispersed in a variety of ways and locations. The cremation occurs in a 'crematorium' consisting of one or more cremator furnaces or cremation 'retorts' for the ashes. A crematorium may be part of chapel or a funeral home, or part of an independent facility or a service offered by a cemetery. Among the different religions practicing cremation, Hinduism and Jainism are notable for not only allowing but also prescribing cremation. The Rig Veda contains a reference to the emerging practice, in RV 10.15.14, where the forefathers "both cremated (*agnidagdhá-*) and uncremated (*ánagnidagdha-*)" are invoked.

## 2.0 Cremation Process

Cremation dates to at least 26,000 years ago in the archaeological record with the Mungo Lake cremation. Cremation in India is first attested in the Cemetery H culture (from ca. 1900 BC), considered the formative stage of Vedic civilization. In cremation, the corpse is burnt open (or) placed in the retort and incinerated at a temperature of 760 to 1150 °C. During the cremation process, a large part of the body (especially the organs) and other soft tissue are vaporized and oxidized due to the heat and the gases are discharged through the exhaust system. The entire process usually takes about two hours. All that remains after cremation are dry bone fragments (mostly calcium phosphates and minor minerals). Their color is usually light gray. They represent very roughly 3.5 per cent of the body's original mass (2.5 per cent in children).

Major reasons for which the people adopt cremation process as compared to burial are:

- Religious Custom
- Very economical use of cemetery space
- Simplified funeral process
- Cost factor (Generally speaking, cremation costs less than traditional burial services)
- Cremated remains can be scattered or buried.

## 3.0 Types of Crematorium

Crematoriums could be classified on a number of ways, predominant of which is based on Fuel. The concept of open burning using firewood is conventionally adopted. However modern cremators operating with electricity, natural gas and diesel exists.

Open burning requires minimum ¼ tonne of fuel wood. As an alternate to this fuel wood burning, Diesel fired Crematoria have been installed in few locations in India to mitigate the environmental pollution. In a diesel crematorium, approximately 30–40 liters of diesel are consumed to cremate one body and the process takes about two hours. Due to continued escalation in the price of petroleum products, the operating cost of diesel-fired cremation makes it an expensive option. Of late, Electrical Crematoria have started replacing the wood fired ones due to the simplicity of operation.

A comparative analysis on the Fuel wood and Electrical Crematorium is given below:

### 3.1 Wood Fired Cremation

Open burning wood (depicted in Figure 20.1) fired cremation systems are widely adopted but however suffer from certain inherent limitations namely,

1. Higher quantity of wood requirement
2. Higher time of cremation
3. Need for supplementary fuels like cow dung cake, tyres etc.
4. Higher cost of cremation
5. Pollution nuisance

In addition to all these, cremation becomes quite a task during rainy season.

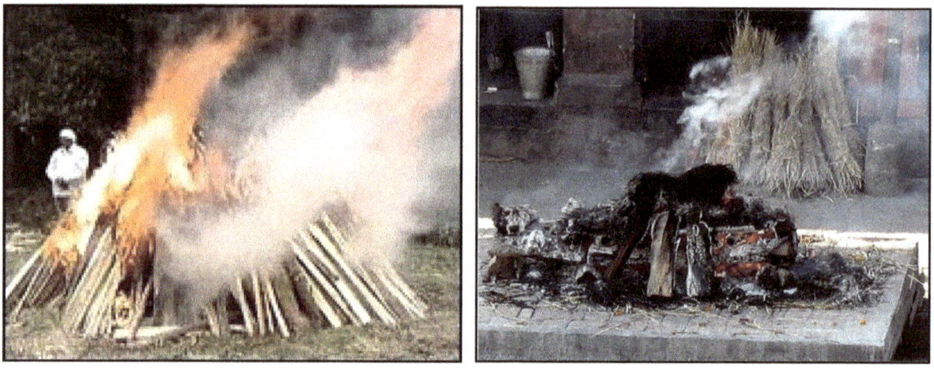

**Figure 20.1: Wood Fired Cremation–Open Burning**

### 3.2 Electrical Crematorium

On the other hand, all the limitations cited above in wood fired open burning have been addressed very effectively when electrical cremation (depicted in Figure 20.2) is resorted to. It has only one major limitation, namely, the cost of operation. It has been estimated that an electric crematorium consumes around 100 kWh of power to cremate a body, which is quite expensive. In addition, the adoption of this technology in rural areas / remote villages is questionable where the power supply would be quite erratic and very unreliable in terms of quality. Affordability is also one of the factors and poor people cannot afford this.

## 4.0 Gasifier Based Crematorium (GBC)

Gasifier based crematorium is an excellent application of Gasification technology to act as a 'trade off' between the high cost electrical crematorium and low cost

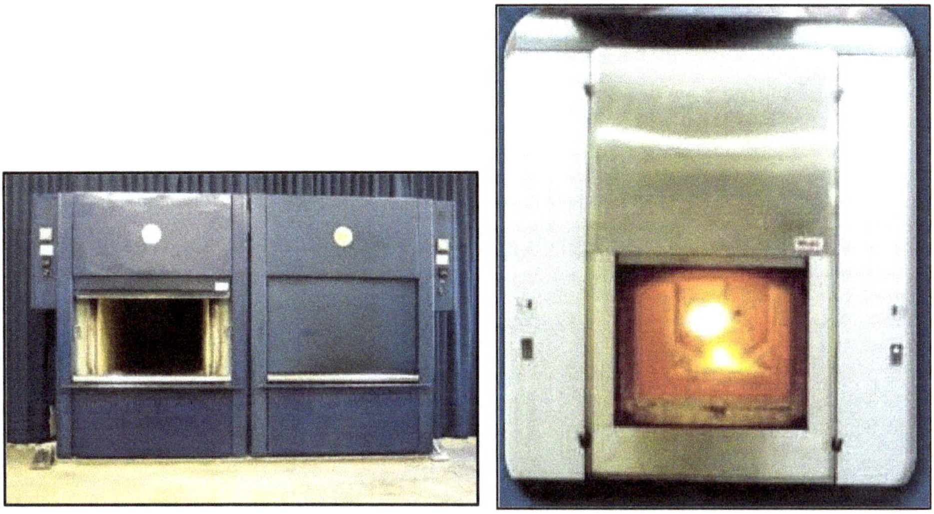

**Figure 20.2: Electrical Crematorium**

firewood based open burning. In simple terms, it offers almost all the advantages of electrical crematorium at the cost of open burning.

Gasification technology is based on a high temperature complex thermo-chemical reactions which converts solid biomass fuel into a more conveniently usable gaseous form (commonly known as Producer Gas) of fuel–with a calorific value of about 4 500 kJ / $nm^3$–which can be used to meet thermal energy requirements.

Gasifier has following major advantages:

- Better conversion (solid to gas) efficiency (above 70 per cent)
- Good turn down ratio–better control on burning (handling gaseous fuel)
- Production of clean flue gas (as gas is burnt completely)
- Fuel versatility
- Limited attendance
- Lower emissions
- Compact
- Better responsive characteristics
- Possible emergency shut down

The induction of gasifier systems for crematoria will improve the energy efficiency (as gaseous fuel is now burned instead of solid wood fuel) thereby saving substantial quantities of fuel wood. Such a system is likely to gain social acceptance also, as the body will be burnt in flames of producer gas obtained through wood gasification, which conforms to the conventional Hindu custom.

The utilities and accessories associated with GBC comprise the following:

1. Gasifier
2. Main Burners

3. Pilot Burner
4. Ash Cooling and Cleaning System
5. Instrumentation and Control Systems
6. Auxiliary Equipment *viz.*, Wood Cutter etc.

## 5.0 Scheme of Operation of GBC

The process flow chart of a gasifier based crematorium is depicted in Figure 20.3.

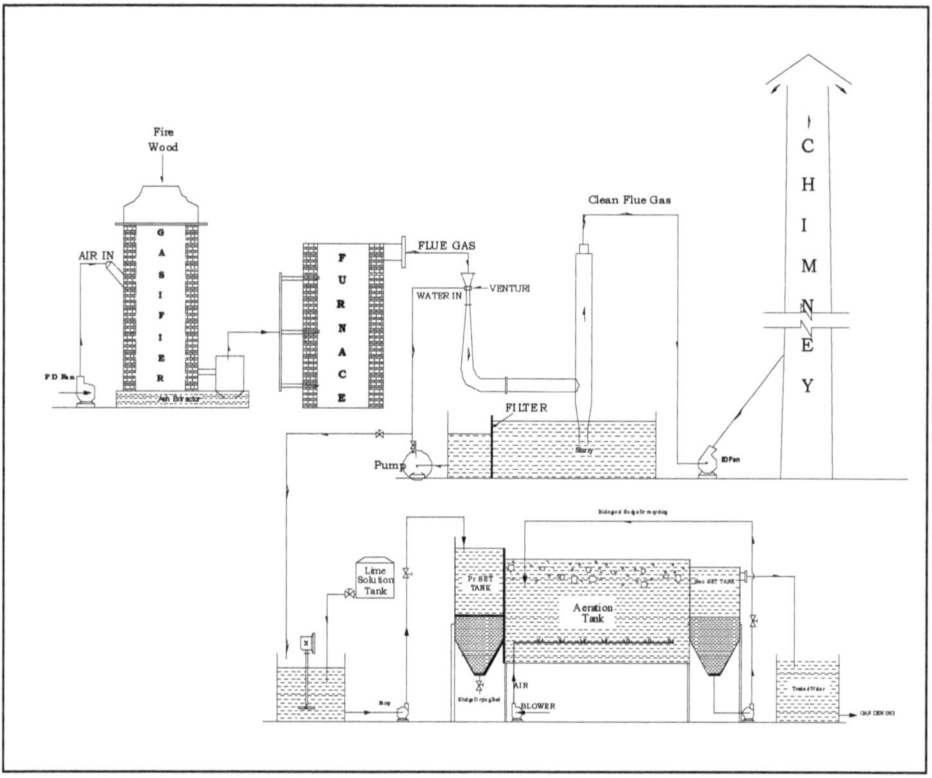


**Figure 20.3: Process Flow Chart of Gasifier Based Crematorium**

The producer gas generated from the gasifier (through gasification of cut to size wood) is sent to the furnace where it is burnt with the aid of requisite oxygen. Although the gas is a low btu one, it is still capable of producing a flame temperature of 1000°C and above. The burning of the gas inside the furnace results in cremation of the body. The temperature inside the furnace would be about 800°C, which is sufficient enough to burn the dead body completely.

The burnt gas–also known as Flue Gas–is then washed in a venturi scrubber arrangement to make it devoid of possible chemical and ash contamination. Venturi scrubber is capable of washing the gas completely to remove contaminants to an extent of 99 per cent and above. The cleaned flue gas is then exited to the atmosphere through an ID fan–Chimney arrangement. The wash water from the venturi system would be periodically sent to the effluent treatment plant for further treatment to bring it to the disposal norms prescribed by PCB.

## 6.0 Ambient Air Quality (AAQ) Notification for Gasifier Operation

Table 20.1 details the norms laid by PCB on concentration of various pollutants in ambient air.

**Table 20.1: AAQ Norms for Gasifier Operation**

| No. | Pollutant | Time Weighted Average | Concentration in Ambient Air: Area Classification | | | |
|---|---|---|---|---|---|---|
| | | | Industrial | Residential, Rural and Others ($\mu g/m^3$) | Sensitive | Measurement Method |
| 1. | $SO_2$ | Annual* | 80 | 60 | 15 | Improved West and Gaeke Method Ultraviolet |
| | | 24 hrs# | 120 | 80 | 30 | Fluorescence |
| 2. | NOx | Annual* | 80 | 60 | 15 | 1. Jacob and Hochheiser modified (Na-Arsenite) Method |
| | | 24 hrs# | 120 | 80 | 30 | 2. Gas Phase Chemiluminescence |
| 3. | SPM | Annual* | 360 | 140 | 70 | High Volume Sampling (Average flow rate not less |
| | | 24 hrs# | 500 | 200 | 100 | than $1.1 m^3$ / min) |
| 4. | RPM (< 10μm) | Annual* | 120 | 60 | 50 | Respirable Particulate |
| | | 24 hrs# | 150 | 100 | 75 | Matter Sampler |
| 5. | Pb | Annual* | 1.0 | 0.75 | 0.5 | AAS method after sampling using EPM 2000 or equivalent |
| | | 24 hrs# | 1.5 | 1.0 | 0.75 | filter paper |
| 6. | CO | 8 hrs* | 5 000 | 2 000 | 1 000 | Non dispersive Infrared |
| | | 1 hr | 10 000 | 4 000 | 2 000 | Spectroscopy |

*: Annual Arithmetic Mean of minimum 104 measurements in a year, taken twice a week, 24 hourly at uniform interval.

#: 24 hourly / 8 hourly values should meet 98 per cent of the time in a year. However 2 per cent of the time, it may exceed but not on two consecutive days.

## 7.0 Economics

Institute for Energy Studies, Anna University Chennai has offered consultancy services to the Commissionerate of Municipal Administration; Government of Tamilnadu (GoTN) for installation of Gasifier based Crematorium in 52 Urban Local Bodies of Tamilnadu. With the grants provided by the GoTN, the payback period of commissioning a GBC–on comparison with opens burning system– works out to less than 5 years at a cremation rate of Rs. 700 per body. However excluding the grants, the simple payback period exceeds 7 years.

## 8.0 Merits of Gasifier Crematorium

The Gasifier Crematorium possesses the following advantages over conventional Open Cremation or Electrical Cremation.

- Wood is the fuel. Thus the sentiments attached with the cremation are fulfilled.
- Requires less than 50 per cent of firewood for cremation otherwise needed in the open burning due to improved / upgraded technology adoption.
- Cremation can be completed in 60–90 minutes as against the conventional time requirement of not less than 4 hours.
- Gasifier cremation can be performed equally efficiently during rainy season also.
- Easy to Operate and calls for nil / minimal maintenance.
- Absolutely environment friendly.
- Compared to Electric Crematorium–that requires 100 kWh of power per body–Gasifier Crematorium requires less than 5 kWh of power / body–a tremendous savings in Electricity requirement.
- Clean and neat surroundings can be ensured through this technology adoption, which is normally not possible in open cremation grounds.

## 9.0 Conclusions

Gasifier based crematorium is a viable technique for cremation of corpses, both technically and economically. The option is very lucrative if electricity or diesel is replaced and is superior to conventional open firing on all fronts. Implementation of this program would lead to savings of about 1 million tones of firewood (associated pollution mitigation with possible CDM benefits under clustering concept). Hence, Government of India may consider this project as a societal development one and may provide Interest free loans to all Urban Local Bodies towards making this project a reality.

## References

1. Census statistics of India (2001) published by Ministry of Home Affairs, Govt of India
2. Mortality country fact sheet – India, Published by WHO, 2006.
3. Wikipedia @ http://www.wikipedia.org/

## Abbreviations

GBC: Gasifier Based Crematorium

PCB: Pollution Control Board

ID: Induced Draught

AAQ: Ambient Air Quality
SPM: Suspended Particulate Matters
RPM: Residual Particulate Matters
GoTN: Government of Tamil Nadu
CDM: Clean Demand Management

# Chapter 21

# *DG Set Combined Heat and Power (CHP) Generation: A Successful Case Study of a Cold Storage*

**G. Subramanyam**
*Director, Siri Energy and Carbon Advisory Services Pvt. Ltd.*
*2nd Floor, Narayan Villa, Opp.Priyadarshini Park, Saroornagar,*
*Hyderabad, Andhra Pradesh, India*
*Email: subramanyam@siriexergy.com*

The following is a successful case study of Cold Storage Unit situated at outskirts of Hyderabad. The Unit started its operations way back in 1987. The company is one of the first and commercial De-Humidification Cold Storage Unit in the state of Andhra Pradesh, exclusively designed for high value hybrid seeds, tamarind, chilies etc. The company has state-of-art pre-cooling facility for export of quick perishables like fruits, vegetables and eggs. The unit needs both power and steam. The electrical power consumption is towards running refrigeration compressor and associated equipment like condenser pumps, chilled water pumps, cooling tower and lighting etc. The steam is used for de-humidification. In cold storage units, the temperature and RH control are most critical in cold storage units, and the moisture is absorbed by in adsorber using silica gel as desiccant. After absorption, the silica gel needs to be regenerated, and steam is used for *Dry Air* regeneration purpose. In *Dry Air* dehumidification, in one cycle-adsorption of moisture takes place and in other cycle-re-activation of silica gel is carried out using hot air. Hot air is generated using steam in the radiator coils.

## Cold Storage Unit

Initially, the unit went for 75HP LT connection with the SEB, which is meant for lighting only. For balance power requirements, 2 x 320 kVA DG sets were installed. For steam generation, 600-kg/hr capacity HSD fired boiler was installed. The unit

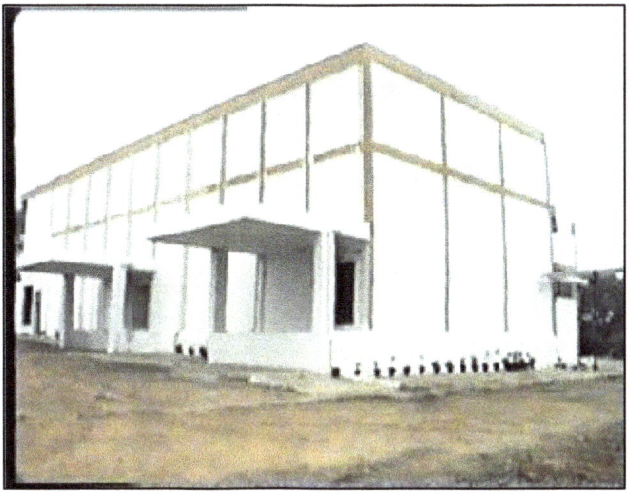

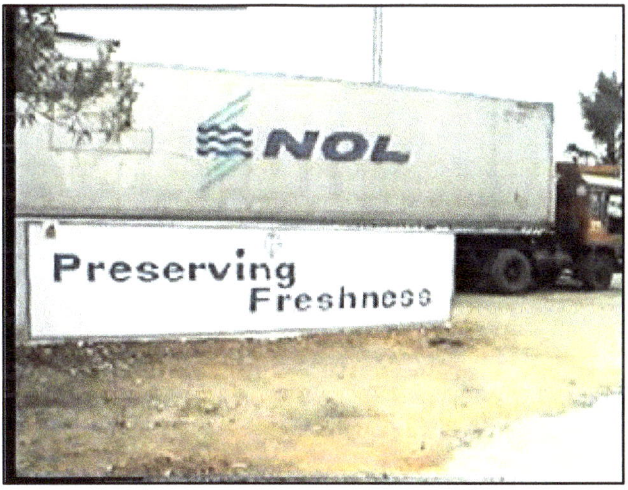


requires around 185 kW of power and 300-350 kg/hr of steam. Initially the unit used to consume about 55-60 lit/hr Diesel in the DG sets and another 22-25 lit/hr Diesel in the boiler. The DG sets were used to operate for 12Hrs/day and the boiler was needed to operate about 10-12 Hrs/day. With this combination, the daily energy cost used to be around Rs.31,416/day, and the annual bill of the order of Rs.115 Lakhs/year. (All the costs are compared at the present rate).

When the cost of Diesel went up exorbitantly, the unit has gone for a gassifier, using rice husk as fuel, by investing Rs.15 Lakhs. The gassifier was supplied by M/s Ankur Scientific Energy Systems, Baroda. The rice husk is partially oxidized, and biogas is generated. The other by-products of the gassifier are ash and tar. The ash has very good market value and is being used by brick manufacturers and tar is sold as a lubricant. One problem with the gassifier is the DG set can't run on completely

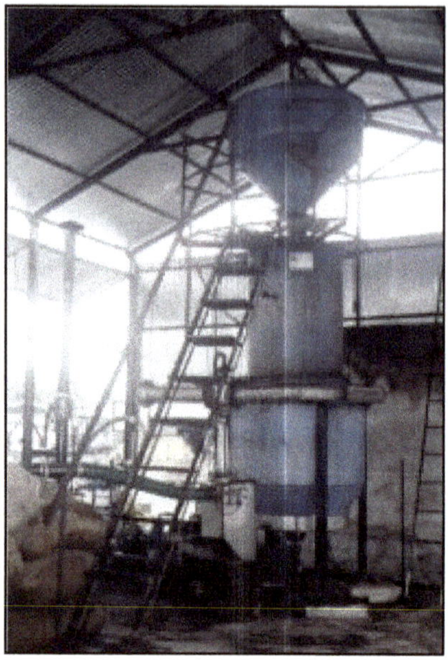

on 100 per cent biogas. A minimum of 30 per cent Diesel and 70 per cent gas is needed. But this unit decided to use 40 per cent Diesel and 60 per cent biogas for running the DG sets. By installing this Gassifier, the Unit has reduced their total energy bill to Rs. 79 Lakhs from the initial Rs. 115 lakhs a savings of by Rs.36 lakhs, a significant reduction of more than 31 per cent.

When these people decided to go for a gassifier, the cost of Rice husk used to be in the range of Rs.700-800/ton. But over a period, when more people started using rice husk, the cost of rice husk almost doubled and is now available in the range of Rs1500-1600/Ton.At this stage, they approached National Productivity Council, to help them to reduce their energy bill further by carrying out Energy Audit.

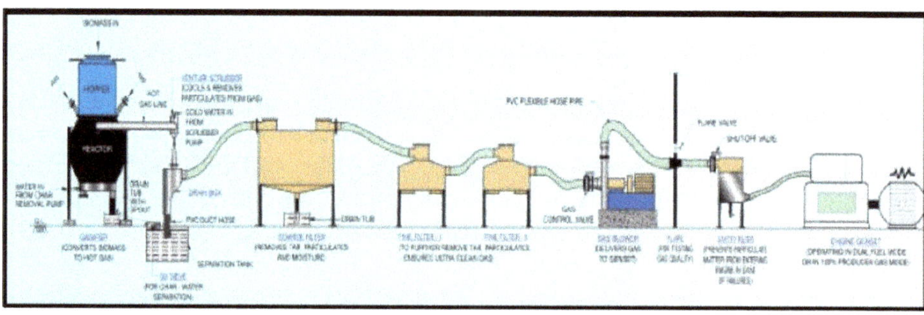


After preliminary visits and taking some trails on DG sets, Boiler and refrigeration system, we thought Vapour Absorption Refrigeration (VAR) system would be best bet to replace the existing load of 100 TR, which is running on vapour compression Ammonia refrigeration system. Since steam is available and in case, if we want more steam, we thought we can generate from biogas. At this stage, two alternatives were worked out. First one is to generate steam by installing Waste Heat Boiler (WHB), another alternative was to generate hot air directly from the Flue gas by installing simple air heater. First of all, why to generate steam and then produce hot air, if we are able to generate directly hot air, so that we can avoid one more inefficiency. But after working out various alternatives, our suggestion of heat recovery from DG set's exhaust gases and generating hot air was well appreciated by the management. Since the DG sets are run continuously we felt this is an ideal case for Combined Heat and Power (CHP) generation.

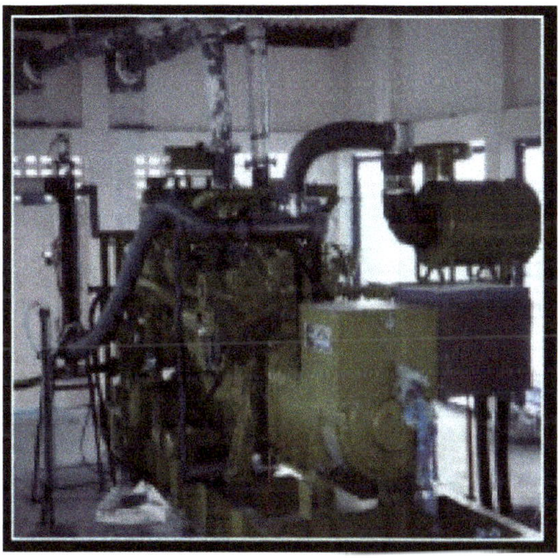

The exhaust temperature of the DG sets used to be in the range of 390-400°C. After detailed measurement of Flue Gas flow, the potential heat available assessed to be around 1,00,000 kcal/h, which is more than sufficient to generate 1500 CFM hot air at 120°C. This perfectly matched the plant requirements. After giving this idea, the unit has implemented the Waste Heat Recovery (WHR) system for their DG sets, in a record time of one month. Interesting part is, based on the heat duty; we have given them a rough estimate of heat transfer area required and the line diagram of the modification. The unit has fabricated the shell and tube heat exchanger on their own inside the factory itself at a cost of only Rs.3.5 lakhs (year 2000 cost) and present cost Rs.6 lakhs. One single heat exchanger, with blower can cater to both the DG sets, with by-pass arrangement. When we gave them, the heat exchanger design, in fact, we ourselves were not confident of our own design. But after seeing the working model, which was giving hot air at 140-150°C, at the exit of the Waste heat recovery system, we are proud that, we can do even designing.

By implementation of the WHR, the existing steam boiler was completely stopped and has become a stand-by boiler. By this the annual bill of the plant has further reduced to Rs.46 Lakhs from the initial Rs. 115 Lakhs. In another words, additional savings of Rs.33 Lakhs, by stopping the boiler alone. This amounts to reduction of

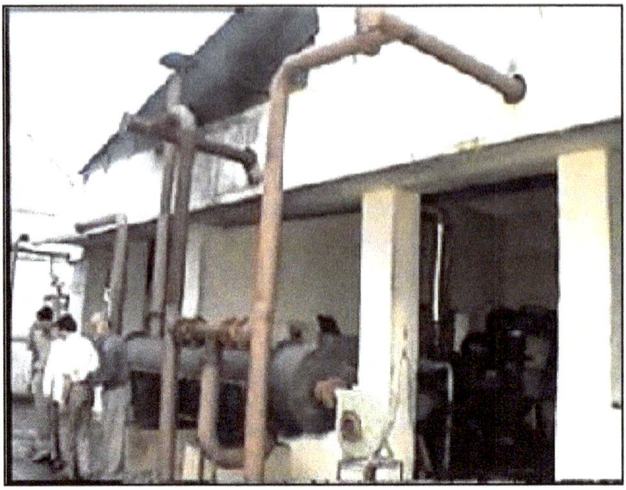

**Modification**

another 42 per cent . The Table 21.1 presents the cost comparison of three different combinations, *i.e.*Case-1) DG set and Boiler, Case-2) DG set with Gassifier and Boiler and Case -3) DG set with Gassifier and WHR system.

**Table 21.1: Cost Benefit Analysis of DG set Cogeneration with Gassifier**

| *Sl.No.* | *Parameter* | *Unit* | *Case -1* Only DG Set | *Case-II* DG set with Gassifier | *Case-III* DG Set with Gassifier and WHR |
|---|---|---|---|---|---|
| 1. | DG set Rating | KVA | 2 x 320 | 2 x 320 | 2 x 320 |
| 2. | Normal operation | ? | 320 | 320 | 320 |
| 3. | Avg. Power Generation | kw | 180 | 180 | 180 |
| 4. | Diesel Consumption in DG set | Lit/hr | 55 | 22 | 22 |
| 5. | Rice husk consumption in Gassifier | kg/hr | 0 | 200 | 200 |
| 6. | DG set F.Gas Exit Temp. | °C | 400 | 400 | 400 |
| 7. | Steam requirement | kg/hr | 300 | 300 | 0 |
| 8. | Diesel consumption in the Boiler | Lit/hr | 22 | 22 | 0 |
| 9. | Total Energy Cost | Rs./day | 31416 | 21552 | 12576 |
| 10. | Annual Energy Bill | Rs. Lakhs | 115 | 79 | 46 |
| 11. | Annual Monetary Savings | Rs. Lakhs | | 36 | 69 |
| 12. | Investment towards Gassifier | Rs. Lakhs | | 15 | 15 |
| 13. | Investment towards Waste Heat Recovery | Rs. Lakhs | | 0 | 6 |
| 14. | Total Investment | Rs. Lakhs | | 15 | 21 |
| 15. | Simple Pay Back Period | Months | | 5 | 4 |

Basis: Cost of Diesel Rs.34/lit and Cost of Rice Husk Rs.1500/ton, Operating Hours = 12/day.

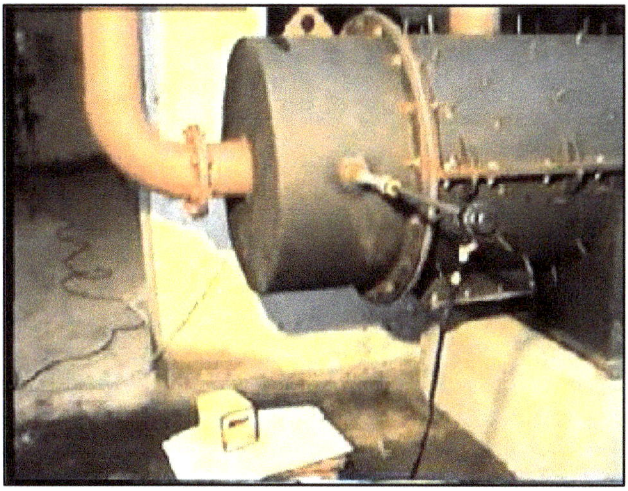

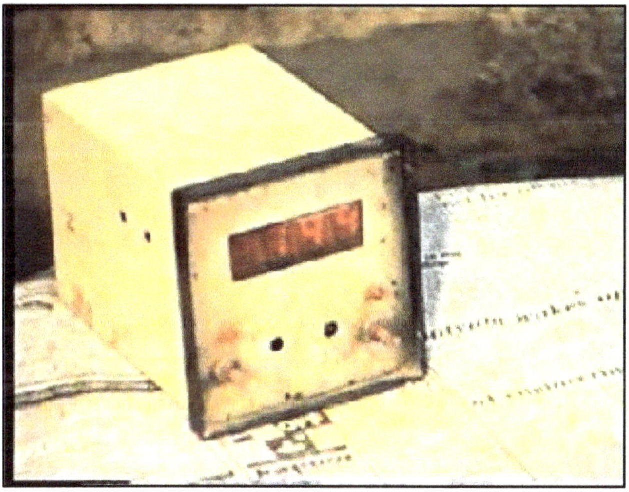

**Temperature Monitoring**

After implementation of the scheme, one problem faced by the unit was the soot deposition and fouling of the heat transfer surfaces. Since the unit is using 60 per cent bio-gas, the problem of soot deposition aggravated with the introduction of additional equipment in the flue gas path. So, the unit decided to operate the DG sets alternately so that they get sufficient time to clean the heat transfer surfaces and for any maintenance requirements.

This is a successful case study of a small cold storage unit, which experimented, by installing a gassifier and also the waste heat recovery system, there by reducing their energy bill from the initial Rs.115 lakhs to Rs.46 lakhs, a savings of Rs.69 lakhs equivalent to 60 per cent over a period. The simple payback period was hardly less than 4 months, really unbelievable. This savings are based on only 12hrs/day of operation. Imagine the savings, if the plant operates for 24 hrs and all the 3 shifts.

This technology can be replicated not only in the similar cold storage units but also for other units, who needs combined heat and power (CHP) simultaneously and urge to reduce their energy bills. The significant other benefits are substantial reduction of $CO_2$ the main green house gas.

The estimated GHG emission reduction amounts to around 600 Ton/year. We thought of converting it to a CDM project, but the CER numbers does not favor in comparison with the transaction costs. If more people impalement similar measure then it will be a good case for bundled CDM project. Is there any taker for this technology! We have a video film to support this idea and seeing is believing.

## Acknowledgements

I want to acknowledge my senior colleagues Mr. D. Pawan Kumar and Mr. R. Virendra of National Productivity Council, for their inspiration and guidance in converting an idea to reality.

www.ingramcontent.com/pod-product-compliance
Ingram Content Group UK Ltd.
Pitfield, Milton Keynes, MK11 3LW, UK
UKHW022213051026
15624UKWH00001B/146

9 788170 359838